Numerical analysis 1999

CHAPMAN & HALL/CRC

Research Notes in Mathematics Series

Submission of proposals for consideration

Suggestions for publication, in the form of outlines and representative samples, are invited by the Editorial Board for assessment. Intending authors should approach one of the main editors or another member of the Editorial Board, citing the relevant AMS subject classifications. Alternatively, outlines may be sent directly to the publisher's offices. Refereeing is by members of the board and other mathematical authorities in the topic concerned, throughout the world.

Preparation of accepted manuscripts

On acceptance of a proposal, the publisher will supply full instructions for the preparation of manuscripts in a form suitable for direct photo-lithographic reproduction. Specially printed grid sheets can be provided. Word processor output, subject to the publisher's approval, is also acceptable.

Illustrations should be prepared by the authors, ready for direct reproduction without further improvement. The use of hand-drawn symbols should be avoided wherever possible, in order to obtain maximum clarity of the text.

The publisher will be pleased to give guidance necessary during the preparation of a typescript and will be happy to answer any queries.

Important note

In order to avoid later retyping, intending authors are strongly urged not to begin final preparation of a typescript before receiving the publisher's guidelines. In this way we hope to preserve the uniform appearance of the series.

CRC Press UK

Chapman & Hall/CRC Statistics and Mathematics
Pocock House
235 Southwark Bridge Road
London SE1 6LY
Tel: 020 7450 7335

D F Griffiths and G A Watson
(Editors)

Numerical analysis 1999

CHAPMAN & HALL/CRC

Boca Raton London New York Washington, D.C.

Library of Congress Cataloging-in-Publication Data

Catalog record is available from the Library of Congress.

Direct all inquiries to CRC Press LLC, 2000 N.W. Corporate Blvd., Boca Raton, Florida 33431.

International Standard Book Number 1-58488-020-1
Printed in the United States of America 2 3 4 5 6 7 8 9 0
Printed on acid-free paper

Contents

Preface

The 18th Dundee Biennial Conference on Numerical Analysis was held at the University of Dundee during the four days 29th June – 2nd July, 1999. It was preceded on the 28th June by the Leslie Fox Prize Meeting, when the six finalists gave their presentations. As usual the standard of talks was impressively high, and the prize was shared by Niles Pierce from the Californian Institute of Technology and Reha Tütüncü from Carnegie Mellon University.

The conference itself was attended by around 200 participants from 32 countries, with just under half the participants coming from outside the UK. The opening talk was the fifth A. R. Mitchell lecture, delivered by Franco Brezzi, of the University of Pavia. The tradition was continued of having a speaker who has made many important contributions to numerical analysis, but also has a long association with Dundee, and with Ron Mitchell. Twelve other invited speakers presented talks and, in addition, there were 128 submitted talks presented in four parallel sessions. This volume contains full versions of all of the invited papers; the titles of all contributed talks given at the meeting, together with the names and addresses of the presenters (correct at the time of the meeting), are also listed here.

We would like to take this opportunity of thanking all the speakers, including the after-dinner speaker David Sloan, from the University of Strathclyde, all those who chaired sessions, and all participants for their contributions. The conference is also indebted to the University of Dundee for making available various university facilities throughout the week, and for the provision of a reception for the participants. Thanks are also due to staff at CRC Press, to Sunil Nair and Dawn Sullivan for their guidance through the pre-publication process and to Maria Jennings for her careful proof-reading of the manuscript.

Finally, we are pleased to announce that the 19th Biennial Conference will be held from 25th–29th June 2001. Information will be posted on the conference web page, which may be accessed at

`http://www.maths.dundee.ac.uk/~naconf/`

These pages also give details of earlier conferences in the series.

D F Griffiths
G A Watson

October 1999

M. AINSWORTH, P. COGGINS AND B. SENIOR

Mixed *hp*-finite element methods for incompressible flow

Abstract Recent progress in the design and analysis of mixed *hp*-finite element methods for the approximation of the Stokes' equations is outlined. The *hp*-version of the finite element method can give exponential rates of convergence on properly designed meshes. In particular, the accurate *approximation* of boundary layers dictates the use of high aspect ratio elements. However, the *stability* of mixed finite elements is highly sensitive to the aspect ratio ρ^{-1} to the extent that the advantages of the enhanced approximation properties may be severely degraded (even if the approximation error is reduced exponentially fast) or even dissipated altogether (in the case of an algebraic rate of convergence). The degeneration of the stability constant is due to a single pressure mode that may be stabilised by using polynomials of sufficiently high degree. Unfortunately, the required degree is prohibitively large for use in practical computations. A practical alternative is to generate high aspect ratio elements through a series of geometric refinements based on a more modest grading factor for which the threshold on the polynomial degree needed to restore stability is met by lower order elements. Following this strategy, the degeneration of the stability constant is ameliorated from $\mathcal{O}(\rho^{-1/2})$ to $\mathcal{O}(|\log \rho|^{1/2})$.

1 Introduction

The steady-state, slow flow of an incompressible fluid in a polygonal domain Ω may be modelled by Stokes' equations. In particular, conservation of momentum implies that the velocity $\mathbf{u}$ and pressure p satisfy

$$-\Delta \mathbf{u} + \nabla p = \mathbf{f} \text{ in } \Omega \tag{1.1}$$

where $\mathbf{f}$ is a source term, while the assumption of incompressibility and conservation of mass mean that

$$\nabla \cdot \mathbf{u} = 0 \text{ in } \Omega. \tag{1.2}$$

For simplicity, we assume that the velocity vanishes on the boundary $\partial\Omega$. It is convenient to reformulate the equations in variational form: find $\mathbf{u} \in \mathbf{H}_0^1(\Omega) =$

$\mathbf{V}$ and $p \in L_0^2(\Omega) = M$, such that

$$\begin{aligned} a(\mathbf{u}, \mathbf{v}) + b(p, \mathbf{v}) &= (\mathbf{f}, \mathbf{v}) \\ b(q, \mathbf{u}) &= 0 \end{aligned} \tag{1.3}$$

for all $\mathbf{v} \in \mathbf{V}$ and $q \in M$. Here, $a(\cdot,\cdot)$ and $b(\cdot,\cdot)$ are bilinear forms defined by

$$a(\mathbf{u}, \mathbf{v}) = \int_\Omega \nabla\mathbf{u} : \nabla\mathbf{v} d\mathbf{x} = \int_\Omega \sum_{j=1}^{2} \nabla u_j \cdot \nabla v_j d\mathbf{x}$$

$$b(q, \mathbf{w}) = -\int_\Omega q\nabla \cdot \mathbf{w} d\mathbf{x}$$

while $(\cdot,\cdot)$ denotes the inner product on $\mathbf{L}^2(\Omega)$. The notations $\mathbf{L}^2(\Omega)$ and $\mathbf{H}_0^1(\Omega)$ are used to denote the product spaces $L^2(\Omega)^2$ and $H_0^1(\Omega)^2$, where the standard notations for Sobolev spaces are adopted [1]. In addition, $L_0^2(\Omega)$ denotes the subspace of $L^2(\Omega)$ consisting of functions with vanishing average value over the domain.

The Galerkin discretisation of the Stokes problem, based on a family of finite dimensional subspaces $\mathbf{V}_N \subset \mathbf{V}$ and $M_N \subset M$, $N \in \mathbb{N}$, consists of finding $\mathbf{u}_N \in \mathbf{V}_N$ and $p_N \in M_N$ such that

$$\begin{aligned} a(\mathbf{u}_N, \mathbf{v}_N) + b(p_N, \mathbf{v}_N) &= (\mathbf{f}, \mathbf{v}_N) \\ b(q_N, \mathbf{u}_N) &= 0 \end{aligned} \tag{1.4}$$

for all $\mathbf{v}_N \in \mathbf{V}_N$ and $q_N \in M_N$.

The selection of a proper combination of velocity–pressure spaces, $\mathbf{V}_N$–M_N, is essential for both the *stability* of the discrete problem and the *accuracy* of the resulting approximation. The notion of what constitutes a suitable combination of spaces is quantified by the *Babuška-Brezzi* condition. This condition may be motivated by examining the well-posedness of the discrete problem by introducing bases $\{\phi_m\}$ and $\{\psi_n\}$ for $\mathbf{V}_N$ and M_N, respectively, and writing

$$\mathbf{u}_N = \vec{u} \cdot \vec{\phi}; \quad p_N = \vec{p} \cdot \vec{\psi} \tag{1.5}$$

where the coefficient vectors $\vec{u}$ and $\vec{p}$ are determined by the linear system

$$\begin{aligned} \mathcal{A}\vec{u} + \mathcal{B}^\top \vec{p} &= \vec{f} \\ \mathcal{B}\vec{u} &= \vec{0} \end{aligned} \tag{1.6}$$

with $\mathcal{A} = [a(\phi_j, \phi_k)]$, $\mathcal{B} = [b(\psi_j, \phi_k)]$ and $\vec{f} = [(\mathbf{f}, \phi_k)]$. The stiffness matrix $\mathcal{A}$ is invertible and the coefficients of the approximate velocity $\vec{u}$ may be eliminated using the first equation. Substituting into the second equation leads to the following equation for the coefficients of the approximate pressure,

$$\mathcal{B}\mathcal{A}^{-1}\mathcal{B}^{\top}\vec{p} = \mathcal{B}\mathcal{A}^{-1}\vec{f}. \tag{1.7}$$

Hence, a necessary and sufficient condition for the existence of a unique solution $\vec{p}$, and consequently $\vec{u}$, is that the matrix $\mathcal{B}\mathcal{A}^{-1}\mathcal{B}^{\top}$ be positive definite. A quantitative version of this condition is to introduce a positive, quantity β_N–the *inf-sup* constant–such that

$$\mathcal{B}\mathcal{A}^{-1}\mathcal{B}^{\top} \geq \beta_N^2 \mathcal{M} \tag{1.8}$$

where $\mathcal{M} = [(\psi_j, \psi_k)]$ is the pressure mass matrix.

A simple computation reveals that for any given non-zero discrete pressure $q = \vec{q} \cdot \vec{\psi}$, the supremum

$$\sup_{\mathbf{v} \in \mathbf{V}_N} \frac{b(q, \mathbf{v})}{|\mathbf{v}|_{1,\Omega}} = \sqrt{\vec{q}^{\top} \mathcal{B}\mathcal{A}^{-1}\mathcal{B}^{\top} \vec{q}} \tag{1.9}$$

and is attained when $\mathbf{v} = \vec{v} \cdot \vec{\phi}$, where

$$\vec{v} \propto \mathcal{A}^{-1}\mathcal{B}^{\top} \vec{q}. \tag{1.10}$$

Thus, the solvability condition given by (1.8) is equivalent to the Babuška-Brezzi condition:

$$\sup_{\mathbf{v} \in \mathbf{V}_N} \frac{b(q, \mathbf{v})}{|\mathbf{v}|_{1,\Omega}} \geq \beta_N \|q\|_{L_2(\Omega)} \quad \forall q \in M_N. \tag{1.11}$$

This condition should be interpreted as a compatibility condition between the discrete spaces $\mathbf{V}_N$ and M_N. In particular, for a given choice of velocity space, the condition means that the pressure space M_N must not be too 'rich', in the sense that it contains pressure modes that cannot be 'seen' by the velocities. For such a *spurious* pressure mode $q \subset M_N$, the quantity $b(q, \mathbf{v})$ vanishes identically meaning that the inf-sup constant vanishes, $\beta_N = 0$. This corresponds to the matrix $\mathcal{B}\mathcal{A}^{-1}\mathcal{B}^{\top}$ being positive *semi*-definite and results in the non-uniqueness of the pressure approximation. Obviously, it is desirable to avoid unstable combinations of spaces $\mathbf{V}_N$-M_N in practice.

Difficulties are not confined to the case when the inf-sup constant vanishes. Whilst this guarantees the existence of a unique solution of the Galerkin discretisation, if the inf-sup constant degenerates, $\beta_N \to 0$, as the discretisation is refined, $N \to \infty$, then the loss of stability may adversely affect the accuracy of the approximation. Pressure modes leading to the degeneration of the inf-sup constant are said to be *weakly spurious*. The accuracy of the Galerkin approximation depends on the balance between the approximation properties of the spaces $\mathbf{V}_N$ and M_N compared with the rate at which the inf-sup constant degenerates. In particular, it may be shown [8, 10, 12] that

$$||\mathbf{u}-\mathbf{u}_N||_{1,\Omega} \leq C\left(\frac{1}{\beta_N}\inf_{\mathbf{v}\in\mathbf{V}_N}||\mathbf{u}-\mathbf{v}||_{1,\Omega}+\inf_{q\in M_N}||p-q||_{0,\Omega}\right) \quad (1.12)$$

$$||p-p_N||_{0,\Omega} \leq \frac{C}{\beta_N}\left(\frac{1}{\beta_N}\inf_{\mathbf{v}\in\mathbf{V}_N}||\mathbf{u}-\mathbf{v}||_{1,\Omega}+\inf_{q\in M_N}||p-q||_{0,\Omega}\right). \quad (1.13)$$

This means that while approximation properties may be improved by refining the spaces, if the inf-sup constant degenerates too rapidly, then the overall accuracy of the Galerkin approximation may be lost due to the presence of the inf-sup constant in the denominator.

With the prospect of exponentially accurate approximation properties, it is tempting to disregard even quite severe algebraic degeneration of the inf-sup condition. However, the above derivation of the Babuška-Brezzi condition shows that the conditioning of the discrete algebraic system (1.6) depends on the value of the inf-sup constant. Consequently, the efficiency and accuracy of the numerical techniques will be adversely affected by a poor inf-sup constant. In conclusion, it is of importance to investigate and control the behaviour of the inf-sup constant as the subspaces are refined.

2 Mixed hp-finite element spaces

Suppose that the domain Ω is partitioned into the union of affine, quadrilateral elements of size h, satisfying the usual conditions for a finite element scheme [11]. Let N be a non-negative integer specifying the polynomial degree of the approximation. The space $\mathcal{Q}_N$ of polynomials of degree N is defined by

$$\mathcal{Q}_N = \operatorname{span}\{x^i y^j : 0 \leq i,j \leq N\},$$

the space $\mathcal{P}_N$ of polynomials of total degree N is defined by

$$\mathcal{P}_N = \operatorname{span}\{x^i y^j : 0 \leq i+j \leq N\}$$

while the *trunk space* is defined by

$$\mathcal{Q}'_N = \mathcal{P}_N \oplus \operatorname{span}\{xy^N, x^N y\}. \quad (2.1)$$

The approximation properties of these spaces over a single element of size h, are given by

$$\inf_{v_N\in\mathcal{Q}_N}||u-v_N||_{1,K} \leq Ch^{\min(m-1,N)}N^{-(m-1)}||v||_{m,K} \quad (2.2)$$

provided that $v \in H^m(K)$. Similarly,

$$\inf_{q_N \in \mathcal{R}_{N-1}} ||p - q_N||_{0,K} \leq C h^{\min(m-1,N)} N^{-(m-1)} ||p||_{m-1,K} \tag{2.3}$$

where $p \in H^{m-1}(K)$, where $\mathcal{R}_N$ is any of the spaces $\mathcal{Q}_N$, $\mathcal{P}_N$ or $\mathcal{Q}'_N$. The same approximation results are valid for finite element spaces over a quasi-uniform partitioning of the domain Ω. Thus, viewed solely from the point of approximation properties, if the discretisation of the velocity is based on polynomials of degree N, it is natural to use a polynomial subspace of degree $N-1$ for the pressure discretisation.

2.1 Elements based on discontinuous pressures

The continuity properties of the subspace $M_N \subset M = L_0^2(\Omega)$ allows the pressure approximation to be sought in the form of a discontinuous, piecewise polynomial. Perhaps the most natural combination is the $\mathcal{Q}_1$-$\mathcal{P}_0$ element (whereby the velocity approximation is based on continuous piecewise bilinear functions in conjunction with discontinuous piecewise constant approximation of the pressures) since it offers balanced approximation orders in the mesh-size h. However, this is the classic example of an unstable combination with vanishing inf-sup constant corresponding to the so-called checker-board spurious pressure mode [10, 12]. However, increasing the order of the spaces to give the $\mathcal{Q}_2$-$\mathcal{P}_1$ element gives balanced approximation orders and inf-sup constant independent of the mesh-size.

In search of a higher order scheme, the obvious combination $\mathcal{Q}_N$-$\mathcal{Q}_{N-1}$ offers balanced approximation orders but is again unstable. The stability may be recovered by reducing the pressure space and using the combination $\mathcal{Q}_N$-$\mathcal{Q}_{N-2}$. This combination is stable, but the inf-sup constant degrades as $\beta_N = \mathcal{O}(N^{-1/2})$ as shown in [14]. In addition, the element gives unbalanced approximation orders for a fixed order N. However, for a spectral element or p-version finite element method [5, 6], whereby the mesh is fixed and convergence is achieved by increasing the polynomial order N, the approximation orders in N are balanced. This element is often used in a spectral method for incompressible flow, despite the fact that the inf-sup constant degrades. Another alternative is to reduce the pressure space to $\mathcal{P}_{N-1}$ rather than $\mathcal{Q}_{N-2}$. The element $\mathcal{Q}_N$-$\mathcal{P}_{N-1}$ represents the best possible state of affairs, offering balanced approximation orders in both the mesh-size h and the degree N of the elements, along with a uniformly bounded inf-sup constant as shown in [7]. These properties make the element attractive for hp-finite element approximation.

2.2 Elements based on continuous pressures

Despite the variational formulation allowing the use of discontinuous pressures, many practitioners prefer to use continuous pressures since the number of degrees of freedom is reduced and the elements are more stable than the corresponding combination based on a discontinuous pressure space.

For instance, the Taylor-Hood element is based on continuous piecewise biquadratic approximation of the velocity in conjunction with continuous bilinear approximation of the pressures. However, whereas the corresponding discontinuous $\mathcal{Q}_2$-$\mathcal{Q}_1$ element is unstable, the Taylor-Hood element is known to be stable [10].

The higher order Taylor-Hood element, $\mathcal{Q}_N$-$\mathcal{Q}_{N-1}$, uses continuous pressures and provides balanced approximation orders. The combination is known [9] to be stable with inf-sup constant independent of the mesh-size h. The dependence on the polynomial order N is an open question, but numerical computations (shown later) support the conjecture that β_N degrades at a rate of $\mathcal{O}(N^{-1/2})$.

Unfortunately, the analogue of the stable $\mathcal{Q}_N$-$\mathcal{P}_{N-1}$ element is unattractive if continuous pressures are used, since enforcing the continuity of the pressures results in the effective order of the pressures being reduced by one degree thereby losing the balanced approximation properties. A natural alternative is to enhance the pressure space to the trunk space $\mathcal{Q}'_{N-1}$ which allows the use of continuous pressures without loss of order. The combination $\mathcal{Q}_N$-$\mathcal{Q}'_{N-1}$ provides balanced approximation and was shown to be stable with inf-sup constant independent of both h and N in [4].

The actual values of the inf–sup constants on a single element, for the $\mathcal{Q}_N$-$\mathcal{P}_{N-1}$ and $\mathcal{Q}_N$-$\mathcal{Q}'_{N-1}$ methods are shown in Figure 2.1. The numerical results confirm the theoretical predictions that both methods are stable with inf-sup constant independent of N. As a matter of fact, for $N \geq 5$, the inf-sup constants for both methods are practically identical. The values of the inf-sup constant on a mesh consisting of four square elements for the higher order Taylor-Hood element and the $\mathcal{Q}_N$-$\mathcal{Q}'_{N-1}$ element with continuous pressures are shown in Figure 2.2. It is observed that the latter method remains uniformly stable while the inf-sup constant for the Taylor-Hood method degenerates as N is increased.

3 Mixed hp-FEM on high aspect ratio elements

The solutions of the full Navier-Stokes equations typically exhibit highly localised features such as boundary layers and singularities at re-entrant corners.

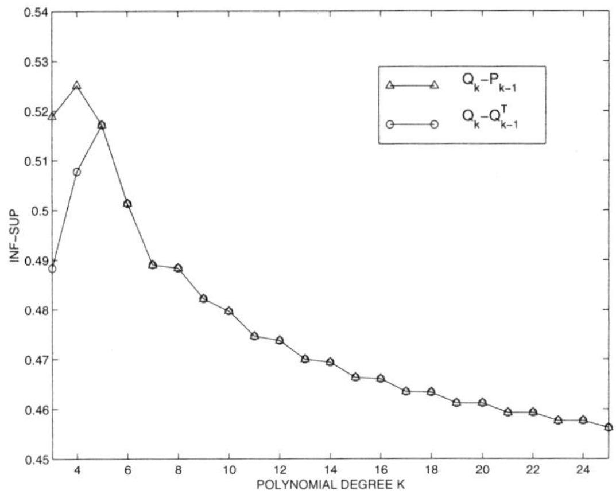

Figure 2.1: Inf-sup constants for the $\mathcal{Q}_N$-$\mathcal{P}_{N-1}$ and $\mathcal{Q}_N$-$\mathcal{Q}'_{N-1}$ methods on a single square element.

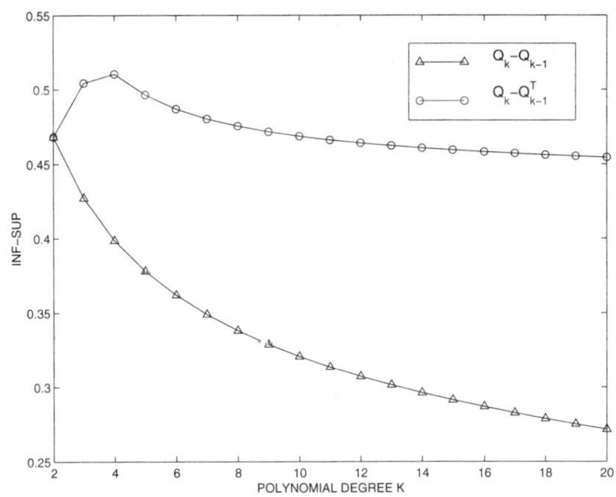

Figure 2.2: Inf-sup constants for the continuous $\mathcal{Q}_N$-$\mathcal{Q}_{N-1}$ and $\mathcal{Q}_N$-$\mathcal{Q}'_{N-1}$ methods on a mesh of four square elements.

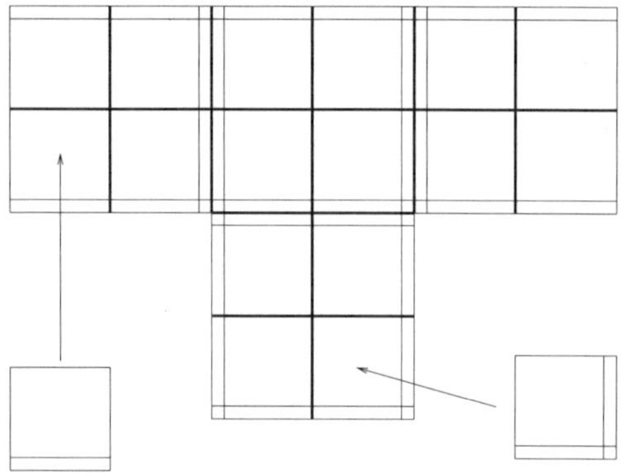

Figure 3.1: Example of a typical affine quadrilateral mesh showing high-aspect ratio elements and decomposition into (a) edge macro-elements and (b) corner macro-elements.

Such features are difficult to approximate accurately using the almost square elements discussed in the previous section. Instead, it is necessary to use elements whose thickness is of the order of the boundary layer, but whose length is of the order of the size of the computational domain. The ratio ρ of these lengths will generally be extremely small. An example of the type of mesh that we have in mind is shown in Figure 3.1. The use of such high aspect ratio elements is beneficial from the viewpoint of approximation properties but has a detrimental effect on the stability properties, exhibited by a degeneration of the inf-sup constant in the limit $\rho \to 0$. Consequently, the danger is that the improved approximation properties might easily be lost due to the poor stability constants.

Schötzau and Schwab [13] developed higher order elements based on the $\mathcal{Q}_N$-$\mathcal{Q}_{N-2}$ element with discontinuous pressures based on decomposing the mesh into edge and corner macro-elements as shown in Figure 3.1. The stability of the elements on the macro-elements may be used to deduce bounds on the stability for the entire mesh.

3.1 Stability on edge macro-elements

The dependence of the inf-sup constant on the aspect ratio ρ for an edge macro-element was analysed in [13] where it was shown that by augmenting the velocity space using higher order polynomial functions on the interior edge removes the dependence on ρ. In particular, it was shown that for the $\mathcal{Q}_N$-$\mathcal{Q}_{N-2}$ family with discontinuous pressures, the inf-sup constant behaves as $\mathcal{O}(N^{-1/2})$ but is independent of the aspect ratio ρ. Thus, the inclusion of additional functions on the interior edge restores the variation of the inf-sup constant to the same behaviour as observed on a regular mesh.

In [2], the question whether the $\mathcal{Q}_N$-$\mathcal{P}_{N-1}$ element of [7] might be used to construct a method on high aspect ratio meshes to obtain an inf-sup constant independent of ρ was considered. The anisotropic tensor product space $\mathcal{Q}_{r,s}$ is defined by

$$\mathcal{Q}_{r,s} = \text{span}\{x^i y^j : 0 \leq i \leq r, 0 \leq j \leq s\}. \tag{3.1}$$

The mixed finite element involved an augmented velocity space $\mathcal{Q}_{(1+\mu)(N+1),N}$, where μ is a fixed positive constant. The precise result [2] is as follows:

Theorem 3.1 *Let ω_e be an edge macro-element as in Figure 3.1(a). Let $\mu > 0$ be a fixed positive constant and define the mixed finite elements on the edge macro-element by*

$$\begin{aligned} \mathbf{V}_N(\omega_e) &= \mathcal{Q}_{(1+\mu)(N+1),N} \\ M_N(\omega_e) &= \mathcal{P}_{N-1}. \end{aligned} \tag{3.2}$$

Then, there exists a positive constant C such that

$$\inf_{q_N \in W_N(\omega_e)} \sup_{\mathbf{v}_N \in \mathbf{V}_N(\omega_e)} \frac{(\nabla \cdot \mathbf{v}_N, q_N)_{\omega_e}}{|\mathbf{v}_N|_{1,\omega_e} \|q_N\|_{0,\omega_e}} \geq C(\log N)^{-1/2} \tag{3.3}$$

where C is independent of N and the aspect ratio of the elements in ω_e.

A similar result holds even if μ is chosen to be zero. However, the dependence on the polynomial degree N would increase from $\log N$ to N. The behaviour of the inf-sup constant using the minimal space $\mathcal{Q}_N$-$\mathcal{P}_{k-1}$ is compared with the enhanced space in Figure 3.2. From a purely theoretical point of view, the inclusion of interior bubbles generated from the parameter μ seems to be necessary if the inf–sup constant is to be bounded independently of N if the modest $(\log N)^{-1/2}$ growth is ignored. However, for moderate values of N the parameter μ may be chosen to be zero for the purposes of practical computation.

3.2 Mixed finite elements on corner macro-elements

The chief source behind the degeneration of the inf-sup constant at high aspect ratios lies in the pressure modes on corner macro-elements as shown in Figure 3.3. It may be shown [2] that the degeneration of the inf-sup constant with

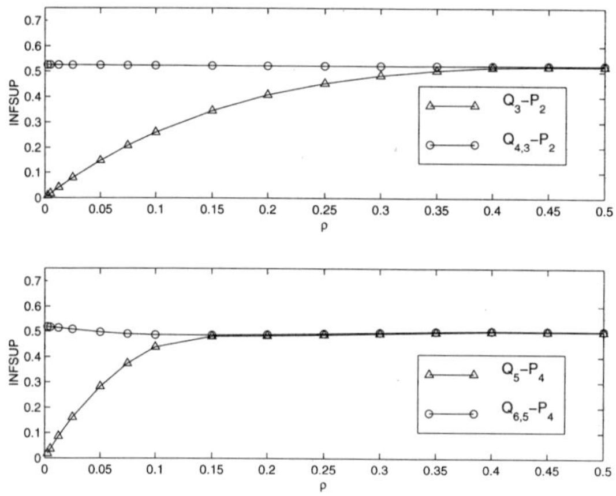

Figure 3.2: Variation of inf-sup constant with aspect ratio on an edge macro-element. The minimal space $\mathcal{Q}_N$-$\mathcal{P}_{k-1}$ degenerates at high aspect ratio while the enhanced space remains robust.

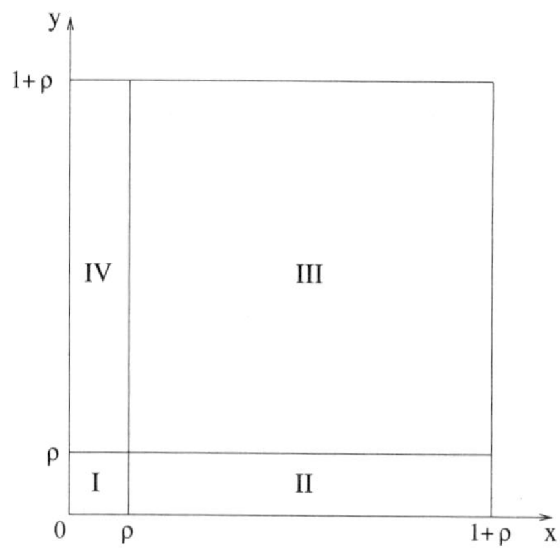

Figure 3.3: A corner macro-element based on four elements.

the aspect ratio ρ is basically due to a *single* pressure mode. This pressure mode $\bar{q}^\star$ is given by

$$\bar{q}^\star \propto \begin{cases} 1 & \text{on } I \\ -\rho^2 & \text{on } III \\ 0 & \text{otherwise} \end{cases} \tag{3.4}$$

and is shown graphically in Figure 3.4.

It may be shown [2] that if the pressure space is depleted by removing this pressure mode, then the resulting mixed finite element is stable with inf-sup constant independent of the aspect ratio. Unfortunately, while this would be a rather satisfactory result from the viewpoint of stability, the approximation properties of the pressure space would be severely degraded in the neighbourhood of re-entrant corners where it is precisely the pressure mode $\bar{q}^\star$ that is needed to adequately approximate the true pressure.

One alternative is to enhance the velocity space so that the spurious mode $\bar{q}^\star$ is stabilised. The question of how this may be achieved most efficiently was considered in [2], where it was shown that the addition of a *single*, additional function $\mathbf{v}^\star$ to the velocity space would restore the stability of the method–even with the troublesome pressure mode $\bar{q}^\star$ present. In principle then, it would seem that the solution is simply augment using the function $\mathbf{v}^\star$. As usual, there is no free lunch, and in this case the price to be paid is that the function $\mathbf{v}^\star$ is a polynomial of 'sufficiently' high degree. The polynomial is shown graphically in Figures 3.4 and 3.5. It transpires [2] that 'sufficiently' high means that degree of the polynomial needed to restore stability is of the order $\mathcal{O}(\rho^{-1/2})$. For highly anisotropic elements, the required degree is excessively large to be of use in practical computation.

However, one consequence of this observation is that as the polynomial order N of the underlying space is increased, the stability of the method with respect to aspect ratio should be restored as the degree N approaches the critical threshold $\mathcal{O}(\rho^{-1/2})$ beyond which the stabilising polynomial $\mathbf{v}^\star$ is automatically present in the space. Thus, we arrive at a perhaps rather surprising conclusion: the stability of the space *improves* as the degree N is *increased*. This is in stark contrast with the usual state of affairs where the stability generally degrades with increasing polynomial degree. The precise statement of these conclusions was proved in [2]:

Lemma 3.1 *For $0 < \rho \ll 1$, let ω_c be the macro-element shown in Figure 3.1(b) and let $\bar{q}^\star$ be the pressure mode defined above. Then, there exists*

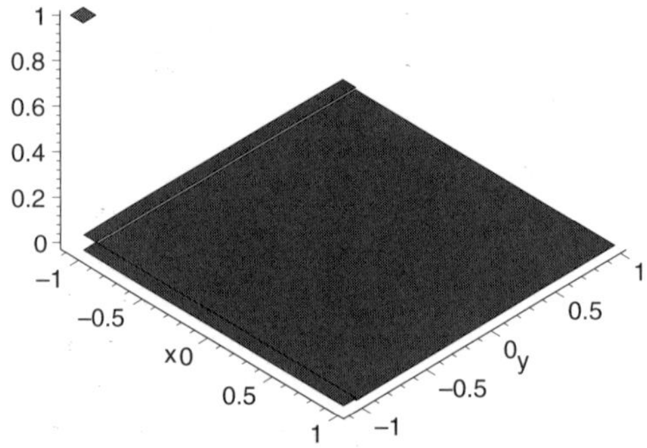

(a) The pressure mode $p^\star$.

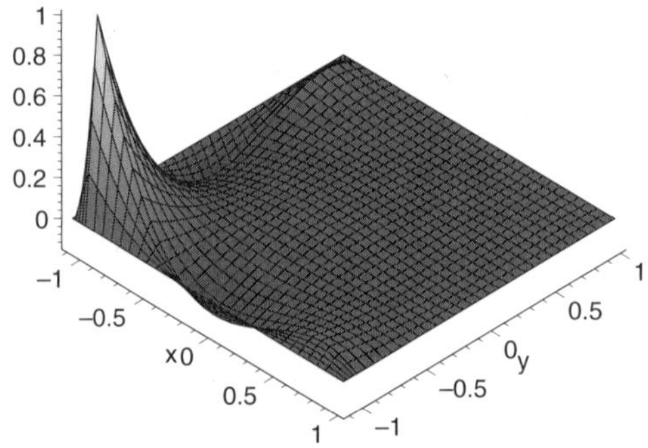

(b) The stabilising velocity function $\mathbf{v}^\star$.

Figure 3.4: Plot showing the pressure mode $p^\star$ responsible for the degeneration of the inf-sup constant on high aspect ratio corner macro-elements. One component of the velocity function $\mathbf{v}^\star$ used to stabilise the pressure mode is shown.

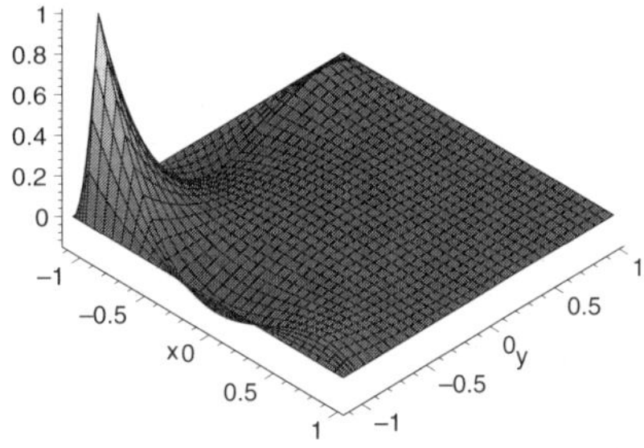

(a) Aspect ratio 10, degree 3.

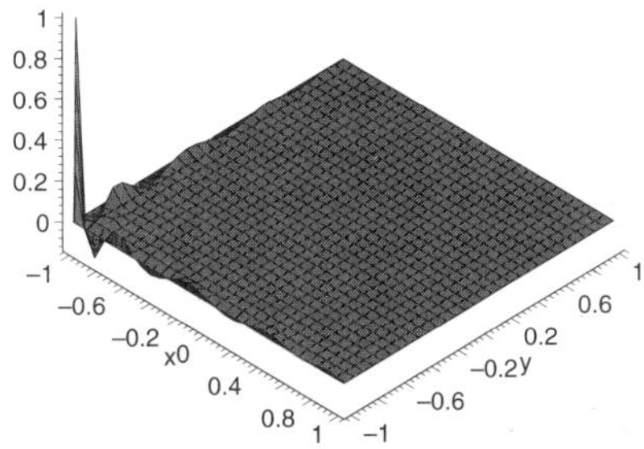

(b) Aspect ratio 100, degree 10.

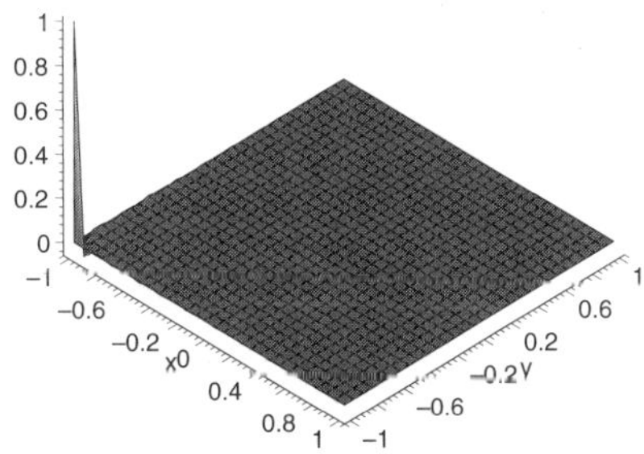

(c) Aspect ratio 100, degree 30.

Figure 3.5: Plots of one component of the function $\mathbf{v}^\star$ used to stabilise the spurious pressure mode. The function is shown on corner macro-elements of varying aspect ratio and varying polynomial degrees.

a polynomial $\mathbf{v}^\star$ *of degree at most* N *such that*

$$\frac{b(\mathbf{v}^\star, \bar{q}^\star)_{\omega_c}}{|\mathbf{v}^\star|_{1,\omega_c}\|\bar{q}^\star\|_{0,\omega_c}} \geq \beta^\star \min(1, N\sqrt{\rho}) \tag{3.5}$$

where $\beta^\star$ *is independent of* ρ *and* N.

This result exhibits the predicted switch in behaviour of the inf-sup constant as the polynomial degree is increased.

A close examination of the proof of this result would reveal that the function $\mathbf{v}^\star$ provides (up to a constant) the best value for the quotient. As a consequence, it follows that the inf-sup behaves as $\beta = \mathcal{O}(\rho^{-1/2})$ for fixed polynomial order N. Therefore, for a fixed order of polynomial, or for polynomials of relatively low degree compared to $\mathcal{O}(\rho^{-1/2})$, the degeneration of the inf-sup constant with the aspect ratio ρ is inevitable.

3.3 Geometrically graded boundary layer meshes

The presence of boundary layers and other localised phenomena can only be efficiently resolved by finite element spaces based on locally refined meshes containing high aspect ratio elements. The discussion shows that unless an extremely high degree polynomial space is used, the inf-sup constant will behave like $\mathcal{O}(\rho^{-1/2})$. This casts doubt over the advisability of using meshes of the type shown in Figure 3.1 where elements of aspect ratio ρ are generated by a single refinement. An alternative possibility is to generate high aspect ratio elements through a *sequence* of refinements using a geometric grading factor σ as shown in Figure 3.6. The number R of geometric refinements needed to produce elements of aspect ratio ρ is given by

$$R = \frac{\log \rho}{\log \sigma}. \tag{3.6}$$

The corner macro-element shown in Figure 3.6 may be regarded as being composed of R four-element corner macro-elements as shown in Figure 3.7. The analysis of a single four-element corner macro-element identified the spurious pressure mode responsible for the degradation of the inf-sup constant to be the function shown in Figure 3.4. The stability on the geometrically generated corner macro-element is dictated by R pressure modes similar to the one shown in Figure 3.4. The plethora of weakly spurious pressure modes may appear to only make a bad situation worse. However, the previous analysis shows that inf-sup constant behaves like $\min(1, N\sqrt{\sigma})$ on each of the four-element macro-elements. The essential point is now apparent: while the overall aspect ratio ρ may be

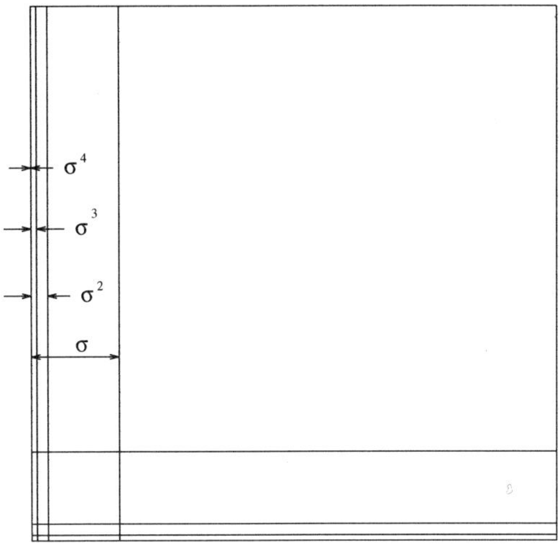

Figure 3.6: A corner macro-element with high aspect ratio elements generated through geometric refinements with grading factor σ.

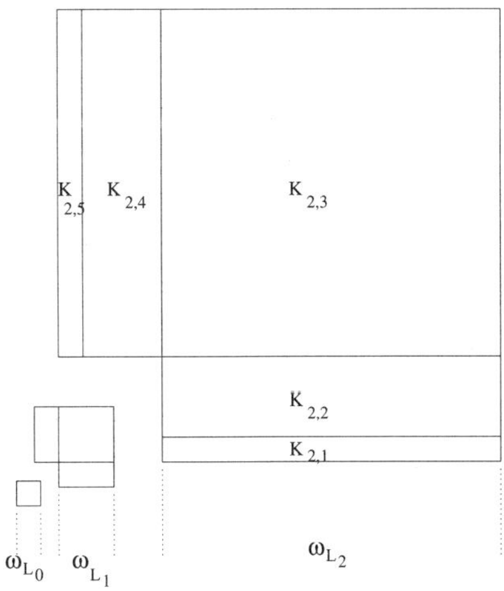

Figure 3.7: Decomposition of a geometrically refined corner macro-element into four-element corner macro-elements.

extremely small, the geometric grading factor σ may be rather more modest, to the extent that the polynomial degree N crosses the threshold $\mathcal{O}(\sigma^{-1/2})$ beyond which stability is restored. Of course, this informal discussion hardly constitutes a proof due to the possibility of additional weakly spurious modes being generated through superposition. Nonetheless, the conclusion is essentially correct as shown by the following result taken from [3]:

Theorem 3.2 *Let μ be a non-negative constant, and define the mixed finite element space based on the combination $\mathcal{Q}_{(1+\mu)(N+1),N}$-$\mathcal{P}_{N-1}$. Then, the inf-sup constant on a corner macro-element generated by R refinements using a geometric grading factor σ satisfies*

$$\beta^\star(N) = \frac{C}{\sqrt{R}}\min(1, N\sqrt{\sigma})\begin{cases} N^{-1/2} & \text{if } \mu = 0 \\ (\log N)^{-1/2} & \text{if } \mu > 0 \end{cases}$$

where $C > 0$ independent of σ, N and R.

The appearance of the factor $\sqrt{R}$ is due to superposition. However, to generate elements of aspect ratio ρ requires $R \propto |\log \rho|$ geometric refinements. The overall dependence of the inf-sup constant on the aspect ratio ρ is no longer $\mathcal{O}(\rho^{-1/2})$ but has been improved to $\mathcal{O}(|\log \rho|^{1/2})$.

The actual numerical values of the inf-sup constants for various configurations is shown in Figures 3.8 and 3.9. It is observed that the inf-sup constant is insensitive to the number of refinements R. In particular, it will be observed that an aspect ratio of 3×10^7 may be achieved in a stable fashion based on five geometric refinements with grading factor 1/20 in conjunction with polynomials of order 5.

3.4 Conclusion

In conclusion, the use of high aspect ratio elements is desirable for the accurate *approximation* of boundary layers. However, the *stability* of mixed finite elements is highly sensitive to the aspect ratio to the extent that the advantages of the enhanced approximation properties may be severely degraded (even if the approximation error is reduced exponentially fast) or even dissipated altogether (in the case of an algebraic rate of convergence). The degeneration of the stability constant is due to a single pressure mode that may be stabilised by using polynomials of sufficiently high degree. Unfortunately, the required degree is prohibitively large for use in practical computations. A practical alternative is to generate high aspect ratio elements through a series of geometric refinements based on a modest grading factor for which increasing the polynomial degree

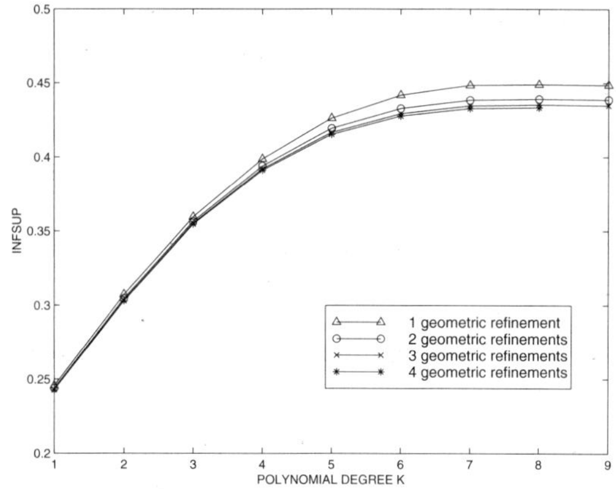

Figure 3.8: Inf–sup constants for the $\mathcal{Q}_{N+1}$-$\mathcal{P}_{N-1}$ method on geometrically refined corner macro-elements, for $R = 1, 2, 3$ and 4. The value of σ was chosen to be 1/50. For all values of R shown, the inf–sup constant increases, at least until the velocity space contains polynomials of degree 7, as predicted by the theory for $\rho = 1/50$.

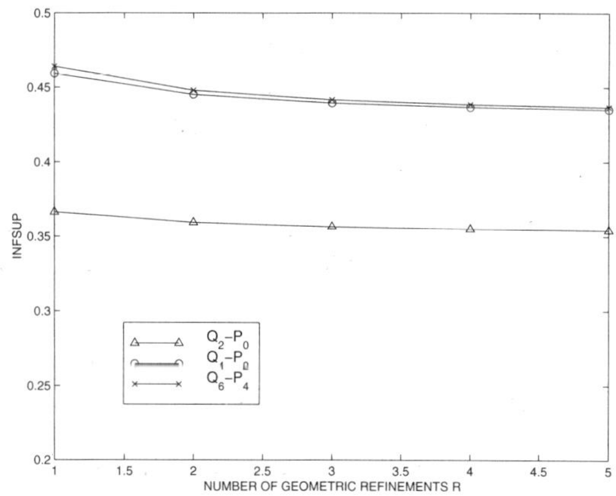

Figure 3.9: Inf–sup constants for the $\mathcal{Q}_{N+1}$-$\mathcal{P}_{N-1}$ method ($N = 1, 3, 5$), for increasing R, the number of geometric refinements. Here σ is chosen to be 1/20, which for $R = 5$, gives an aspect ratio of the order 3×10^{-7}. There is no appreciable observed degeneration of the inf–sup constant with respect to R.

beyond the threshold at which the effect of the grading factor is stabilised. With such a strategy, the degeneration of the stability constant is ameliorated from $\mathcal{O}(\rho^{-1/2})$ to $\mathcal{O}(|\log \rho|^{1/2})$.

References

[1] R.A. Adams. *Sobolev Spaces*, volume 65 of *Pure and Applied Mathematics Series*. Academic Press, New York, 1978.

[2] M. Ainsworth and P.W. Coggins. The stability of mixed hp-finite element methods for Stokes flow using high aspect ratio elements. In Preparation.

[3] M. Ainsworth and P.W. Coggins. The stability of mixed hp-finite element methods for Stokes flow using geometrically graded high aspect ratio elements. In Preparation.

[4] M. Ainsworth and P.W. Coggins. A uniformly stable family of mixed hp-finite elements with continuous pressures for incompressile flow. In Preparation.

[5] I. Babuška and M. Suri. The p and h-p versions of the finite element method. Basic principles and properties. *SIAM Rev.*, 36:578–632, 1994.

[6] C. Bernardi and Y. Maday. Spectral methods. In P.G. Ciarlet and J.L. Lions, editors, *Handbook of Numerical Analysis: Techniques of Scientific Computing*, volume V, pages 209–482. Elsevier Science Publishers B.V. (North-Holland), Amsterdam, 1997.

[7] C. Bernardi and Y. Maday. Uniform inf-sup conditions for the spectral discretisation of the Stokes problem. *Math. Methods Appl. Sci.*, 9(3):395–414, 1999.

[8] S.C. Brenner and L.R. Scott. *The Mathematical Theory of Finite Element Methods*. Springer-Verlag, New York, 1994.

[9] F. Brezzi and R.S. Falk. Stability of higher order Taylor-Hood methods. *SIAM J. Numer. Anal.*, 28(3):581–590, 1991.

[10] F. Brezzi and M. Fortin. *Mixed and Hybrid Finite Element Methods.* Springer-Verlag, Berlin, 1991.

[11] P.G. Ciarlet. *The Finite Element Method for Elliptic Problems.* North–Holland, Amsterdam, 1978.

[12] V. Girault and P.A. Raviart. *Finite Element Methods for Navier Stokes Equations*, volume 5 of *Springer Series in Computational Mathematics*. Springer-Verlag, Berlin, 1986.

[13] C. Schwab and D. Schötzau. Mixed hp-finite element methods on anisotropic meshes. *Math. Methods Appl. Sci.*, 8(5):787–820, 1998.

[14] R. Stenberg and M. Suri. Mixed *hp* finite element methods for problems in elasticity and Stokes flow. *Numer. Math.*, 72:367–389, 1996.

Acknowledgement

The support of P. Coggins through an EPSRC research studentship is gratefully acknowledged.

Mark Ainsworth
Department of Mathematics
Strathclyde University
26 Richmond Street
Glasgow G1 1XH Scotland
`M.Ainsworth@strath.ac.uk`

Patrick Coggins and Bill Senior
Department of Mathematics
University of Leicester
Leicester LE1 7RH, England
`pwc2@mcs.le.ac.uk`, `wcs1@mcs.le.ac.uk`

R. BECKER, H. KAPP AND R. RANNACHER

Adaptive finite element methods for optimization problems

Abstract We present a new approach to error control and mesh adaptivity in the numerical solution of optimal control problems governed by elliptic differential equations. The indefinite boundary value problems obtained by the Lagrangian formalism are discretized by the Galerkin finite element method. The mesh adaptation is driven by residual-based a posteriori error estimates which are derived by global duality arguments. This general approach facilitates control of the error with respect to any quantity of physical interest. In discretizing an optimization problem it seems natural to control the error with respect to the given cost functional. In this way, the computed solution can directly be used in weighting the cell residuals in the a posteriori error estimate. This approach has the features of model reduction as used in optimal control of complex dynamical systems. For illustration, we present some results of test computations for simple model problems in optimal control of super-conductivity.

1 Introduction

In this note we describe a new adaptive finite element method for optimal control problems governed by elliptic partial differential equations. The control problem is treated by the classical Lagrange formalism yielding the Euler-Lagrange equations as first-order optimality conditions. This indefinite system of partial differential equations for the state variable u, the Lagrange multiplier λ and the control variable q is discretized by a standard finite element method. In this approach the set of admissible solutions is also discretized and we generally obtain inadmissible states. Since discretization in partial differential equations is expensive, at least for praxis-relevant models, the question of how this "model reduction" affects the quality of the optimization result is crucial for an economical computation. The need for adaptive error control is therefore evident.

For a posteriori error estimation, we apply the general optimal-control approach developed in [3] and [4] for Galerkin finite element discretizations of partial differential equations. In this method a posteriori estimates for the error with respect to arbitrary functional output are obtained via duality arguments. In these estimates local cell residuals of the computed solution are weighted by certain derivatives of the dual solution which describe the dependence of the error on variations of the local mesh size. In general these weights have to be approximated by numerically solving the corresponding dual problem. This results in a feed-back process by which successively more and more accurate error bounds and economical meshes are generated.

In applying this approach to optimization problems it seems natural to base the error control on the given cost functional. In this particular case the corresponding dual solution is given in terms of the state variable, Lagrangian multiplier and control variable. Hence, the evaluation of the a posteriori error estimates does not require extra work for solving a dual problem and a posteriori error estimation is almost for free. This observation leads to a particularly simple and efficient strategy for mesh adaptation in the finite element discretization of optimal control problems. We provide the theoretical background for this approach and discuss some computational tests.

2 An optimization model problem

We consider optimal control problems of the form

$$J(u,q) \;\to\; \min!, \qquad A(u,q) = f, \tag{2.1}$$

where A is a partial differential operator relating a state variable u and a control variable q, and $J(\cdot,\cdot)$ is a cost functional. As a prototypical model case, we consider a problem of boundary control in super–conductivity ("Ginzburg-Landau model"; see Du, Gunzburger and Peterson [6], and Ito and Kunisch [9]). The state equations are

$$\begin{aligned} -\Delta u + s(u) &= f \quad \text{on } \Omega, \\ \partial_n u = q \ \text{ on } \Gamma_C, \quad \partial_n u &= 0 \ \text{ on } \partial\Omega \setminus \Gamma_C, \end{aligned} \tag{2.2}$$

defined on an open bounded domain with boundary $\partial\Omega$. The control q acts on the boundary component Γ_C while the observations $u_{|\Gamma_O}$ are taken on a component Γ_O; see Figure 2.1. The cost functional is defined by

$$J(u,q) = \tfrac{1}{2}||u - u_O||^2_{\Gamma_O} + \tfrac{1}{2}\alpha||q||^2_{\Gamma_C} \tag{2.3}$$

with a prescribed boundary function u_O and a regularization parameter $\alpha \geq 0$. The zero-order term may be linear $s(u) = u$ or nonlinear of the form $s(u) = u^3 - u$.

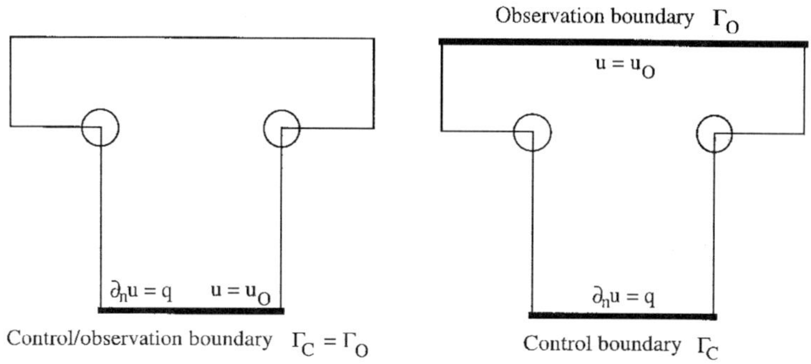

Figure 2.1: Configuration of the boundary control model problem on a T-domain (Ginzburg-Landau model): Configuration 1 (left), Configuration 2 (right).

We employ the Lagrangian formalism with the Lagrange function

$$L(u,\lambda,q) = J(u,q) - (-\Delta u + s(u) - f, \lambda)_\Omega + (\partial_n u - q, \lambda)_{\Gamma_C}, \tag{2.4}$$

involving the adjoint variable λ. The corresponding Euler-Lagrange equations, expressed in weak form, read as follows:

$$(u,\psi)_{\Gamma_O} + (\nabla\psi, \nabla\lambda)_\Omega + (s'(u)\psi, \lambda)_\Omega = (u_O, \psi)_{\Gamma_O} \quad \forall \psi \in H, \tag{2.5}$$
$$(\nabla u, \nabla\pi)_\Omega + (s(u),\pi)_\Omega - (q,\pi)_{\Gamma_C} = (f,\pi)_\Omega \quad \forall \pi \in H, \tag{2.6}$$
$$-(\chi,\lambda)_{\Gamma_C} + \alpha(q,\chi)_{\Gamma_C} = 0 \quad \forall \chi \in Q. \tag{2.7}$$

Here, the natural function space for the state variable u and the adjoint variable λ is the Sobolev space $H := H^1(\Omega)$, while the control q is determined in the Lebesgue space $Q := L^2(\Gamma_C)$. The stationary points of this saddle-point problem yield possible extreme points of the optimization problem. The system (2.5) - (2.7) can be written in matrix form as follows:

$$\begin{bmatrix} C & A^T + N'(u) & 0 \\ A + \widetilde{N}'(u) & 0 & -B \\ 0 & -B^T & \alpha S \end{bmatrix} \begin{bmatrix} u \\ \lambda \\ q \end{bmatrix} = \begin{bmatrix} u_O \\ f \\ 0 \end{bmatrix}, \tag{2.8}$$

where $\widetilde{N}'(u) := N(0) + \int_0^1 N'(tu)\,dt$, and the operators C, A, B, S, $N(\cdot)$, and $N'(\cdot)$ have their obvious meaning. For solving this saddle–point problem, we use a Newton iteration which is defined on the continuous level in the function space $H \times H \times Q$. The Newton steps are discretized by an adaptive Galerkin finite element method using the conforming bilinear Q_1 element for all unknowns u, λ and q. The mesh adaptation is driven by residual-based a posteriori error estimates for the nonlinear problem. The linearized algebraic problems are solved by a GMRES method with multigrid preconditioning; for algorithmic details, we refer to [1], [2] and [10].

3 Galerkin finite element discretization

The Galerkin finite element discretization of the saddle–point problem (2.5) - (2.7) uses subspaces $H_h \subset H$ and $Q_h \subset Q$ of piecewise polynomial functions defined on regular decompositions $\mathbb{T}_h = \cup_{T\in\mathbb{T}_h}\{T\}$ of the domain Ω into cells T (triangles or quadrilaterals); see Brenner and Scott [5]. Here, we think of bilinear shape functions. In order to avoid unnecessary complications due to curved boundaries, we suppose the domain Ω to be polygonal. We use the notation $h_T := diam(T)$ and $h_\Gamma := diam(\Gamma)$ for the width of a cell $T \in \mathbb{T}_h$ or a cell edge $\Gamma \subset \partial\Omega$. In order to ease local mesh refinement and coarsening, hanging nodes are allowed but at most one per edge; see Figure 3.1. The

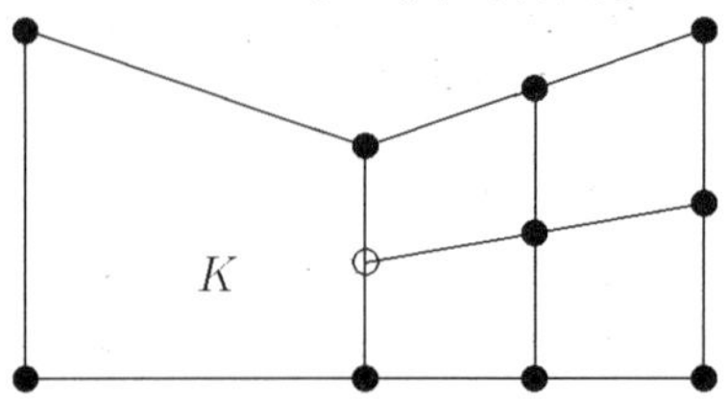

Figure 3.1: Quadrilateral mesh with "hanging nodes".

corresponding degrees of freedom are eliminated by interpolation in order to keep the discretization conforming. For simplicity, we assume that the space Q_h of discrete controls is given by the traces of the finite element functions of V_h. This is not necessary for our results but simplifies notation.

The discrete problems seek $\{u_h, \lambda_h, q_h\} \in H_h \times H_h \times Q_h$, such that

$$(u_h, \psi_h)_{\Gamma_O} + (\nabla\psi_h, \nabla\lambda_h)_\Omega + (s'(u_h)\psi_h, \lambda_h)_\Omega = (u_O, \psi_h)_{\Gamma_O} \quad \forall\psi_h \in H_h, \tag{3.1}$$

$$(\nabla u_h, \nabla\pi_h)_\Omega + (s(u_h), \pi_h)_\Omega - (q_h, \pi_h)_{\Gamma_C} = (f, \pi_h)_\Omega \quad \forall\pi_h \in H_h, \tag{3.2}$$

$$-(\chi_h, \lambda_h)_{\Gamma_C} + \alpha(q_h, \chi_h)_{\Gamma_C} = 0 \quad \forall\chi_h \in Q_h. \tag{3.3}$$

For this approximation, we have the following theoretical a priori estimate for the errors $e_u := u - u_h$, $e_\lambda := \lambda - \lambda_h$, and $e_q := q - q_h$ (see Gunzburger and Hou [8], and also [2]):

$$||\nabla e_u||_\Omega + ||e_q||_{\Gamma_C} + ||\nabla e_\lambda||_\Omega \\ \leq c\big\{||\nabla(u-\psi_h)||_\Omega + ||\nabla(\lambda-\pi_h)||_\Omega + ||q-\chi_h||_{\Gamma_C}\big\},$$

for arbitrary approximations $\psi_h,\ \pi_h \in H_h$ and $\chi_h \in Q_h$.

4 A posteriori error estimation

4.1 A model problem

We give a brief introduction to a posteriori error estimation using duality techniques. For illustration, we consider the simple model problem

$$-\Delta u = f \quad \text{in } \Omega, \qquad u = 0 \quad \text{on } \partial\Omega. \tag{4.1}$$

The variational formulation of this problem seeks $u \in H = H_0^1(\Omega)$ such that

$$(\nabla u, \nabla\phi)_\Omega = (f,\phi)_\Omega \quad \forall \phi \in H. \tag{4.2}$$

The discrete approximations $u_h \in H_h$ are determined by the Galerkin equations

$$(\nabla u_h, \nabla\phi_h)_\Omega = (f,\phi_h)_\Omega \quad \forall \phi_h \in H_h. \tag{4.3}$$

The essential feature of this approximation scheme is the "Galerkin orthogonality" of the error $e := u - u_h$,

$$(\nabla e, \nabla\phi_h)_\Omega = 0, \quad \phi_h \in H_h. \tag{4.4}$$

Next, we seek to derive a posteriori error estimates. Let $J(\cdot)$ be a *linear* "error functional" defined on H and $z \in H$ the solution of the corresponding dual problem

$$(\nabla\phi, \nabla z)_\Omega = J(\phi) \quad \forall \phi \in H. \tag{4.5}$$

Taking $\phi = e$ in (4.5) and using the Galerkin orthogonality, cell-wise integration by parts (observing that $z_{|\partial\Omega} = 0$) results in the error representation

$$\begin{aligned} J(e) &= (\nabla e, \nabla z)_\Omega = (\nabla e, \nabla(z-\phi_h))_\Omega \\ &= \sum_{T\in\mathbb{T}_h} \big\{(-\Delta u + \Delta u_h, z-\phi_h)_T - (\partial_n u_h, z-\phi_h)_{\partial T}\big\} \\ &= \sum_{T\in\mathbb{T}_h} \big\{(f + \Delta u_h, z-\phi_h)_T - \tfrac{1}{2}(n\cdot[\nabla u_h], z-\phi_h)_{\partial T\setminus\partial\Omega}\big\}, \end{aligned} \tag{4.6}$$

where $[\nabla u_h]$ denotes the jump of ∇u_h across the interelement boundary, and $\phi_h \in H_h$ is an arbitrary approximation. In this relation the factors $z - \phi_h$ may be viewed as weights expressing the sensitivity of the error quantity $J(e)$ with respect to changes of the equation residuals $(f + \Delta u_h)_{|T}$ and jump residuals $n\cdot[\nabla u_h]_{|\partial T}$. From the error identity (4.6), we infer the following a posteriori error estimate

$$|J(e)| \le \eta(u_h) := \sum_{T\in\mathbb{T}_h} h_T^2 \,\{\rho_T^{(u)}\omega_T^{(z)} + \rho_{\partial T}^{(u)}\omega_{\partial T}^{(z)}\}, \tag{4.7}$$

with the cell residuals and weights

$$\rho_T^{(u)} := h_T^{-1}\|f + \Delta u_h\|_T, \rho_{\partial T}^{(u)} := \tfrac{1}{2}h_T^{-3/2}\|n\cdot[\nabla u_h]\|_{\partial T\setminus\partial\Omega},$$
$$\omega_T^{(z)} := h_T^{-1}\|z - \phi_h\|_T, \omega_{\partial T}^{(z)} := h_T^{-1/2}\|z - \phi_h\|_{\partial T\setminus\partial\Omega}.$$

In view of the local approximation properties of finite elements, there holds

$$\omega_T^{(z)} + \omega_{\partial T}^{(z)} \le c_I h_T^2 \max_T |\nabla^2 z|\,. \tag{4.8}$$

We remark that in a finite difference discretization of the model problem (4.1) the corresponding "influence factors" behave like $\omega_T^{(z)} \approx \max_T |z|$.

In practice the weights $\omega_T^{(z)}$, $\omega_{\partial T}^{(z)}$ have to be determined computationally. Let $z_h \in H_h$ be the Galerkin approximation of z defined by

$$(\nabla\phi_h, \nabla z_h)_\Omega = J(\phi_h) \quad \forall \phi_h \in V_h. \tag{4.9}$$

In view of the estimate (4.8), we can approximate

$$\omega_T^{(z)} + \omega_{\partial T}^{(z)} \approx c_I h_T^2 \max_T |\nabla_h^2 z_h|, \tag{4.10}$$

where $\nabla_h z_h$ is a suitable difference quotient approximating $\nabla^2 z$. The interpolation constant is usually in the range $c_I \approx 0.1 - 1$ and can be determined by calibration. Alternatively, we may construct from $z_h \in H_h$ a patchwise biquadratic interpolation $I_h^2 z_h$ and replace $z - \phi_h$ in the weights by $I_h^2 z_h - z_h$. This gives an approximation which is free of any interpolation constant. The quality of these approximations for the model problem has been analysed in [4].

Next, we apply the foregoing argument for deriving an energy-norm error estimate. This is intended to prepare for our analysis of the error in approximating optimal control problems. The solution $u \in H$ of (4.1) is characterized by the property

$$L(u) = \tfrac{1}{2}\|\nabla u\|_\Omega^2 - (f, u)_\Omega \;\rightarrow\; \min! \quad \text{on } H. \tag{4.11}$$

Further, we note that

$$L(u) - L(u_h) = \tfrac{1}{2}\|\nabla u\|_\Omega^2 - (f,u)_\Omega - \tfrac{1}{2}\|\nabla u_h\|_\Omega^2 + (f,u_h)_\Omega$$
$$= -\tfrac{1}{2}\|\nabla u\|_\Omega^2 - \tfrac{1}{2}\|\nabla u_h\|_\Omega^2 + (\nabla u, \nabla u_h)_\Omega = -\tfrac{1}{2}\|\nabla e\|_\Omega^2.$$

Hence, energy-error control means control of the error with respect to the "energy functional" $L(\cdot)$. This can be accomplished by using the linear error functional

$$J(\phi) = -\tfrac{1}{2}(\nabla e, \nabla\phi)_\Omega$$

in the corresponding dual problem (4.5). Obviously, the dual solution is then simply given by $z = -\frac{1}{2}e$. Accordingly, the general a posteriori error estimate (4.7) takes the particular form

$$\|\nabla e\|_\Omega^2 \le \sum_{T\in\mathbb{T}_h} h_T^2 \{\rho_T^{(u)}\omega_T^{(u)} + \rho_{\partial T}^{(u)}\omega_{\partial T}^{(u)}\}, \tag{4.12}$$

with the weights $\omega_T^{(u)} = h_T^{-1}\|u-\phi_h\|_T$ and $\omega_{\partial T}^{(u)} = h_T^{-1/2}\|u-\phi_h\|_{\partial T\setminus\partial\Omega}$. Then, using the local approximation estimate

$$\inf_{\phi_h\in H_h}\Big(\sum_{T\in\mathbb{T}_h}\{h_T^{-2}\|u-\phi_h\|_T^2 + h_T^{-1}\|u-\phi_h\|_{\partial T}^2\}\Big)^{1/2} \le c_I\|\nabla e\|_\Omega, \tag{4.13}$$

we conclude from (4.12) that

$$\|\nabla e\|_\Omega^2 \le c_I\Big(\sum_{T\in\mathbb{T}_h} h_T^4\{\rho_T^{(u)2} + \rho_{\partial T}^{(u)2}\}\Big)^{1/2}\|\nabla e\|_\Omega.$$

This implies the standard residual based energy norm a posteriori error estimate (e.g., Verfürth [12]):

$$|L(u) - L(u_h)| = \tfrac{1}{2}\|\nabla e\|_\Omega^2 \le \tfrac{1}{2}c_I^2\sum_{T\in\mathbb{T}_h} h_T^4\{\rho_T^{(u)2} + \rho_{\partial T}^{(u)2}\}. \tag{4.14}$$

Below, we will see that the peculiar relation $z = -\frac{1}{2}e$ for the dual solution corresponding to the "energy functional" $L(\cdot)$ follows from a general principle which can be used also for the discretization of the optimal control problem described above.

4.2 A general paradigm for a posteriori error estimation

Next, we consider an abstract situation. Let $L(u)$ be a twice differentiable functional on some Hilbert space V, e.g., the energy functional related to the Poisson problem or the Lagrangian functional defined for the Ginzburg-Landau model. For its first and second differentials at u, we use the notation $L'(u;\cdot)$ and $L''(u;\cdot,\cdot)$, respectively. Notice that $L''(u;\cdot,\cdot)$ is symmetric. We seek stationary points $u \in V$ of $L(\cdot)$,

$$L'(u;\phi) = 0 \quad \forall \phi \in V. \tag{4.15}$$

Corresponding approximations $u_h \in V_h$ are defined in finite dimensional subspaces $V_h \subset V$ by the Galerkin equations

$$L'(u_h;\phi_h) = 0 \quad \forall \phi_h \in V_h. \tag{4.16}$$

Let $J(\cdot)$ be a functional chosen for measuring the error $e = u - u_h$. Then,

$$J(u) - J(u_h) = \int_0^1 J'(u_h + te; e)\, dt, \tag{4.17}$$

$$L'(u;\cdot) - L'(u_h;\cdot) = \int_0^1 L''(u_h + te; e, \cdot)\, dt, \tag{4.18}$$

leads us to consider the "dual problem"

$$\int_0^1 L''(u_h + te; \phi, z)\, dt = \int_0^1 J'(u_h + te; \phi)\, dt \quad \forall \phi \in V, \tag{4.19}$$

which is assumed to have a solution $z \in V$. Then, taking $\phi = e$ in (4.19) and using the Galerkin equation (4.16) results in the error identity

$$J(u) - J(u_h) = L'(u;z) - L'(u_h;z) = -L'(u_h; z - \phi_h), \tag{4.20}$$

with arbitrary $\phi_h \in V_h$. In general, this error representation cannot be evaluated since the left-hand side as well as the right-hand side in the dual problem (4.19) depend on the unknown solution u. The simplest way of approximation is to replace u by u_h, which yields the perturbed dual problem

$$L''(u_h;\phi,\tilde{z}) = J'(u_h;\phi) \quad \forall \phi \in V. \tag{4.21}$$

Controlling the effect of this perturbation on the accuracy of the resulting error estimate may be a delicate task and depends strongly on the particular problem under consideration. Our own experiences with different types of applications (e.g., the Navier-Stokes equations and models in elasto-plasticity) indicate that this problem is not critical as long as the solution to be computed is stable. The crucial problem is the approximation of the perturbed dual solution by solving a discretized dual problem

$$L''(u_h;\phi_h,\tilde{z}_h) = J'(u_h;\phi_h) \quad \forall \phi_h \in V_h. \tag{4.22}$$

So far, the derivation of the error representation (4.20) did not use that the variational equation (4.15) stems from an "energy functional". In fact it can be used for much more general situations; see the surveys given in Eriksson et al. [7] and in [11]. It seems natural to control the error $e = u - u_h$ with respect to the given "energy" functional $L(\cdot)$. Observing that $L'(u;\phi) = 0$, it follows by integration by parts that

$$\int_0^1 L'(u_h + te;\phi)\,dt = -\int_0^1 L''(u_h + te; e,\phi)t\,dt = -\int_0^1 L''(u_h + te;\phi, e)t\,dt.$$

Hence, in this case the dual problem (4.19) takes the special form

$$\int_0^1 L''(u_h + te;\phi, z)\,dt = -\int_0^1 L''(u_h + te;\phi, e)t\,dt \quad \forall \phi \in V. \tag{4.23}$$

If the functional $L(\cdot)$ is quadratic or in the general case by linearization $u \to u_h$, we obtain the perturbed dual problem

$$L''(u_h;\phi,\tilde{z}) = -\tfrac{1}{2}L''(u_h;\phi,e) \quad \forall \phi \in V, \tag{4.24}$$

with the solution $\tilde{z} = -\frac{1}{2}e$. The resulting a posteriori error estimate has the form

$$|L(u) - L(u_h)| \approx \inf_{\phi_h \in V_h} |L'(u_h;\tilde{z} - \phi_h)| = \inf_{\phi_h \in V_h} \tfrac{1}{2}|L'(u_h;\tilde{u} - \phi_h)|. \tag{4.25}$$

In the ideal case of a quadratic functional $L(\cdot)$ linearization is not required and this error bound becomes exact. Here, again the quantity

$$\tilde{z} - \phi_h = -\tfrac{1}{2}e - \phi_h = \tfrac{1}{2}(u - \psi_h)$$

has to be approximated as described above by using the computed solution $u_h \in H_h$.

We emphasize that in this particular case the evaluation of the a posteriori error estimate with respect to the "energy functional" does not require the explicit solution of the dual problem. This abstract reasoning can be taken as guide-line for systematically deriving a posteriori error estimates in concrete situations. In the following, we will carry this out for the optimal control problem described in Section 2.

4.3 A posteriori error analysis for the optimization problem

Now, we want to apply the formalism of the previous section to the particular situation of the boundary control problem described in Section 2. The functional of interest is the Lagrangian functional of the optimal control problem,

$$L(v) = J(u,q) + (\nabla u, \nabla\lambda)_\Omega + (s(u) - f, \lambda)_\Omega - (q,\lambda)_{\Gamma_C},$$

defined for triples $v = \{u, \lambda, q\}$ in the Hilbert space $V := H \times H \times Q$. We recall the equations for stationary points $v = \{u, \lambda, q\} \in V$ of $L(\cdot)$ in slightly rearranged form:

$$\begin{aligned} (\psi, u - u_O)_{\Gamma_O} + (\nabla\psi, \nabla\lambda)_\Omega + (\psi, s'(u)\lambda)_\Omega &= 0 \quad \forall\psi \in H, && (4.26)\\ (\nabla u, \nabla\pi)_\Omega + (s(u) - f, \pi)_\Omega - (q,\pi)_{\Gamma_C} &= 0 \quad \forall\pi \in H, && (4.27)\\ (\lambda - \alpha q, \chi)_{\Gamma_C} &= 0 \quad \forall\chi \in Q. && (4.28) \end{aligned}$$

The corresponding discrete approximations $v_h = \{u_h, \lambda_h\, q_h\}$ are determined in the finite element space $V_h = H_h \times H_h \times Q_h \subset V$ by

$$\begin{aligned} (\psi_h, u_h - u_O)_{\Gamma_O} + (\nabla\psi_h, \nabla\lambda_h)_\Omega + (\psi_h, s'(u_h)\lambda_h)_\Omega &= 0 \quad \forall\psi_h \in H_h, && (4.29)\\ (\nabla u_h, \nabla\pi_h)_\Omega + (s(u_h) - f, \pi_h)_\Omega - (q_h,\pi_h)_{\Gamma_C} &= 0 \quad \forall\pi_h \in H_h, && (4.30)\\ (\lambda_h - \alpha q_h, \chi_h)_{\Gamma_C} &= 0 \quad \forall\chi_h \in Q_h. && (4.31) \end{aligned}$$

Following the formalism of the previous section, we seek to estimate the error $e = \{e_u, e_\lambda, e_q\}$ with respect to the Lagrangian functional $L(\cdot)$. The corresponding linearized dual problem

$$L''(u_h; \phi, \tilde{z}) = -\tfrac{1}{2} L''(u_h; \phi, e) \quad \forall\phi \in V, \tag{4.32}$$

then has the solution $\tilde{z} = -\frac{1}{2}\{e_u, e_\lambda, e_q\}$. Hence, we neither have to build this dual problem nor do extra work solving it. This leads us to the following result.

Theorem 4.1 *For the finite element discretization of the variational equation (2.5) - (2.7), there holds the a posteriori error relation*

$$\begin{aligned} |J(u,q) - J(u_h,q_h)| \approx& \sum_{\Gamma\subset\partial\Omega} h_\Gamma^2 \{\rho_\Gamma^{(\lambda)}\omega_\Gamma^{(u)} + \rho_\Gamma^{(u)}\omega_\Gamma^{(\lambda)}\} + \sum_{\Gamma\subset\Gamma_C} h_\Gamma^2 \rho_\Gamma^{(q)}\omega_\Gamma^{(q)} \\ &+ \sum_{T\in\mathbb{T}_h} h_T^2 \{\rho_T^{(u)}\omega_T^{(\lambda)} + \rho_{\partial T}^{(u)}\omega_{\partial T}^{(\lambda)} + \rho_T^{(\lambda)}\omega_T^{(u)} + \rho_{\partial T}^{(\lambda)}\omega_{\partial T}^{(u)}\}, \end{aligned} \tag{4.33}$$

with the cell residuals and weights

$$
\begin{aligned}
\rho_\Gamma^{(\lambda)} &= h_\Gamma^{-3/2}\|u_h - u_O + \partial_n\lambda_h\|_\Gamma, \quad \Gamma\subset\Gamma_O, & \omega_\Gamma^{(u)} &= h_\Gamma^{-1/2}\|u-\psi_h\|_\Gamma,\\
\rho_\Gamma^{(\lambda)} &= h_\Gamma^{-3/2}\|\partial_n\lambda_h\|_\Gamma, \quad \Gamma\subset\partial\Omega\backslash\Gamma_O, & &\\
\rho_\Gamma^{(u)} &= h_\Gamma^{-3/2}\|\partial_n u_h - q_h\|_\Gamma, \quad \Gamma\in\Gamma_C, & \omega_\Gamma^{(\lambda)} &= h_\Gamma^{-1/2}\|\lambda-\pi_h\|_\Gamma,\\
\rho_\Gamma^{(u)} &= h_\Gamma^{-3/2}\|\partial_n u_h\|_\Gamma, \quad \Gamma\subset\partial\Omega\backslash\Gamma_C, & &\\
\rho_\Gamma^{(q)} &= h_\Gamma^{-3/2}\|\lambda_h - \alpha q_h\|_\Gamma, & \omega_\Gamma^{(q)} &= h_\Gamma^{-1/2}\|q-\chi_h\|_\Gamma,
\end{aligned}
$$

$$
\begin{aligned}
\rho_T^{(u)} &= h_T^{-1}\|\Delta u_h - s(u_h) + f\|_T, & \omega_T^{(\lambda)} &= h_T^{-1}\|\lambda-\pi_h\|_T,\\
\rho_{\partial T}^{(u)} &= \tfrac{1}{2}h_T^{-3/2}\|n\cdot[\nabla u_h]\|_{\partial T\backslash\partial\Omega}, & \omega_{\partial T}^{(\lambda)} &= h_T^{-1/2}\|\lambda-\pi_h\|_{\partial T\backslash\partial\Omega},\\
\rho_T^{(\lambda)} &= h_T^{-1}\|\Delta\lambda_h - s'(u_h)\lambda_h\|_T, & \omega_T^{(u)} &= h_T^{-1}\|u-\psi_h\|_T,\\
\rho_{\partial T}^{(\lambda)} &= \tfrac{1}{2}h_T^{-3/2}\|n\cdot[\nabla\lambda_h]\|_{\partial T\backslash\partial\Omega}, & \omega_{\partial T}^{(u)} &= h_T^{-1/2}\|u-\psi_h\|_{\partial T\backslash\partial\Omega},
\end{aligned}
$$

where $\psi_h,\ \pi_h \in H_h$ *and* $\chi_h \in Q_h$ *are arbitrary approximations. If the Lagrangian functional* $L(\cdot)$ *is quadratic this error relation yields a true upper bound.*

Proof In the present case, there holds

$$
\begin{aligned}
L(v) - L(v_h) &= J(u,q) + (\nabla u, \nabla\lambda)_\Omega + (s(u) - f, \lambda)_\Omega - (q,\lambda)_{\Gamma_C}\\
&\quad - J(u_h,q_h) - (\nabla u_h,\nabla\lambda_h)_\Omega - (s(u_h) - f,\lambda_h)_\Omega + (q_h,\lambda_h)_{\Gamma_C}\\
&= J(u,q) - J(u_h,q_h),
\end{aligned}
$$

since $\{u,\lambda,q\}$ and $\{u_h,\lambda_h,q_h\}$ satisfy the equations (4.27) and (4.30), respectively. Hence, error control with respect to the Lagrangian functional $L(\cdot)$ and the cost functional $J(\cdot)$ is equivalent. Now, the general error identity (4.25) implies that

$$
|J(u,q) - J(u_h,q_h)| \approx \inf_{\phi_h\in V_h} |L'(v_h; v-\phi_h)|, \tag{4.34}
$$

where $v_h = \{u_h,\lambda_h,q_h\}$ and $v = \{u,\lambda,q\}$. Notice that this relation is an identity if the functional $J(\cdot)$ is quadratic. From (4.29) – (4.31), we see that

$$
\begin{aligned}
L'(v_h; v-\phi_h) &= (u_h - u_O, u-\psi_h)_{\Gamma_O} + (\nabla(u-\psi_h),\nabla\lambda_h)_\Omega\\
&\quad + (u-\psi_h, s'(u_h)\lambda_h)_\Omega + (\nabla u_h, \nabla(\lambda-\pi_h))_\Omega\\
&\quad + (s(u_h) - f, \lambda-\pi_h)_\Omega - (q_h,\lambda-\pi_h)_{\Gamma_C} + (\lambda_h - \alpha q_h, q-\chi_h)_{\Gamma_C}.
\end{aligned}
$$

Splitting the global integrals into the contributions from each single cell $T \in \mathbb{T}_h$ and each cell edge $\Gamma \subset \partial\Omega$, respectively, and integrating locally by parts yields

$$\begin{aligned}
L'(v_h; v - \phi_h) = & \sum_{\Gamma \subset \Gamma_O} (u_h - u_O + \partial_n \lambda_h, u - \psi_h)_\Gamma + \sum_{\Gamma \subset \partial\Omega \setminus \Gamma_O} (\partial_n \lambda_h, u - \psi_h)_\Gamma \\
& + \sum_{\Gamma \subset \Gamma_C} (\partial_n u_h - q_h, \lambda - \pi_h)_\Gamma + \sum_{\Gamma \subset \partial\Omega \setminus \Gamma_C} (\partial_n u_h, \lambda - \pi_h)_\Gamma \\
& + \sum_{\Gamma \subset \Gamma_C} (\lambda_h - \alpha q_h, q - \chi_h)_\Gamma \\
& + \sum_{T \in \mathbb{T}_h} \big\{(-\Delta u_h + s(u_h) - f, \lambda - \pi_h)_T + \tfrac{1}{2}(n\cdot[\nabla u_h], \lambda - \pi_h)_{\partial T \setminus \partial\Omega}\big\} \\
& + \sum_{T \in \mathbb{T}_h} \big\{(u - \psi_h, -\Delta \lambda_h + s'(u_h)\lambda_h)_T + \tfrac{1}{2}(u - \psi_h, n\cdot[\nabla \lambda_h])_{\partial T \setminus \partial\Omega}\big\}.
\end{aligned}$$

From this the asserted relation follows by applying the Hölder inequality. □

In the computational examples presented in the next section, we evaluate the weights by using the local interpolation estimate

$$h_T^{-1}\|\phi - I_h\phi\|_T + h_T^{-1/2}\|\phi - I_h\phi\|_{\partial T} \le c_I h_T^2 \max_T |\nabla^2 \phi|. \tag{4.35}$$

The derivatives $\nabla^2 u$, $\nabla^2 \lambda$ and $\partial_s^2 q$ are then approximated by corresponding difference quotients of the computed solution $\{u_h, \lambda_h, q_h\}$.

Below, we will compare the performance of the dual–weighted error estimator (4.33) with more traditional error indicators. Control of the error in the "energy norm" of the state equation alone leads to the a posteriori error indicator

$$\eta_E(u_h) := c_I \sum_{T \in \mathbb{T}_h} h_T^3 \rho_{\partial T}^{(u)2} + c_I \sum_{\Gamma \subset \partial\Omega} h_\Gamma^3 \rho_\Gamma^{(u)2}, \tag{4.36}$$

with the cell residuals $\rho_{\partial T}^{(u)}$ and $\rho_\Gamma^{(u)}$ as defined above. Further, incorporating error control for the adjoint equation results in

$$\eta_E(u_h, \lambda_h) := c_I \sum_{T \in \mathbb{T}_h} h_T^3 \{\rho_{\partial T}^{(u)2} + \rho_{\partial T}^{(\lambda)2}\} + c_I \sum_{\Gamma \subset \partial\Omega} h_\Gamma^3 \{\rho_\Gamma^{(u)2} + \rho_\Gamma^{(\lambda)2}\}. \tag{4.37}$$

Both ad-hoc criteria aim at satisfying the state equation and the adjoint equation uniformly with good accuracy. However, this concept seems questionable since it does not take into account the sensitivity of the cost functional with respect to the local perturbations introduced by discretization. Capturing these dependencies is the particular feature of our approach. In practice, one may use a simplified version of the error estimator (4.33) which incorporates only the essential jump-residual terms and the residuals along the observation and control boundary:

$$\eta_\omega(u_h,\lambda_h,q_h) := \sum_{T\in\mathbb{T}_h} h_T^3\,\rho_{\partial T}^{(u)}\rho_{\partial T}^{(\lambda)} + \sum_{\Gamma\subset\partial\Omega} h_\Gamma^2\{\rho_\Gamma^{(\lambda)}\omega_\Gamma^{(u)} + \rho_\Gamma^{(u)}\omega_\Gamma^{(\lambda)}\}. \tag{4.38}$$

The performance of these different error indicators is compared by an example in the next section.

Strategies for mesh adaptation

Finally, we briefly describe strategies for mesh adaptation on the basis of a posteriori error estimates of the kind

$$\eta := \sum_{T\in\mathbb{T}_h} \eta_T, \tag{4.39}$$

with certain cell indicators, e.g. $\eta_T := h_T^3\rho(u_h)_{\partial T}\rho(\lambda_h)_{\partial T}$. We aim at achieving a prescribed tolerance TOL for the quantity $J(u)$ and the number of mesh cells N which measures the complexity of the computational model. Usually the admissible complexity is constrained by some maximum value $N_{\max}$.
i) Error balancing strategy: We cycle through the mesh and equilibrate the local error indicators η_T according to $\eta_T \approx TOL/N$. This process requires iteration with respect to the number of mesh cells N and results in $\eta \approx TOL$.
ii) Fixed fraction strategy: We order the cells according to the size of η_T and refine a certain percentage (say 30%) of cells with largest η_T, or those cells which contribute to a certain percentage of the value of $\eta(u_h)$. A certain part of cells with small η_T may be coarsened. By this strategy, one can achieve a prescribed rate of increase of N (or keep it constant as may be desirable in nonstationary computations).

5 Numerical results

We perform computations for the Ginzburg-Landau model described in Section 2 on a series of locally refined meshes. On each mesh, the Euler-Lagrange equations are discretized by the Galerkin finite element method as described above using piecewise bilinear shape functions for both the state and adjoint variables u and λ, while the traces on Γ_C of the bilinear shape functions form the control space Q_h. Then, the resulting discrete systems are solved iteratively and new meshes are generated on the basis of a posteriori error estimators. In all cases, the weights are evaluated by using difference approximation as described in the previous section with interpolation constants set to $c_I = 0.1$. The mesh refinement follows the "fixed fraction" strategy. The *effectivity index*

$$I_{eff} := \frac{|J(u_{ref}, q_{ref}) - J(u_h, q_h)|}{\eta(u_h, q_h)},$$

is used as a measure for the quality of our a posteriori error estimator. The reference solutions are obtained on adaptively generated meshes with about $2 \cdot 10^5$ cells.

5.1 A "forward" test case

First, we want to illustrate the difference between traditional "energy error control" and our functional-oriented "dual-weighted error control" by considering the following linear primal test example:

$$\begin{aligned} -\Delta u + u = 0 \quad \text{on } \Omega, \\ \partial_n u = q \quad \text{on } \Gamma_C, \quad \partial_n u = 0 \quad \text{on } \partial\Omega \setminus \Gamma_C; \end{aligned} \tag{5.1}$$

see Configuration 2 in Figure 2.1. The boundary control is frozen as $q \equiv 0.0503455$ (taken from an optimization result). The corresponding discrete equations read

$$(\nabla u_h, \nabla \psi_h)_\Omega + (u_h, \psi_h)_\Omega = (q, \psi_h)_{\Gamma_C} \quad \forall \psi_h \in V_h. \tag{5.2}$$

We control the error $e = u - u_h$ with respect to the *quadratic* observation functional

$$J(u) = \tfrac{1}{2}\|u - u_O\|^2_{\Gamma_O}.$$

The corresponding dual solution $z \in V$ is obtained by solving the corresponding system (2.5) - (2.7) with frozen boundary function q and linearized right-hand side $J'(u_h; \psi)$. The resulting a posteriori error bound is

$$|J(u) - J(u_h)| \le \eta_\omega(u_h) := \sum_{T \in \mathbb{T}_h} h_T^2 \{\rho_T^{(u)} \omega_T^{(z)} + \rho_{\partial T}^{(u)} \omega_{\partial T}^{(z)}\} + \sum_{\Gamma \in \partial\Omega} h_\Gamma^2 \rho_\Gamma^{(u)} \omega_\Gamma^{(z)}, \tag{5.3}$$

with cell residuals and weights defined as above. The asymptotic correctness of this error estimator is demonstrated in Table 5.1.

Table 5.1: Effectivity index of the dual–weighted error estimator $\eta_\omega(u_h)$ applied to the linear primal model problem; $E(u_h) := |J(u) - J(u_h)|$.

N	1376	5840	22544	57104	84368
$E(u_h)$	1.64e-05	4.17e-06	1.01e-06	3.5e-07	2.49e-07
I_{eff}	0.81	0.91	0.92	0.95	0.88

We compare the dual–weighted error estimator with the traditional energy-norm error estimator which in this case reads as follows:

$$||\nabla e||_\Omega^2 \leq \eta_E(u_h) := c_I \sum_{T \in \mathbb{T}_h} h_T^4 \{\rho_T^{(u)2} + \rho_{\partial T}^{(u)2}\} + c_I \sum_{\Gamma \in \partial\Omega} h_\Gamma^4 \rho_\Gamma^{(u)2}, \tag{5.4}$$

with the notation as introduced above. Clearly, small $||\nabla e||_\Omega$ implies small $E(u_h)$, but not vice versa. Hence, mesh adaptation based on the energy-error estimator may result in overly refined meshes. This is clearly seen in Figure 5.1.

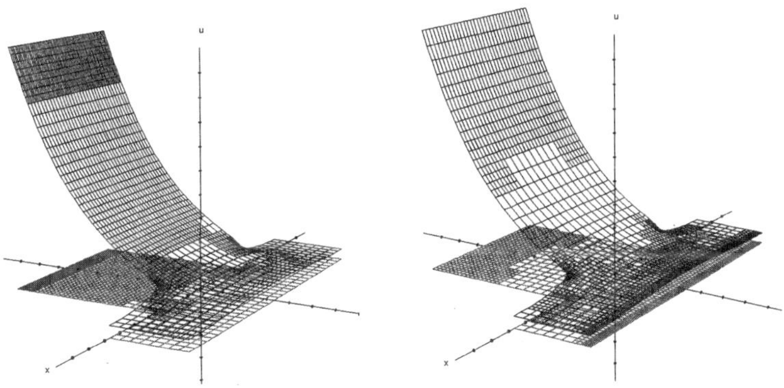

Figure 5.1: Meshes obtained by the error estimators $\eta_E(u_h)$ (left) and $\eta_\omega(u_h)$ (right) with $N \sim 5000$ cells in both cases; the graph of the solution is strongly scaled up.

Obviously, the energy-error estimator puts too much emphasis on refining at the reentrant corners which is obviously less important for achieving good accuracy along the observation boundary Γ_O. In contrast to that, the dual–weighted error estimator provides a better balance between resolving the corner singularities and the neighbourhood of Γ_O. This results in a higher mesh economy as shown by the corresponding error plots in Figure 5.2. This demonstrates the

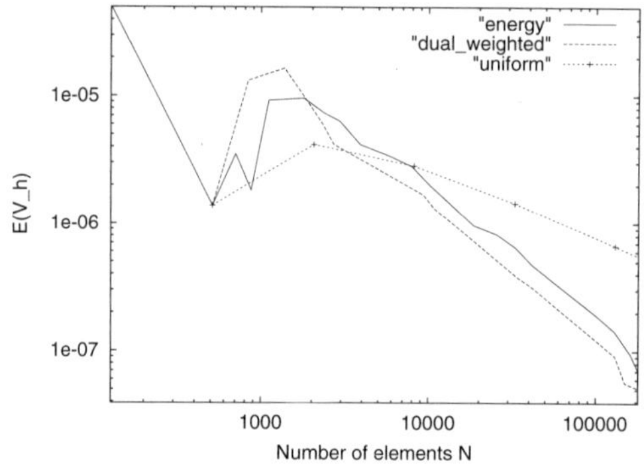

Figure 5.2: Efficiency of meshes generated by the error estimators $\eta_E(u_h)$ (solid line) and $\eta_\omega(u_h)$ (dashed line), and by uniform refinement (crosses) (*log/log* scale).

value of capturing the sensitivities inherent to the problem under consideration. This effect will become even more pronounced in solving the optimal control problem.

5.2 Optimal control test cases

Next, we consider the optimal control problem for the *nonlinear* Ginzburg-Landau model with $s(u) = u^3 - u$ and $f \equiv 0$. The solution approach is as described in the preceding sections.

Test for Configuration 1

We begin with Configuration 1 shown in Figure 2.1. The cost functional is chosen with $u_O = \sin(0.19x)$ and $\alpha = 0$. This test case represents an extreme situation since the observation $u_{|\Gamma_O}$ is evaluated right at the control boundary Γ_C , i.e., the information relevant for the optimization problem does not have to pass through the domain. Hence, the corner singularities do not much affect the optimization process and should therefore not induce mesh refinement. This is accomplished by our dual–weighted error estimator as shown by Table 5.2. In this case, the heuristic energy-error indicators (4.36) and (4.37) produce inefficient meshes as seen in Figures 5.3, 5.4 and 5.5. In fact, in (4.38) the boundary indicators are dominant over the interior indicators.

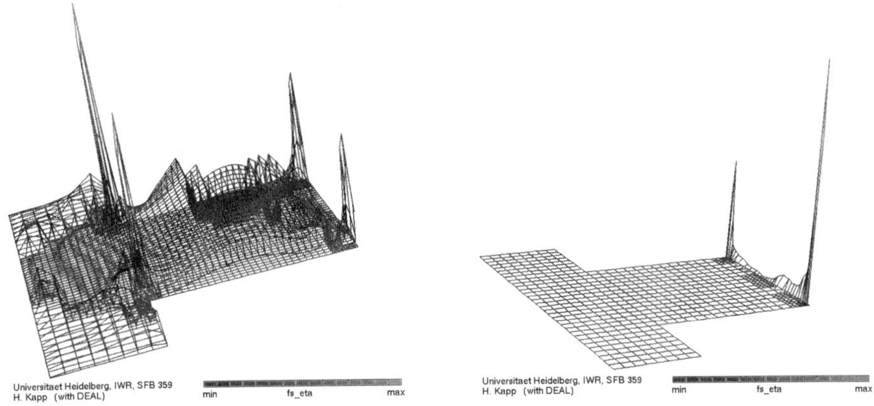

Figure 5.3: Configuration 1: Size of cell indicators η_T in the error estimators $\eta_E(u_h)$ (left) and $\eta_\omega(u_h, \lambda_h, q_h)$ (right).

Table 5.2: Configuration 1: Effectivity index of the dual-weighted error estimator $\eta_\omega(u_h, \lambda_h, q_h)$ (reference value $J(u,q) = 0.0119467...$).

N	596	1616	5084	8648	15512
$E(v_h)$	0.0002555	0.0002375	8.22e-05	4.21e-05	3.99e-05
I_{eff}	0.34	0.81	0.46	0.29	0.43

In Figure 5.5, we compare the efficiency of the meshes generated by the dual–weighted error estimator and the energy-error indicator. The first one yields significantly more economical meshes; the value 0.011948 of the cost function is obtained with only $N \sim 3500$ cells compared to the $N \sim 100000$ cells needed by the energy-norm indicator.

Test for Configuration 2

For the last test, we take the observations as $u_O \equiv 1$ and set the regularization parameter to $\alpha = 0.1$. Now, depending on the nonlinearity $s(u)$, there may be several stationary points of the Euler-Lagrange equations. By varying the starting values for the Newton iteration, we can approximate these solutions. A trivial solution corresponds to $u \equiv u_O$ and is actually the global minimum of the cost functional. The two other computed solutions correspond to a local

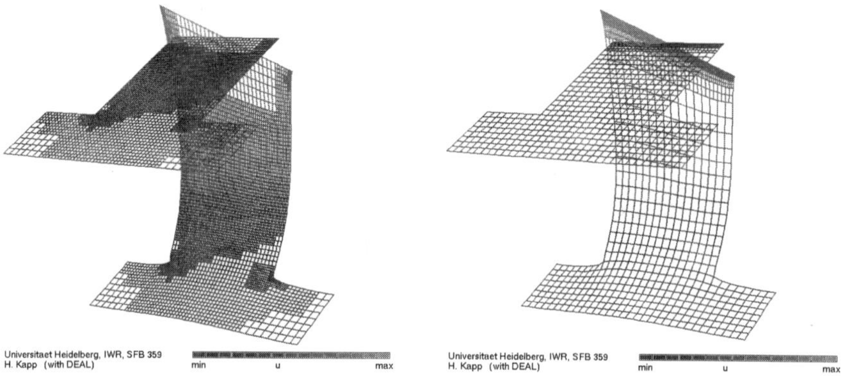

Figure 5.4: Configuration 1: Comparison of discrete solutions obtained by the error estimators $\eta_E(u_h)$ (right, $N \sim 4800$ cells) and $\eta_\omega(u_h, \lambda_h, q_h)$ (left, $N \sim 5000$ cells).

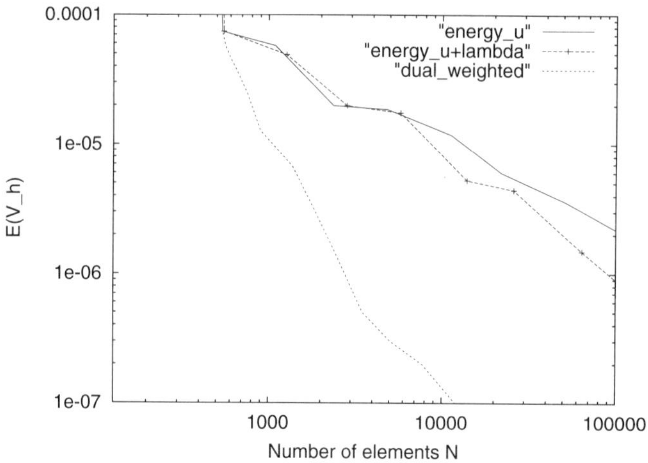

Figure 5.5: Configuration 1: Comparison of efficiency of meshes generated by the error indicators $\eta_E(u_h)$ (solid line), $\eta_E(u_h, \lambda_h)$ (crosses) and $\eta_\omega(u_h, \lambda_h, q_h)$ (dashed line).

minimum and a local maximum. The effectivity of the dual–weighted error estimator for computing the local minimum is shown in Table 5.3.

Table 5.3: Configuration 2: Effectivity index of the weighted error estimator $\eta_\omega(u_h, \lambda_h, q_h)$ for computing the local minimum (reference value $J(u,q) = 0.04888934625\ldots$).

N	512	15368	27800	57632	197408
$E(v_h)$	9.29e-05	8.14e-07	4.86e-07	2.31e-07	4.58e-08
I_{eff}	1.32	0.56	0.35	0.42	0.32

Next, Figure 5.6 shows the distribution of the local cell indicators η_T for the two error estimators; the corresponding results are shown in Figures 5.7 and 5.8. Obviously, the weighted error estimator induces a stronger refinement along the observation and control boundaries which seems more relevant for the optimization process than resolving the corner singularities. However, the total contribution of the interior cell indicators is dominant over that of the boundary indicators. This explains why the gain in efficiency (about 25%) of the dual-weighted error estimator over the energy-norm indicator is less significant here compared to the previous example.

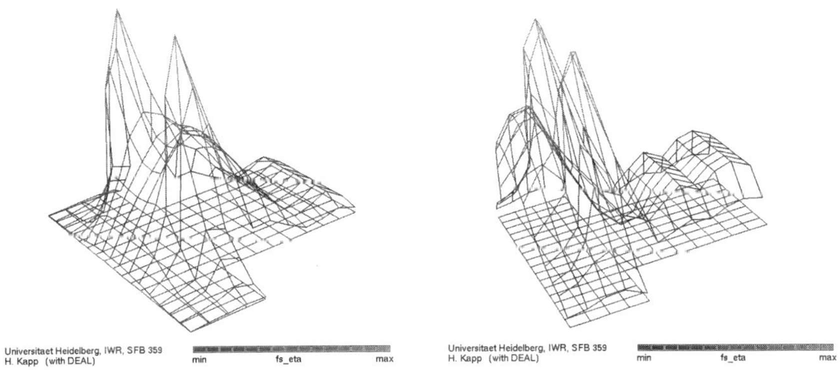

Figure 5.6: Configuration 2: Distributions of the cell indicators η_T in the error estimators $\eta_E(u_h)$ (left) and $\eta_\omega(u_h, \lambda_h, q_h)$ (right).

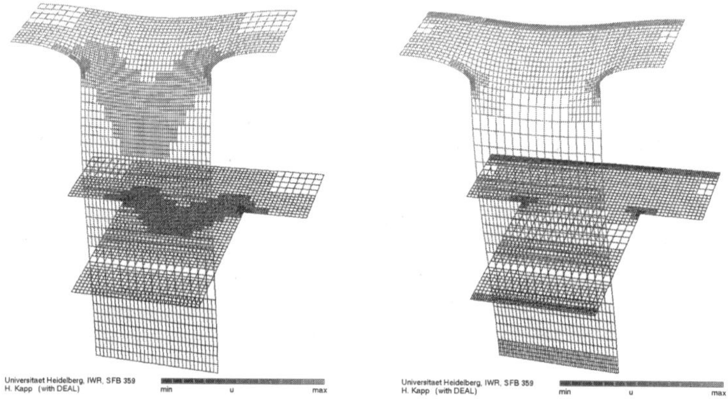

Figure 5.7: Configuration 2: Comparison of discrete solutions obtained by the error estimators $\eta_E(u_h)$ (right, $\sim$ 3300 cells) and $\eta_\omega(u_h, \lambda_h, q_h)$ (left, $\sim$ 3000 cells).

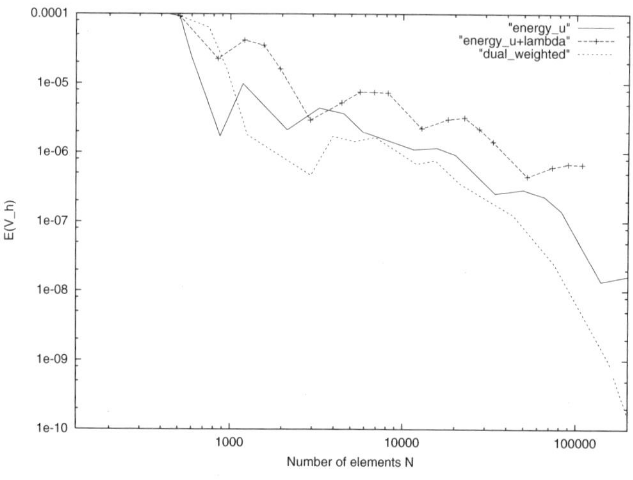

Figure 5.8: Configuration 2: Comparison of efficiency of meshes generated by the error estimators $\eta_E(u_h)$ (solid line) and $\eta_\omega(u_h, \lambda_h, q_h)$ (dashed line)+.

References

[1] R. Becker and H. Kapp, *Optimization in PDE Models with Adaptive Finite Element Discretization*, Proc. ENUMATH'97, Heidelberg, Sept.28 - Oct.3, 1997, World Scientific Publishers, Singapore, 1998.

[2] R. Becker, H. Kapp and R. Rannacher, *Adaptive Finite Element Methods for Optimal Control of Partial Differential Equations: Basic Concepts*, Preprint 98-55, SFB 359, University of Heidelberg, Nov. 1998, submitted for publication.

[3] R. Becker and R. Rannacher, *Weighted A Posteriori Error Control in FE Methods*, ENUMATH'95, Paris, Sept. 18-22, 1995, Proc. ENUMATH'97 (H. G. Bock et al., eds.), pp. 621–637, World Scientific Publishers, Singapore, 1998.

[4] R. Becker and R. Rannacher, A feed-back approach to error control in finite element methods: Basic analysis and examples, *East-West J. Numer. Math* 4, 237-264 (1996).

[5] S. C. Brenner and R. Scott, *The Mathematical Theory of Finite Element Methods*, Springer-Verlag, Berlin, 1994.

[6] Q. Du, M. D. Gunzburger and J. S. Peterson, Analysis and approximation of the Ginzburg-Landau model of superconductivity, *SIAM Rev.* 34, 54-81 (1992).

[7] K. Eriksson, D. Estep, P. Hansbo and C. Johnson, *Introduction to adaptive methods for differential equations*, Acta Numerica 1995 (A. Iserles, ed.), pp. 105-158, Cambridge University Press, 1995.

[8] M. D. Gunzburger and L. S. Hou, Finite dimensional approximation of a class of constrained nonlinear optimal control problems, *SIAM J. Contr. Opt.* 34, 1001-1043 (1996).

[9] K. Ito and K. Kunisch, Augmented Lagrangian-SQP methods for nonlinear optimal control problems of tracking type, *SIAM J. Contr. Opt.* 34, 874-891 (1996).

[10] H. Kapp, *Adaptive Galerkin Finite Element Methods for Optimal Control of Partial Differential Equations*, Doctorate thesis, Institute of Applied Mathematics, University of Heidelberg, in preparation.

[11] R. Rannacher, *Error control in finite element computations*, Proc. NATO-Summer School "Error Control and Adaptivity in Scientific Computing" Antalya (Turkey), Aug. 1998 (H. Bulgak and C. Zenger, eds.), pp. 247–278, NATO Science Series, Series C, Vol. 536, Kluwer, Dordrecht, 1999.

[12] R. Verfürth, *A Review of A Posteriori Error Estimation and Adaptive Mesh-Refinement Techniques*, Wiley/Teubner, New York-Stuttgart, 1996.

Acknowledgement

The authors acknowledge support by the German Research Association (DFG) through the SFB 359 "Reactive Flows, Diffusion and Transport", University of Heidelberg.

Roland Becker
Hartmut Kapp
Rolf Rannacher
Institute of Applied Mathematics
University of Heidelberg
INF 293, D-69120 Heidelberg, Germany
`roland.becker@iwr.uni-heidelberg.de`
`hartmut.kapp@iwr.uni-heidelberg.de`
`rannacher@iwr.uni-heidelberg.de`
`http://gaia.iwr.uni-heidelberg.de`

P. Blomgren, T. F. Chan, P. Mulet, L. Vese and W. L. Wan

Variational PDE models and methods for image processing

Abstract We describe some numerical techniques for the total variation image restoration method, namely some preconditioning issues and modular solvers for the solution of the Euler-Lagrange equations when constraints are present. We also highlight extension of this technique to handle color images, and to reduce the staircasing effect. Finally, we present segmentation and active contour techniques based on the Mumford-Shah model.

1 Introduction

Variational PDE (Partial Differential Equation) methods have emerged recently as alternatives to traditional statistical and fast transform-based methodologies in image processing. Examples are the Perona-Malik anisotropic diffusion model [27], the total variation restoration method by Rudin, Osher and Fatemi [29], the axiomatic derived fundamental PDE due to Alvarez, Guichard, Lions and Morel [1], the Mumford-Shah segmentation model [25] and various active contour models for object detection. These PDE models offer a systematic way to treat geometric properties of images, as well as to properly handle singularities (e.g., edges), by using PDE concepts, such as gradients, jumps, diffusion, curvature and level sets. Also, these models treat images as continuous functions. Through these models, the vast arsenal of PDE and CFD (Computational Fluid Dynamics) techniques can be brought to bear on image processing problems. The methods are often variational, the associated Euler-Lagrange equations give the PDE model.

On the other hand, these models are usually highly nonlinear and require new computational techniques for their effective solution, including optimization and linearization techniques, as well as level set methods for tracking interfaces, preconditioners, regularization.

The main purpose of this paper is to survey some recent work in our research group on using PDE-based techniques in image restoration, segmentation and active contours. The classical algorithms, which are mainly based on least squares, are not usually appropriate for edge recovery. Instead, we follow a variational formulation based on the minimization of the total variation norm subject to some noise constraints. We will present efficient numerical methods as well as extensions of TV to beyond gray-scale images.

In Section 2 we briefly review the total variation restoration model and some numerical methods for the solution of the Euler-Lagrange equations. We also discuss modular solvers for the solution of these equations when constraints are present. We will next present two types of preconditioning techniques for the iterative solution of the linearized Euler-Lagrange equations, one based on fast transforms and the other on multigrid techniques. In Section 3, we extend the total variation restoration technique to vector images (including color), and discuss techniques to lessen the tendency of TV to over-sharpening smooth images. Finally, in Sections 4 and 5 we present some techniques for image segmentation and active contours based on the Mumford-Shah model.

2 Numerical methods

In this section, we focus on efficient numerical methods to overcome the difficulties due to the highly nonlinearity of the mathematical models, often combined with constraints. We use linearization techniques and primal-dual methods, as well as modular solvers for constrained restoration, preconditioners based on fast transform and multigrid.

2.1 Total variation restoration

Let us denote by u and z the real and observed images, respectively, both defined in a region Ω. The model of degradation we assume is $\mathbb{K}u + n = z$, where n is a Gaussian white noise and $\mathbb{K}$ is a (known) linear *blur operator*.

The problem $\mathbb{K}u = z$, for a compact operator $\mathbb{K}$, is ill-posed, so we regularize it by two related formulations:

(1) its Tikhonov regularization ([34]), which consists in the solution of the variational problem:

$$\min_u \alpha\, R(u) + \frac{1}{2}||\mathbb{K}u - z||^2_{\mathcal{L}^2}, \tag{2.1}$$

for some regularization functional R which measures the irregularity of u and a coefficient α suitably chosen to balance the tradeoff between a good fit to the data and a *regular* solution, and;

(2) the constrained variational problem:

$$\min_u \alpha\, R(u) \text{ subject to } \frac{1}{2}||\mathbb{K}u - z||^2_{\mathcal{L}^2} = \frac{1}{2}\sigma^2, \tag{2.2}$$

where σ^2 is the variance of the noise, which we assume to be known. If λ is the Lagrange multiplier corresponding to (2.2) and $\alpha = \frac{1}{\lambda}$, then both problems are equivalent. Examples of regularization functionals are: $R(u) = ||u||^2, ||\Delta u||^2, ||\nabla u \cdot \nabla u||^2$. The associated Euler-Lagrange equations are linear, but they are not usually suitable for edge recovery.

In [29], the *Total Variation norm*: $R(u) = TV(u) = \int_\Omega |\nabla u|\, dx\, dy$ is proposed as regularization functional. It does not penalize discontinuities in u and thus allows a better edge recovery. Its Euler-Lagrange equation is:

$$0 = -\alpha \nabla \cdot \left(\frac{\nabla u}{|\nabla u|} \right) + \mathbb{K}^* \left(\mathbb{K}u - z \right) = g(u), \tag{2.3}$$

with homogeneous Neumann boundary conditions, where $\mathbb{K}^*$ is the adjoint operator of $\mathbb{K}$. Since this equation is not well defined at points where $\nabla u = 0$, a commonly used technique is to replace $|\nabla u|$ by $\sqrt{|\nabla u|^2| + \beta}$, for a small positive parameter β. The minimization problem is convex and therefore has a unique solution. Then the computational task is to solve (2.3) efficiently.

The main difficulty in solving equation (2.3) is the linearization of the highly nonlinear divergence term. A number of methods have been proposed to solve (2.3). Rudin, Osher and Fatemi [29] used a time marching scheme to compute the steady state of the parabolic equation $u_t = -g(u)$ with initial condition $u = z$. This method can be slowly convergent due to stability constraints. Vogel and Oman [38] proposed the following fixed point iteration to solve the Euler-Lagrange equation: $u^0 = z$, solve for u^{k+1}:

$$-\alpha \nabla \cdot \left(\frac{\nabla u^{k+1}}{|\nabla u^k|} \right) + \mathbb{K}^* \left(\mathbb{K}u^{k+1} - z \right) = 0. \tag{2.4}$$

This method is robust but only linearly convergent (see [20, 11, 4] for convergence proofs).

Due to the high nonlinearity of (2.3), Newton's method has an extremely small domain of convergence for small β. So it is natural to use a continuation procedure, starting with a large value of β and gradually reducing it to the desired value. Although this method is locally quadratically convergent, the choice of the sequence of subproblems to solve is crucial for its efficiency, and we have not succeeded in finding a fully satisfactory selection procedure, although some heuristics can be used, see [15].

A better linearization ([21]) can be achieved by introducing a new variable $\vec{w} = \frac{\nabla u}{|\nabla u|}$ (representing the unit normal to level curves of u), replacing (2.3) by the following equivalent system:

$$\begin{aligned} -\alpha\nabla\cdot\vec{w}+\mathbb{K}^*(\mathbb{K}u-z) &= 0 \\ \vec{w}|\nabla u|-\nabla u &= 0, \end{aligned} \tag{2.5}$$

and then linearizing it by Newton's method. In practice, this method is globally convergent and the local convergence rate is quadratic. Although we do not have a complete theory yet, we believe that the key to its success is that (2.5) is more *globally linear* than (2.3). Our method is related to the recent work by Andersen, Christiansen, Conn and Overton on a primal-dual algorithm for minimizing a sum of Euclidian norms [3].

2.2 Modular solvers for constrained image restoration

In many cases it is quite straightforward to design schemes for solving the corresponding unconstrained problem

$$\min_{u\in X(\Omega)} R(u)+\frac{\lambda}{2}||\mathbb{K}u-z||_2^2, \tag{2.6}$$

which is just (2.1) multiplied by $\lambda=\frac{1}{\alpha}$. If chosen correctly, λ is the *Lagrange multiplier* corresponding to the constraint in the constrained formulation (2.2). Solving the fully constrained problem involves simultaneously identifying the correct λ, as well as solving the *Euler-Lagrange equations* associated with (2.6).

In this section, we present a highly modular approach to solving the constrained problem. Given an efficient solver, $u\leftarrow S(u,\lambda)$, for the unconstrained problem (with λ given), we construct a constrained solver $(u,\lambda)\leftarrow\hat{S}(u,\lambda)$. No knowledge of the regularizing functional $R(u)$ or the existing solver $S(u,\lambda)$ is required.

Approaches for the solution of the constrained problem (2.2) already exist: Rudin, Osher and Fatemi [29] used the projected gradient method of Rosen [28] in the setting of an explicit time-marching algorithm. Chan, Golub and Mulet [21] introduced a fully constrained primal-dual approach, which is quadratically convergent. While quite successful, these approaches are non-modular: the constraints and the regularizing functional are strongly coupled in the solver. Nonmodularity in itself is not an argument against any method. However, it does make it hard to adapt the algorithm to modifications to the regularity functional, and/or the constraints. In our modular approach, we can combine any solver for the unconstrained problem with any set of constraints.

Let $G(u,\lambda)=-\nabla_u R(u)+\lambda\mathbb{K}^*(\mathbb{K}u-z)$, and $N(u,\lambda)=||\mathbb{K}u-z||^2-|\Omega|\sigma^2$. The first order necessary conditions for optimality of the constrained problem

(2.2) can be written as a coupled nonlinear system

$$\begin{bmatrix} G(u,\lambda) \\ N(u,\lambda) \end{bmatrix} = 0. \tag{2.7}$$

Assuming that the solution exists, and the Jacobian is regular enough, we apply a Newton iteration; noticing that $N_\lambda = 0$, we block eliminate (see Keller [24]), and get

$$\begin{bmatrix} I & G_u^{-1}G_\lambda \\ N_u & 0 \end{bmatrix} \begin{bmatrix} \delta u \\ \delta\lambda \end{bmatrix} = - \begin{bmatrix} G_u^{-1}G \\ N \end{bmatrix}. \tag{2.8}$$

The key to the modular solver lies in the fact that the quantities $w = G_u^{-1}G$, and $v = G_u^{-1}G_\lambda$ can be approximated by calls to the unconstrained solver S. w is simply the Newton update for the fixed-λ problem, $S(u,\lambda) - u$. v can be approximated by another call to the solver: $v = G_u^{-1}G_\lambda = -u_\lambda = -S(u,\lambda)_\lambda \approx -S_\lambda$, provided S is sufficiently contractive, i.e., $||S_u|| \ll 1$. In particular, if the solver S is a Newton solver, this approximation is exact, since $S_u = 0$. In practice, all reasonably fast solvers are contractive enough to make this approximation work. Finally, we use a finite difference approximation to S_λ, so that $v = \frac{S(u,\lambda)-S(u,\lambda+\epsilon)}{\epsilon}$, where $|\epsilon| \ll 1$.

Using these approximations, and eliminating the (2,1)-block gives

$$\begin{bmatrix} I & v \\ 0 & -N_u v \end{bmatrix} \begin{bmatrix} \delta u \\ \delta\lambda \end{bmatrix} = - \begin{bmatrix} -w \\ N + N_u w \end{bmatrix}. \tag{2.9}$$

Summary: MODULAR ALGORITHM (1 iteration)

1. Compute $w = S(u^n, \lambda^n) - u^n$.
2. Compute $v = [\epsilon]^{-1} [S(u^n, \lambda^n) - S(u^n, \lambda^n + \epsilon)]$.
3. Compute $\delta\lambda = [N_u v]^{-1}(N + N_u w)$, where $N_u = \mathbb{K}^* (\mathbb{K}u - z)$.
4. Compute $\delta u = w - v\,\delta\lambda$.
5. Update $\begin{cases} u^{n+1} & = & u^n + \delta u \\ \lambda^{n+1} & = & \lambda^n + \delta\lambda. \end{cases}$

Hence, each iteration requires two calls to the unconstrained solver, S. The solution obtained from the first call can be used as an initial guess for the second call, thus speeding up this call considerably.

To illustrate the efficiency and the robustness of the modular algorithm, we apply the modular solver in three different settings: *(i)* Gray-scale denoising

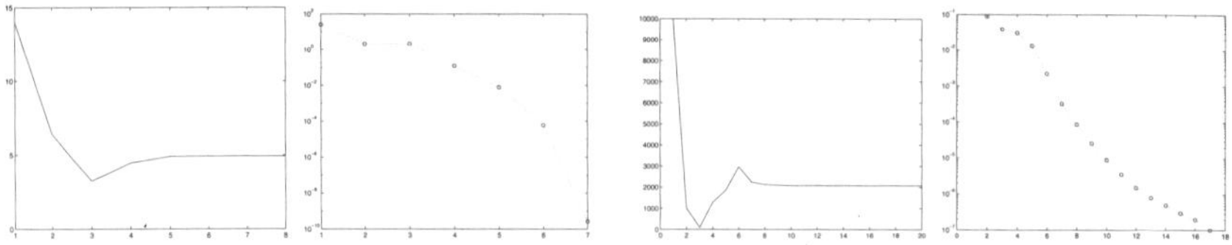

(a) Gray-scale reconstruction. (b) Color reconstruction.

Figure 2.1: **Gray-scale denoising** — (a): LEFT: The regularization parameter, λ, as a function of the number of iterations. RIGHT: The residual of the constraint, e.g., $N(u, \lambda)$ as a function of the iterations. As expected, convergence is quadratic. **Color denoising** — (b): LEFT: λ. RIGHT: $N(u, \lambda)$, as expected, convergence is linear.

Figure 2.2: The true, degraded and recovered images.

using a primal-dual solver S, $R(u) = TV(u)$. In this application we expect the modular solver to be quadratically convergent, which is confirmed by the data, see Figure 2.1a. *(ii)* RGB-color image denoising using a lagged diffusivity fixed point solver (see [36]), $R(\mathbf{u}) = TV_{n,m}(\mathbf{u})$. Here the modular solver inherits the linear convergence, see Figure 2.1b. *(iii)* Reconstruction of an image degraded by both noise and blur (nontrivial $\mathbb{K}$), using a fixed point algorithm with cosine transform-based preconditioning (see [12]), $R(u) = TV(u)$. Again, the convergence is linear, and the recovered result quite remarkable (Figure 2.2).

As indicated by the presented applications, the modular approach is easy to implement in different settings; we just plug in the unconstrained solver. This facilitates experimentation and evaluation of different regularization models. Finding the correct Lagrange multiplier is the only way to make fair comparisons between models.

The approach is very robust. In most cases the modular solver converges for a wide range of initial guesses for the Lagrange multiplier (see Figure 2.3a).

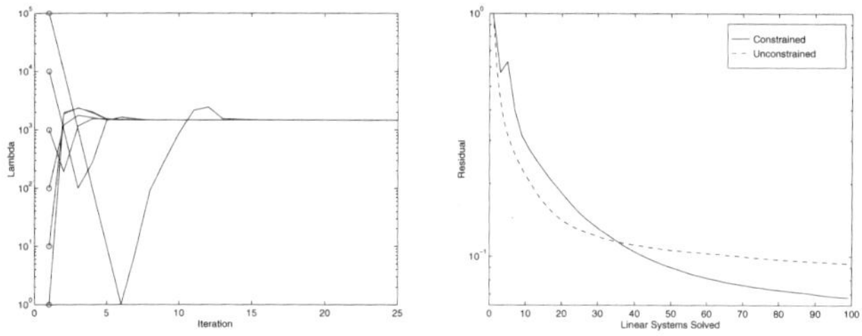

(a) Lambda convergence. (b) Work comparison.

Figure 2.3: (a): The modular solver is very robust, here the correct Lagrange multiplier is identified for initial guesses spread over five orders of magnitude. (b): Comparison of the residual for the constrained, and unconstrained solvers. Notice that the x-axis shows the total number of linear systems solved (the actual work), not the number of iterations.

The modular solver is efficient. In practice we can compute a solution to the constrained problem in the same amount of time it takes to compute the unconstrained problem (see Figure 2.3b).

2.3 Preconditioning

Fast transform-based preconditioners

In a typical efficient iterative solution of (2.3), one has to precondition linear operators of the form $A_{u^k} = \mathbb{K}^*\mathbb{K} + \alpha L_{u^k}$ where $\mathbb{K}$, a Toeplitz-type matrix, corresponds to the blur and L_{u^k} corresponds to the TV regularization. If no blur is present, i.e., $\mathbb{K}$ is the identity, then standard techniques in the PDE field can be applied: inclomplete factorizations, multigrid, domain decomposition, etc. When $\mathbb{K}$ is a general Toeplitz-type matrix the choice is more difficult.

Vogel and Oman [35, 37] have recently proposed using a "product" preconditioner which allows $\mathbb{K}^*\mathbb{K}$ and L_{u^k} to be preconditioned separately. An alternative preconditioner is the cosine transform-based preconditioner [12]. The motivation comes from the fact that cosine transform preconditioner is "good" for solving Toeplitz system [23] and gives exact approximation to discrete Laplacian with Neumann boundary condition.

Let us denote C_n to be the n-by-n discrete cosine transform matrix. If δ_{ij}

is the Kronecker delta, then the (i, j)th entry of C_n is given by

$$\sqrt{\frac{2-\delta_{i1}}{n}} \cos\left(\frac{(i-1)(2j-1)\pi}{2n}\right), \quad 1 \le i, j \le n, \tag{2.10}$$

see Sorensen and Burrus [31]. For any $n^2 \times n^2$ matrix A_{nn}, the optimal level-2 cosine transform preconditioner $c(A_{nn})$ is defined to be the minimizer of the Frobenius norm $||B_{nn} - A_{nn}||_F$ over all matrices B_{nn} that can be diagonalized by the 2D discrete cosine transform matrix $C_n \otimes C_n$, [17].

The construction cost for $c(A)$ is $\mathcal{O}(n^4)$ operations for general A, but can be reduced to $O(n^2 \log n)$ operations in case of band or Toeplitz structure. Therefore, to avoid the $\mathcal{O}(n^4)$ operations cost, a cosine transform preconditioner for $A = \mathbb{K}^* \mathbb{K} + \alpha L_{u^k}$ can be defined as:

$$M = c(\mathbb{K})^* c(\mathbb{K}) + \alpha c(L_{u^k}).$$

Note that M has the eigendecomposition

$$M = (C_n \otimes C_n)(\Lambda_{\mathbb{K}}^* \Lambda_{\mathbb{K}} + \alpha \Lambda_L)(C_n \otimes C_n)^t$$

where $\Lambda_{\mathbb{K}}$ and Λ_L are the eigen-matrices of $c(\mathbb{K})$ and $c(L_{u^k})$, respectively. Hence, in each PCG (Preconditioned Conjugate Gradient) iteration, $M^{-1}v$ can be done in $O(n^2 \log n)$ operation by the 2D Fast Cosine Transform (FCT) algorithm. We remark that by exploiting the Toeplitz structure of $\mathbb{K}$ and by noting that L_{u^k} is a banded matrix, $c(\mathbb{K})$ and $c(L_{u^k})$ can be both constructed in $O(n^2 \log n)$ operations.

The numerical results from [12] indicate that the condition number of the preconditioned system $\kappa(M^{-1}A)$ is $O(n^{0.22})$ while $\kappa(A)$ is $O(n^{1.8})$ without preconditioning. Hence, the preconditioner can significantly speed up the convergence rate of the CG method. Furthermore, numerical experiments indicate that the cosine transform preconditioner is better and more robust than the product preconditioner over a wide range of blurs and α.

Multigrid preconditioners

We now consider the use of multigrid (MG) methods for solving (2.3).

We recall from Subsection 2.3 the linear systems obtained after discretization:

$$(\alpha L + \mathbb{K}^* \mathbb{K})u = f,$$

where L corresponds to the (elliptic) differential operator and $\mathbb{K}$ corresponds to the convolution operator.

Multigrid methods [8, 22] have been known to be very effective for preconditioning the elliptic operator L. The convergence rates are often shown to be independent of the mesh size. For the convolution operator $\mathbb{K}$, circulant [14, 16, 32] and fast transform (FT) [12, 13, 14] preconditioners have been used with success (see Subsection 2.3). Our idea is to combine the benefits of both under the multigrid framework. Specifically, we use FT-based preconditioned conjugate gradient (PCG) as a smoother for MG.

In the following, we focus on two one-dimensional cases: *(i)* $L = I$ (identity), *(ii)* $L = -\Delta$ (Laplacian operator). Preliminary results for the case $L = \nabla(\frac{\nabla u}{|\nabla u|})$ (the divergence operator from the TV-minimization) will be discussed afterwards.

The Case $L = I$. Common relaxation methods fail to smooth the errors because eigenvectors of $L = \alpha I + \mathbb{K}$ corresponding to small eigenvalues are highly oscillatory whereas those corresponding to large eigenvalues are smooth. As a result, relaxation methods do not remove the high frequency errors.

Based on the observations that FT preconditioners are effective for clustering the eigenvalues of L around one, and that CG efficiently annihilates error components corresponding to clusters of eigenvalues, we use PCG with FT preconditioners as smoother for MG. Our numerical results show that the high frequency vectors are concentrated at the cluster for the preconditioned linear system, which are effectively damped away by the cosine transform PCG smoother.

Consequently, faster MG convergence is obtained from the PCG smoothing. We shall now introduce the following notations: the smoother used is indicated in the brackets: PCG(Cos) denotes the cosine transform PCG smoother. We use two presmoothing and no postsmoothing steps. For large α, MG with PCG as smoother is about as efficient as PCG alone, taking into account the two smoothing steps in each MG iteration. But for small α, MG is significantly better, especially for small h.

The Case $L = -\Delta$. The situation of $L = -\alpha\Delta + \mathbb{K}$ is more complicated. The regularization term $\alpha\Delta$ does not simply shift the spectrum, but alters it. For extreme values of α, the eigenvectors of L resemble those of Δ or $\mathbb{K}$. For intermediate α, it is a mixture, but the precise nature of the mixing is not known.

From our numerical results, as expected, we see that MG with relaxation methods as smoother deteriorates by decreasing α since L approaches the convolution operator $\mathbb{K}$. As before, MG(PCG(Cos)) shows better performance over PCG(Cos) alone for small values of α and h.

MG for TV deblurring. MG with PCG as smoother does not work well in this case. We improve its performance by using it as a preconditioner for CG. Since MG with PCG as smoother is a nonstationary preconditioner, we replace the PCG smoother by a relaxation method applied to the cosine transform preconditioned system.

We made a test to see the convergence of PCG using MG as preconditioner for a one-dimensional image restoration problem. For each mesh size h, we use the *optimal* α_{opt} for L so that the recovered image has the specified SNR. Our results indicate that PCG(MG(GS+Cos)) is most effective followed by PCG(MG(GS)). They are not sensitive to α and deterioration with smaller h is slow. However, we have not come up with an efficient implementation for these two methods. Besides, PCG(MG(J+Cos)) shows a degradation over PCG(MG(GS+Cos)), similar to that of the ordinary GS over Jacobi.

In conclusion, even though we feel we have successfully adapted MG for solving (2.3) for the cases of $L = I$ and $L = -\Delta$, the TV-case remains a challenging problem.

3 Extended models

In this section, we extend our total variation reconstruction techniques to color images and to other TV-like functionals based on interpolation, to reduce the staircasing effect.

3.1 Vector valued images: color TV

Given the success of Total Variation restorations for single-channel gray-scale images, it is natural to seek an extension to multiple-channel vector-valued images, e.g., color, and other forms of multispectral data. In Blomgren and Chan [7] we proposed the following extension of the TV-norm to vector-valued functions:

Definition 1 (The Multidimensional $\mathrm{TV}_{\mathrm{n,m}}(\Phi)$ Norm.)
For any function $\Phi : \mathbb{R}^n \to \mathbb{R}^m$, we define:

$$\mathrm{TV}_{\mathrm{n,m}}(\Phi) \stackrel{\mathrm{def}}{=} \sqrt{\sum_{i=1}^{m} [\mathrm{TV}(\Phi^i)]^2}. \tag{3.1}$$

The extension preserves two desirable properties of single-channel TV: *(i)* it does not penalize against edges in the image; and *(ii)* it is rotationally invariant in the image space. Moreover, it reduces to the gray-scale TV-norm when $m = 1$.

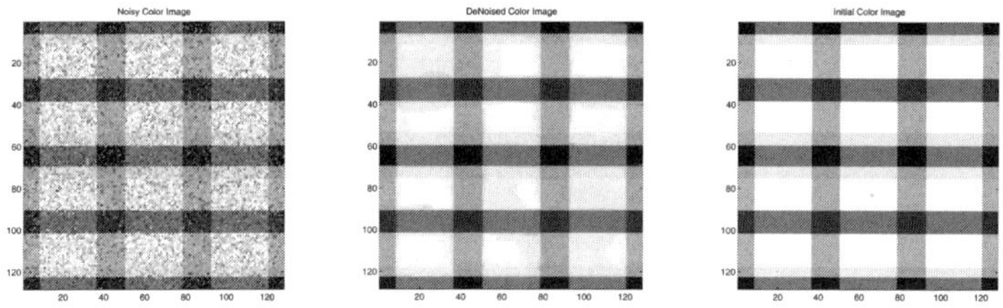

Figure 3.1: Left-to-right: The noisy (SNR = 4.0 dB), denoised and noise-free color images. [This is a color image, please refer to the file *monster.stanford.edu/~blomgren/Research.*]

The norm can be applied to restore color, and other vector-valued images. Given a noisy vector-valued image Φ^0, we are interested in finding the image Φ which minimizes

$$\min_{\Phi \in \mathrm{BV}(\Omega)} \mathrm{TV}_{\mathrm{n,m}}(\Phi) \text{ subject to } \left\| \mathbb{K}\Phi - \Phi^0 \right\|_2^2 = |\Omega|\sigma^2, \tag{3.2}$$

where σ^2 is the variance of the noise. The corresponding *Euler-Lagrange* equations are:

$$\frac{\mathrm{TV}(\Phi_i)}{\mathrm{TV}_{\mathrm{n,m}}(\Phi)} \nabla \circ \left(\frac{\nabla \Phi_i}{||\nabla \Phi_i||} \right) - \lambda \mathbb{K}^* \left(\mathbb{K}\Phi_i - \Phi_i^0 \right) = 0, \tag{3.3}$$

where λ is the associated *Lagrange multiplier*, which determines the correct tradeoff between data-fidelity and smoothing of the solution. See Section 2.2 for more details on determining λ, and Strong and Chan [33] for a discussion of the relation between λ and the minimum scale of recoverable objects.

Solutions of (3.3) can be computed by methods similar to those introduced for gray-scale images, see Section 2.

We show two examples of denoising using the $\mathrm{TV}_{\mathrm{n,m}}$ norm.

Example 1 (Figure 3.1). We created a synthetic 2D image in RGB space, i.e., $\Phi : \mathbb{R}^2 \to \mathbb{R}^3$; Gaussian "white" noise with a *Signal-to-Noise Ratio*, SNR, of 4.0 dB was added to the image, and we ran an explicit time marching scheme, very similar to the one introduced in [29]. As can be observed, the reconstruction is extremely good; the edges are crisply recovered in the correct locations.

Example 2 (Figure 3.2). We denoise (SNR = 3.0 dB) a 128×128 color "Lena" image. We observe that the reconstruction does not smear edges, and objects are recovered in correct locations. However, a human observer may say that the noisy image looks better than the denoised image; this is due to some very advanced image processing and pattern matching in the visual cortex.

Figure 3.2: Left-to-right: The noisy (SNR = 3.0 dB), denoised, and noise-free image. [This is a color image, please refer to the file *monster.stanford.edu/~blomgren/Research.*]

3.2 Reducing the staircasing effect

Images restored using total variation regularization exhibit some artifacts which in 2D manifest themselves as "blockiness," and in 1D as "staircasing" effects (see Figure 3.3b). In 1D, in regions where noise breaks monotone transitions into nonmonotone transitions, the restoration will produce a "step" since straightening the function would increase the fitting error, but not decrease the total variation. The dynamics of the process are more complicated in two dimensions, but the effect is similar — the reconstruction creates small piecewise constant patches. While this blockiness does not constitute a problem for computer vision applications, since no strong false edges are introduced, the restoration is not pleasing to the human eye. In this section, we present some ideas aimed at reducing this staircasing effect.

Functional interpolation

We present here an extension of the total variation norm, based on a functional interpolation between the TV function space, $W^{1,1}$, and the H^1-seminorm space, $W^{1,2}$. The goal is to attain the best of both worlds: sharp edge capturing, and smooth transition in regions without edges, see Figure 3.3. Near an edge we use the $W^{1,1}$ space to capture edges in a TV-like fashion, and in flat regions we use the $W^{1,2}$ space to get an H^1-like reconstruction. In between these regions we use a fractional norm $W^{1,p}$, with $p \in (1,2)$, where p is determined by the image gradient, i.e.,

$$R_{p(x)}(u) = \int_\Omega |\nabla u|^{p(|\nabla u|)}\, d\mathbf{x}, \tag{3.4}$$

where p monotonically decreases from 2 (when $|\nabla u| = 0$), to 1 (as $|\nabla u| \nearrow \infty$).

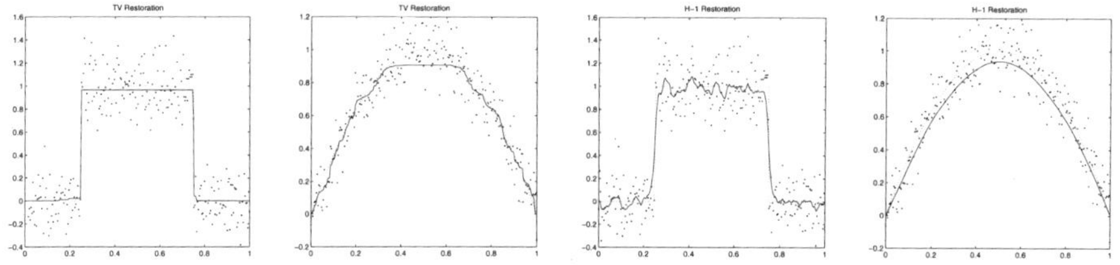

(a) (b) (c) (d)

Figure 3.3: The two leftmost figures show TV restorations of a "box" and a "parabola". The two rightmost figures show H^1-seminorm reconstructions of the same signals. Notice that total variation is very well suited for reconstruction of the piecewise constant image, but suffers from staircasing for the smooth image. The characteristics for the H^1-seminorm is complementary: it suffers from a ringing effect (Gibbs-like phenomena) for the piecewise constant images, but is very well suited for reconstruction of the smooth image.

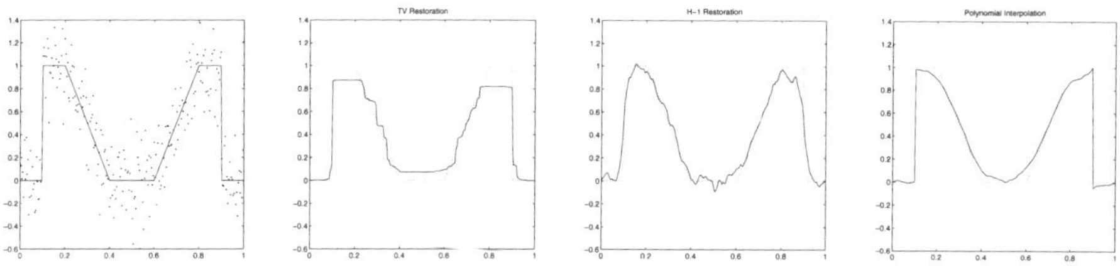

Figure 3.4: Sample restoration using the interpolation norm. Left-to-right: initial+noisy signal, TV, H^1 and polynomial interpolation (PIP)-norm restoration. Notice how the PIP-norm captures the edges and gives smooth ramps.

We have experimented with various transition models $p(|\nabla u|)$: rational functions, polynomials, linear transitions, thresholds, etc. For some details see Blomgren [6, Chapter 4]. Presently we restrict our attention to the case where p is a third order polynomial interpolating between 0 and $|\nabla u|_{\max}$ (the maximal gradient on the discrete grid). Empirically, we have found this polynomial interpolation norm (PIP-norm) to produce good results. Figure 3.4 shows a sample restoration using this p. Notice that the edges are recovered as well, if not better than, for the TV norm, and the recovered ramp-region is much smoother than both the TV- and H^1-semi norms: there is no staircasing and no Gibbs-like phenomena.

We conclude this section by showing some applications of the PIP-norm to

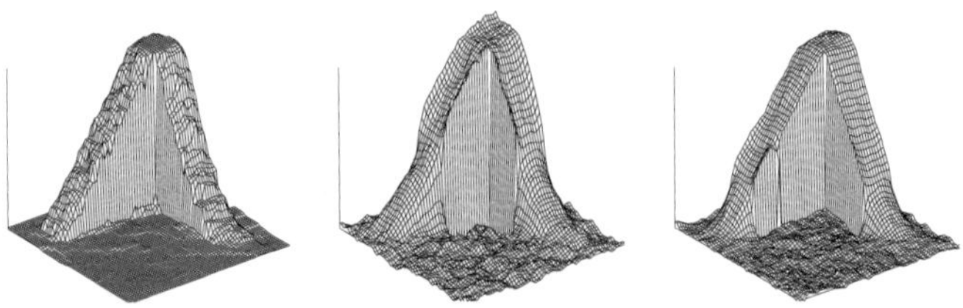

(a) TV (b) PIP (c) $s = 0.5$

Figure 3.5: Comparison of (a) total variation and (b) PIP-interpolation restorations of two disconnected ramps, one linear, the other second order. This example identifies a deficiency of the interpolation approach — when the magnitude of an edge is not large enough to put it in the "TV-region" of the norm, it will be smoothed out. (c) By changing the region of transition of the polynomial, we can determine the level at which smoothing of low-intensity edges sets in.

two-dimensional images. Figure 3.5 shows the impact of TV and PIP reconstructions, from an *SNR = 10 dB* noisy image, of a dual ramp function (one part linear, the other part quadratic). As expected, the PIP reconstruction does not suffer from the staircasing effect, but it has two weaknesses compared with TV: *(i)* The flat region contains more small oscillations — this is not visible at all when rendered as an intensity image, and does not impact edge and object detection. *(ii)* In a region where the magnitude of the edge is small (the bottom part of the ramps), the edge is regularized by a norm closer to H^1, and is thus smoothed. This is a serious drawback, but can be remedied by a rescaling, or tightening of the transition region:

$$\begin{cases} p(x) = ax^3 + bx^2 + cx + d & x < s \cdot |\nabla u|_{\max} \\ p(0) = 2, \quad p'(0) = 0 & \\ p(s \cdot |\nabla u|_{\max}) = 1, \quad p'(s \cdot |\nabla u|_{\max}) = 0 & \\ p(x) = 1 & x >= s \cdot |\nabla u|_{\max}, \end{cases}$$

where $s \in [0, 1]$. The optimal choice of s will depend on the magnitude of objects in the images, and is thus application dependent. Figure 3.5c illustrates how a rescaling by $s = 0.5$ improves the reconstruction.

4 Reduced Mumford-Shah models for segmentation

An important problem in image reconstruction is: given an initial image u_0 (noisy and/or blurry), find an optimal approximation u of u_0, and a set of curves S, such that u is smooth outside S but has jumps or edges only across S.

To solve this problem, Mumford and Shah [25] proposed to minimize the following functional, depending on u and S:

$$F^{MS}(u,S) = \int_{\Omega\setminus S} (\alpha|\nabla u|^2 + \beta(u-u_0)^2)dxdy + |S|,$$

where $\Omega \subset I\!R^2$ is open and bounded, $u_0 : \overline{\Omega} \to I\!R$, $|S|$ denotes the length of S, and $\alpha, \beta > 0$ are parameters. The set S can be considered as S_u, the set of jumps of u.

In practice, it is not easy to solve this problem. Therefore, Ambrosio and Tortorelli [2] proposed an approximation to this problem, replacing the set S_u representing the edges by an edge-function v, equal with one almost everywhere, except in a small neighborhood of S_u. This approximation is:

$$F_\rho^{AT}(u,v) = \int_\Omega \Big[\rho|\nabla v|^2 + \alpha\Big(v^2|\nabla u|^2 + \frac{(v-1)^2}{4\alpha\rho}\Big) + \beta|u-u_0|^2\Big]dxdy. \quad (4.1)$$

If (u_ρ, v_ρ) minimizes F_ρ^{AT}, then $u_\rho \to u$ and $v_\rho \to 1$ in $L^2(\Omega)$ as $\rho \to 0$, and v_ρ is less then one only in a small neighborhood of S_u, which shrinks as $\rho \to 0^+$ ([2]).

Following the idea of Ambrosio-Tortorelli, Shah [30] proposed a more anisotropic approximation:

$$F_\rho^S(u,v) = \int_\Omega \Big[\rho|\nabla v|^2 + \alpha\Big(v^2|\nabla u| + \frac{(v-1)^2}{4\alpha\rho}\Big) + \beta|u-u_0|\Big]dxdy,$$

as a Total Variation minimization variant.

In [18], we show that these two approximations, consisting in solving a system formed by two equations in two unknowns (u,v), can be practically reduced to (nonconvex) approximations with only one equation and one unknown, the image u. In addition, the edge-function v has an explicit representation. Our model can simultaneously be used for denoising and edge detection. We next explain the reduced models.

The system associated with the minimization of $G_\rho^{AT}(u,v)$ is:

$$\begin{cases} \beta u - \alpha\nabla(v^2\nabla u) = \beta u_0 \\ -\triangle v + \frac{1+4\alpha\rho|\nabla u|^2}{4\rho^2}\Big(v - \frac{1}{1+4\alpha\rho|\nabla u|^2}\Big) = 0. \end{cases} \quad (4.2)$$

We can see from the second equation of (4.2) that v is in fact a smoothing of $\frac{1}{1+4\alpha\rho|\nabla u|^2}$. We want to reduce this system to only one equation with only one unknown, while still being able to extract the edges. We see that we cannot simplify further or drop the first equation of (4.2) in u: the parameter α has to be strictly positive, because first of all, we have to regularize u to remove noise. Our main new idea is to drop the term $-\triangle v$ from the second equation in v of (4.2). Then this equation in v has the explicit solution $v = v_\rho(|\nabla u|) = \frac{1}{1+4\alpha\rho|\nabla u|^2}$. Let us replace it in $F_\rho^{AT}(u,v)$, but without the regularizing term $\int_\Omega \rho|\nabla v|^2$. We obtain after computation the functional:

$$F_\rho^{at}(u) = \int_\Omega \left(\alpha \frac{|\nabla u|^2}{1+4\alpha\rho|\nabla u|^2} + \beta|u-u_0|^2\right) dxdy.$$

In this way, the following new problem:

$$\begin{cases} \inf_u \left\{ \, F_\rho^{at}(u) \, \right\} & \text{(restoration for } u\text{)} \\ v_\rho(|\nabla u|) = \frac{1}{1+4\alpha\rho|\nabla u|^2} & \text{(edge-strength function)} \end{cases} \tag{4.3}$$

is our reduced model for the Ambrosio-Tortorelli approximation (4.1). Then we no longer have to solve two coupled PDE's, but only one, from the minimization of $F_\rho^{at}(u)$, v being expressed as an explicit function of the gradient of u.

In the same way, the system associated to the minimization of $F_\rho^S(u,v)$ can be reduced to the minimization of the functional:

$$F_\rho^s(u) = \int_\Omega \left(\alpha \frac{|\nabla u|}{1+4\alpha\rho|\nabla u|} + \beta|u-u_0|\right) dxdy,$$

and the following new problem:

$$\begin{cases} \inf_u \left\{ \, F_\rho^{s}(u) \, \right\} & \text{(restoration for } u\text{)} \\ v_\rho(|\nabla u|) = \frac{1}{1+4\alpha\rho|\nabla u|} & \text{(edge-strength function)} \end{cases} \tag{4.4}$$

is our reduced model for the Shah approximation, with only one PDE in one unknown u, v being expressed as a function of $|\nabla u|$.

Analogous to these new approximations, we may consider a more general model, as follows:

$$\begin{cases} \inf_u F_\rho(u) = \int_\Omega \left(\alpha \frac{|\nabla u|^p}{1+4\alpha\rho|\nabla u|^p} + \beta|u-u_0|^p\right) dxdy, \\ v(|\nabla u|) = \frac{1}{1+4\alpha\rho|\nabla u|^p} \quad \text{(edge-strength function).} \end{cases} \tag{4.5}$$

Removing the regularizing term in v from the A-T and S approximations, we will see that the edges are better preserved. Our reduced approximations are nonconvex, but the numerical finite differences schemes are unconditionally stable (see [5]) and the results are very satisfactory. For more details we refer the reader to [18].

In our reduced models, we use for any $p \geq 1$, the L^2 norm of $(u - u_0)$, instead of the L^p norm in (4.5). We also modify the Shah approximation again replacing the L^1 norm of $(u - u_0)$ by the L^2 norm.

We end this section by presenting numerical results and comparisons for two images. We first choose the parameters for the synthetic image, from Figure 4.1, where we know the original image. Then we use exactly the same corresponding parameters for the real image, from Figure 4.2. We see that our reduced models produce better reconstruction, and with sharper edges.

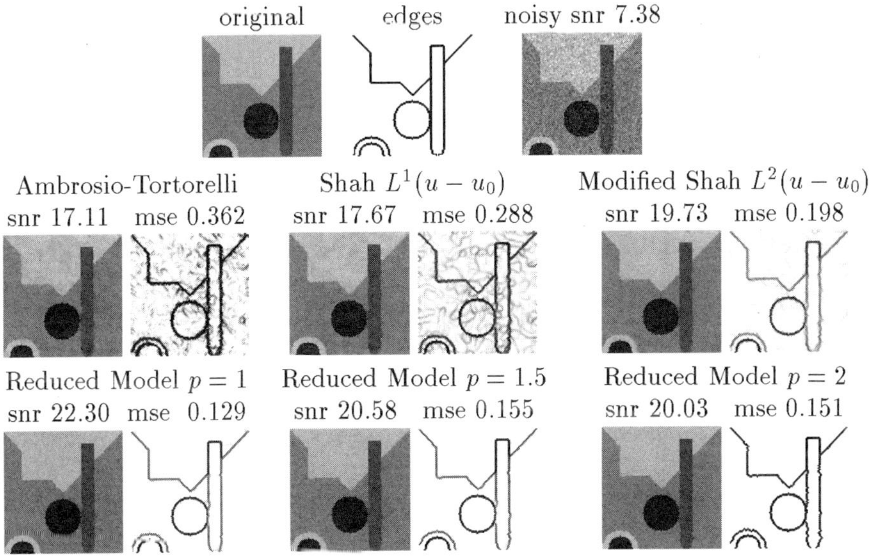

Figure 4.1: Comparisons on the synthetic image. Top: original image u with edges v, and noisy version. Middle: results (u, v) obtained with the Ambrosio-Tortorelli and Shah approximations. Bottom: results (u, v) obtained with our reduced models for $p = 1; 1.5; 2$. Removing the regularization in v, we not only reduced the system to only one equation and one unknown, but also the results have sharper edges, both in the image u and the edge-strength function v. Moreover, the reduced models produce better reconstruction (see the Signal-to-Noise Ratio, SNR, and the Mean Square Error, MSE).

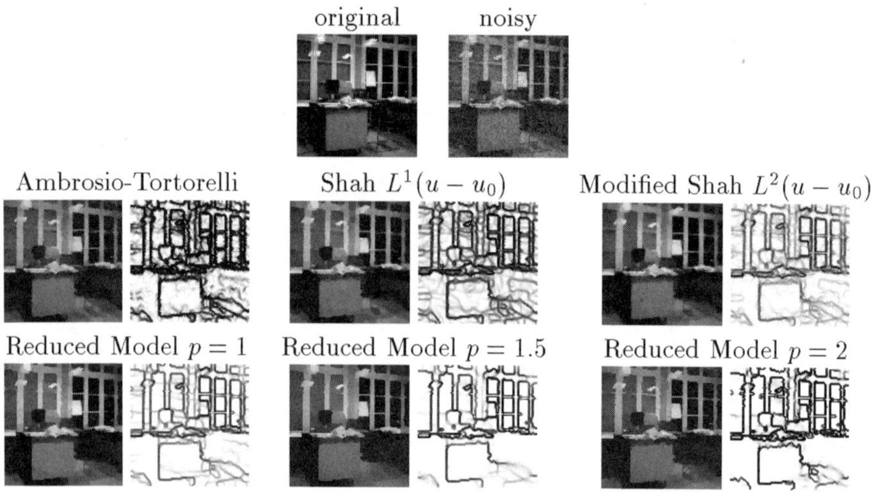

Figure 4.2: Comparisons on the real image. Top: original image u and noisy version. Middle: results (u, v) obtained with the Ambrosio-Tortorelli and Shah approximations. Bottom: results (u, v) obtained with our reduced models for $p = 1; 1.5; 2$. Again, the reduced models produce better reconstruction, and with sharper edges.

5 Active contour models without edges

The basic idea in active contour models or snakes is to evolve a curve, subject to constraints from a given image, in order to detect objects in that image. For instance, starting with a curve around the object to be detected, the curve moves toward its interior normal, and has to stop on the "boundary" of the object.

In problems of curve evolution as active contours or snakes, the level set technique (see [26]) has been used extensively, because it allows for automatic changes in the topology of the evolving curve. To formulate our model, we will use the same technique.

Let $\Omega \subset I\!R^2$ be a bounded open set and $I : \overline{\Omega} \rightarrow I\!R$ be a given image. Usually, $\overline{\Omega}$ is a rectangle in the plane and I takes values between 0 and 255. The level set function $\phi : I\!R^2 \rightarrow I\!R$ is a Lipschitz function. The evolving curve will be represented by the zero-level set of ϕ, such that ϕ has opposite signs on each side of the curve.

Before introducing our model, we recall two more "classical" active contour models based on the mean curvature motion and level set techniques ([26]). In these two models, a "stopping" function g depending on the gradient $|\nabla I|$ of the image I has to be known. For instance, g can be:

$$g(|\nabla I|) = \frac{1}{1 + |\nabla(G_\sigma * I)|^2},$$

where G_σ is a Gaussian kernel to smooth the image I. Basically, we require $\lim_{t\to\infty} g(t) = 0$.

The first geometric model [9] is given by:

$$\begin{cases} \frac{\partial \phi}{\partial t} = g(|\nabla I|)|\nabla \phi| \operatorname{div}\Big(\frac{\nabla \phi}{|\nabla \phi|}\Big) + \mu g(|\nabla I|)|\nabla \phi| \text{ in } [0, \infty[\times\Omega, \\ \phi(0, x, y) = \phi_0(x, y) \text{ in } \Omega \end{cases} \tag{5.1}$$

where μ is a parameter.

The second model, closely related to the previous one, is the geodesic active contour model [10], given by:

$$\begin{cases} \frac{\partial \phi}{\partial t} = |\nabla \phi| \operatorname{div}\Big(g(|\nabla I|)\frac{\nabla \phi}{|\nabla \phi|}\Big) + \mu g(|\nabla I|)|\nabla \phi| \text{ in } [0, \infty[\times\Omega, \\ \phi(0, x, y) = \phi_0(x, y) \text{ in } \Omega. \end{cases} \tag{5.2}$$

In these two models, the curve moves toward the edges of the object, where it stops, because $g \approx 0$ on the contours. But in many cases in practice, g is not exactly zero on the edges, and the curve may pass through the boundary. Also, because the stopping function g depends on the gradient of the image I, these models can detect only edges or boundaries defined by gradient. On the other hand, our proposed model can detect both edges with or without gradient, or very smooth edges.

In our model, we minimize the length of the curve and/or the area inside the curve but, in order to stop the curve on the boundary of the desired object, we use Mumford-Shah segmentation techniques: we also minimize the differences between I and the averages of I "inside" and "outside" the curve. The set S from the Mumford-Shah functional will be the zero-level set of ϕ, and the solution u will be a function taking only two values, namely:

$$u = \begin{cases} \text{average}(I) \text{ on } \phi > 0 \\ \text{average}(I) \text{ on } \phi < 0. \end{cases}$$

We denote by c_1 and c_2 two constants, representing the averages of I inside and outside the evolving curve (the averages of I on $\{\phi > 0\}$ and on $\{\phi < 0\}$, respectively, choosing ϕ to be positive "inside" its zero-level curve and negative "outside").

Using these notations, our model is as follows: we introduce the energy $F(\phi, c_1, c_2)$, defined by:

$$F(\phi, c_1, c_2) = \mu_1 \text{length}\{\phi = 0\} + \mu_2 \text{area}\{\phi \geq 0\} + \lambda\Big(\int_{\phi\geq 0} |I - c_1|^2 + \int_{\phi<0} |I - c_2|^2\Big),$$

where μ_1, μ_2 and λ are positive parameters, and we consider the minimization problem:

$$\inf_{\phi, c_1, c_2} F(\phi, c_1, c_2). \tag{5.3}$$

Using the Heaviside function H defined by $H(x) = 1$ if $x \geq 0$ and $H(x) = 0$ if $x < 0$, and the one-dimensional Dirac measure δ concentrated in 0 and defined by $\delta(x) = \frac{d}{dx}H(x)$ in the sense of distributions, the energy F can be expressed in the following way:

$$\begin{aligned} F(\phi, c_1, c_2) = {} & \mu_1 \int_\Omega \delta(\phi)|\nabla\phi| dxdy + \mu_2 \int_\Omega H(\phi) dxdy \\ & + \lambda\Big(\int_\Omega |I - c_1|^2 H(\phi) + \int_\Omega |I - c_2|^2 (1 - H(\phi))\Big). \end{aligned}$$

Keeping ϕ fixed and minimizing the energy $F(\phi, c_1, c_2)$ with respect to the constants c_1 and c_2, we express these constants function of ϕ as:

$$c_1(\phi) = \frac{\int_\Omega I \cdot H(\phi)}{\int_\Omega H(\phi)}, \quad c_2(\phi) = \frac{\int_\Omega I(1 - H(\phi))}{\int_\Omega (1 - H(\phi))}.$$

We see that c_1 is the average of I in $\{\phi \geq 0\}$ and c_2 is the average of I in $\{\phi < 0\}$.

To solve the problem numerically, we use the Euler-Lagrange equations already derived for c_1 and c_2, for fixed ϕ. To obtain the associated Euler-Lagrange equation in ϕ, we use slightly regularized versions of H and δ. Let H_ε and $\delta_\varepsilon = H'_\varepsilon$ be such regularizations.

The associated evolution problem, parameterizing the descent direction by time, is:

$$\begin{cases} \frac{\partial\phi}{\partial t} = \delta_\varepsilon(\phi)\Big[\mu_1 \text{div}\Big(\frac{\nabla\phi}{|\nabla\phi|}\Big) - \mu_2 - \lambda\Big((I - c_1)^2 - (I - c_2)^2\Big)\Big] \text{ in } \Omega, \\ \phi(t, x, y) = \phi_0(x, y) \text{ in } \Omega, \quad \frac{\delta_\varepsilon(\phi)}{|\nabla\phi|}\frac{\partial\phi}{\partial n} = 0 \text{ on } \partial\Omega, \\ c_1(\phi) = \frac{\int_\Omega I \cdot H(\phi) dxdy}{\int_\Omega H(\phi) dxdy}, \quad c_2(\phi) = \frac{\int_\Omega I(1-H(\phi)) dxdy}{\int_\Omega (1-H(\phi)) dxdy}. \end{cases}$$

We solve this problem using an implicit finite differences scheme. We will not go into the details of the numerical scheme. We refer the reader to [19].

We end this section and the chapter with numerical results using our active contour model without edges (see Figure 5.1). These examples show in particular all the abilities of the model: detecting edges with or without gradient, automatic change of topology, detecting contours on real and very noisy images.

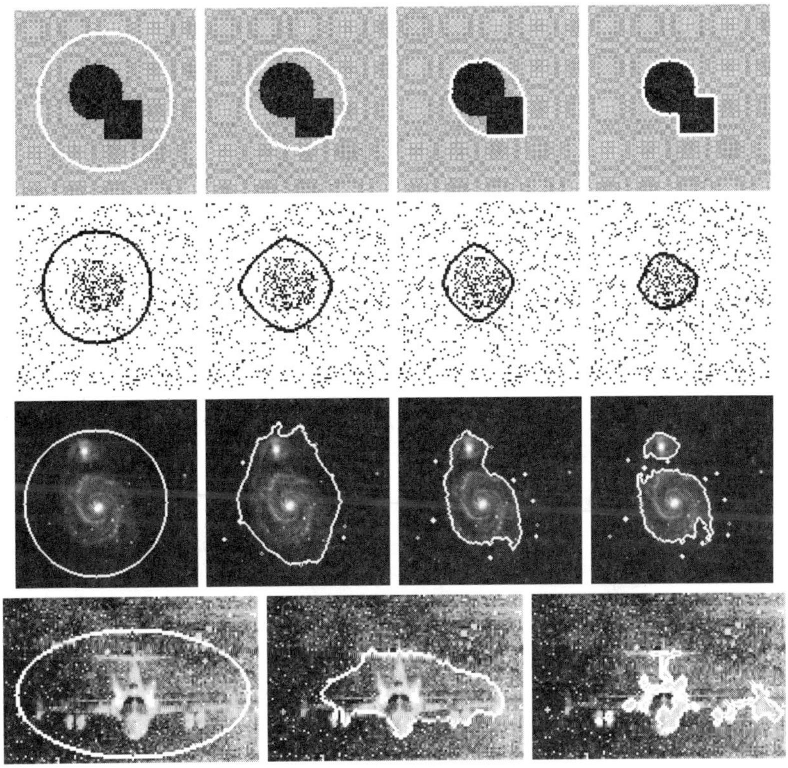

(1)
(2)
(3)
(4)

Figure 5.1: Numerical results using our model "without edges". The model has the abilities of detecting edges with (1) or without (2) gradient, automatical change of topology with blurred edges (3) and detecting edges in real and very noisy images (4). Classical models cannot detect the edges in example (2).

References

[1] L. Alvarez, F. Guichard, P.-L. Lions, and J.-M. Morel. Axioms and fundamental equations of image processing. *Arch. Rat. Mech. Anal.*, 123:199–257, 1993.

[2] L. Ambrosio and V. Tortorelli. On the approximation of free discontinuity problems. *Bollettino U.M.I. (7)*, 6-B:105–123, 1992.

[3] K. Andersen, E. Christiansen, A. R. Conn, and M. L. Overton. An efficient primal-dual interior-point method for minimizing a sum of euclidian norms. *Technical Report TR1998-769, Computer Science Department, Courant Institute,* 1999, to appear in *SIAM J. Sci. Comput.*

[4] G. Aubert, M. Barlaud, L. Blanc-Féraud, and P. Charbonnier. Deterministic edge-preserving regularization in computed imaging. *Technical Report 94-01, Informatique, Signaux et Systèmes de Sophia Antipolis*, 1994.

[5] G. Aubert and L. Vese. A variational method in image recovery. *SIAM J. Num. Anal.*, 34(5):1948–1979, October 1997.

[6] P. Blomgren. *Total Variation Methods for Restoration of Vector Valued Images.* Ph.D. thesis, UCLA CAM Report 98-30, Mathematics Department, Los Angeles, 1998.

[7] P. Blomgren and T. F. Chan. Color TV: total variation methods for restoration of vector valued images. *IEEE Trans. Image Proc.*, 7(3):304–309, March 1998.

[8] A. Brandt. Multi-level adaptive solutions to boundary-value problems. *Math. Comp.*, 31:333–390, 1977.

[9] V. Caselles, F. Catté, T. Coll, and F. Dibos. A geometric model for active contours in image processing. *Numer. Math.*, 66:1–31, 1993.

[10] V. Caselles, R. Kimmel, and G. Sapiro. Geodesic active contours. *Int. J. Comput. Vis.*, 22:61–79, 1997.

[11] A. Chambolle and P.-L. Lions. Image recovery via total variation minimization and related problems. *Technical Report 9509, CEREMADE, 95.*

[12] R. Chan, T. Chan, and C. Wong. Cosine transform based preconditioners for total variation minimization problems in image processing. In

S. Margenov and P. Vassilevski, editors, *Iterative Methods in Linear Algebra, II*, volume 3 of *IMACS Series in Computational and Applied Math.*, pages 311–329. IMACS, Rutgers University, New Brunswick, NJ, 1996.

[13] R. Chan, K. Ng, and C. Wong. Sine transform based preconditioners for symmetric Toeplitz systems. *Linear Algebra Appls.*, 232:237–260, 1996.

[14] R. Chan and M. Ng. Conjugate gradient methods for Toeplitz systems. *SIAM Rev.*, 38:427–482, 1996.

[15] R. H. Chan, T. F. Chan, and H. M. Zhou. Continuation Methods for Total Variation Denoising Problems. In F.T. Luk, editor, *SPIE Proceedings of Advanced Signal Processing Algorithms*, pages 314–325, San Diego, CA, 1995.

[16] T. Chan. An optimal circulant preconditioner for Toeplitz systems. *SIAM J. Sci. Stat. Comput.*, 9:766–771, 1988.

[17] T. Chan and J. Olkin. Circulant preconditioners for Toeplitz-block matrices. *Numer. Algorithms*, 6:89–101, 1994.

[18] T. Chan and L. Vese. Variational image restoration & segmentation models and approximations. UCLA CAM Report 97-47, Mathematics Department, Los Angeles, 1997.

[19] T. Chan and L. Vese. Active contours without edges. UCLA CAM Report 98-53, Mathematics Department, Los Angeles, 1998. To appear in the *IEEE Trans. Image Process.*

[20] T. F. Chan and P. Mulet. On the convergence of the lagged diffusivity fixed point method in image restoration. *SIAM J. Numer. Anal.*, 36(2):354–367, 1999.

[21] T. F. Chan, G. Golub, and P. Mulet. A nonlinear primal-dual method for TV-based image restoration. In M. Berger, R. Deriche, I. Herlin, J. Jaffre, and J. Morel, editors, *ICAOS '96, 12th International Conference on Analysis and Optimization of Systems: Images, Wavelets, and PDEs, Paris, June 26–28, 1996*, number 219 in Lecture Notes in Control and Information Sciences, pages 241–252, 1996.

[22] W. Hackbusch. *Multi-grid Methods and Applications*. Springer-Verlag, Berlin, 1985.

[23] T. Kailath and V. Olshevsky. Displacement structure approach to discrete-trigonometric-transform based preconditioners of G. Strang and T. Chan types. In *Italian Journal Calcolo, devoted to the Proceedings of the International Workshop on "Toeplitz Matrices, Algorithms and Applications", Cortona, Italy,* September 1996.

[24] H. B. Keller. Numerical solution of bifurcation and nonlinear eigenvalue problems. In P. Rabinowitz, editor, *Applications of Bifurcation Theory,* pages 359–384. Academic Press, New York, 1977.

[25] D. Mumford and J. Shah. Optimal approximation by piecewise smooth functions and associated variational problems. *Comm. Pure Appl. Math.,* 42:577–685, 1989.

[26] S. Osher and J. A. Sethian. Fronts propagating with curvature-dependent speed: algorithms based on hamilton-jacobi formulation. *J. Comput. Phys.,* 79:12–49, 1988.

[27] P. Perona and J. Malik. Scale-space and edge detection using anisotropic diffusion. *IEEE Trans. Pattern Anal. Mach. Intell.,* 12:629–639, 1990.

[28] J. G. Rosen. The gradient projection methods for nonlinear programming, part II, nonlinear constraints. *J. Soc. Indust. Appl. Math,* 9:514, 1961.

[29] L. I. Rudin, S. Osher, and E. Fatemi. Nonlinear total variation based noise removal algorithms. *Physica D,* 60:259–268, 1992.

[30] J. Shah. A common framework for curve evolution, segmentation and anisotropic diffusion. In *Proc. of I.E.E.E. Conf. on Computer vision and pattern recognition,* pages 136–142, 1996.

[31] H. Sorensen and C. Burrus. Fast DFT and convolution algorithms. In S. K. Mitra and J. F. Kaiser, editors, *Handbook for Digital Signal Processing,* John Wiley, New York, 1993.

[32] G. Strang. A proposal for Toeplitz matrix calculations. *Stud. Appl. Math.,* 74:171–176, 1986.

[33] D. M. Strong and T. F. Chan. Relation of regularization parameter and scale in total variation based image denoising. In *IEEE Workshop on Mathematical Methods in Biomedical Image Analysis,* 1996.

[34] A. N. Tikhonov and V. Y. Arsenin. *Solutions of Ill-Posed Problems.* John Wiley, New York, 1977.

[35] C. Vogel and M. Oman. A fast, robust algorithm for total variation based reconstruction of noisy, blurred images. *IEEE Trans. Image Process.*, 7:813–824, 1998.

[36] C. R. Vogel and M. E. Oman. Fast total variation-based image reconstruction. In *Proceedings of the ASME Symposium on Inverse Problems*, 1995.

[37] C. R. Vogel and M. E. Oman. Fast total variation-based image reconstruction. In *Proceedings of the 1995 ASME Design Engineering Conferences*, volume 3, pages 1009–1015, 1995.

[38] C. R. Vogel and M. E. Oman. Iterative methods for total variation denoising. *SIAM J. Sci. Stat. Comput.*, 17:227–238, 1996.

Peter Blomgren
Stanford University
Mathematics Department
Building 380, MC 2125
Stanford, CA 94305
`blomgren@math.stanford.edu`

Tony F. Chan and Luminita Vese
UCLA Mathematics Department
405 Hilgard Ave.
Los Angeles, CA 90095-1555
{`chan,lvese`}`@math.ucla.edu`

Pep Mulet
Departamento Matematica Aplicada
Universidad Valencia
Dr. Moliner 50
46100 Burjassot (Valencia)
Spain
`mulet@uv.es`

Win Lok Wan
SCCM, Room 282, Gates Building 2B
Stanford University
Stanford, CA 94305-9025
`wan@sccm.stanford.edu`

F. Brezzi

Interacting with the subgrid world

Abstract In a number of applications, subgrid scales cannot be neglected. Sometimes, they are just a spurious by-product of a discretised scheme that lacks the necessary stability properties. In other cases, they are related to physical phenomena that actually take place on a very small scale but still have an important effect on the solution. We discuss here an attempt to recover information on the subgrid scales, by trying to simulate their effects on the computable ones.

This paper is dedicated to my friend Ron Mitchell.

1 Introduction

There are essentially two types of subgrid phenomena that must be taken into account in modern numerical simulation.

One of them occurs when the discretisation lacks the necessary stability properties. This is often due to the fact that the numerical scheme does not treat in a proper way the smallest scales allowed by the computational grid. As a consequence, they appear as abnormally amplified in the final numerical results. Most types of numerical instabilities are produced in this way, as it can be easily confirmed by checking the beautiful review on numerical instabilities reported in [18]. In the last decade it has become clear that several attempts to recover stability, in these cases, could be interpreted as a way of improving the simulation of the effects of the smallest scales on the larger ones. By doing that, the small scales can be *seen* by the numerical scheme and therefore be kept under control.

A second type of subgrid phenomena is related, instead, with actual physical effects that take place on a scale which is often much smaller than the smallest one representable on the computational grid. These effects have however a strong impact on the larger scales, and cannot be neglected without jeopardizing the overall quality of the final solution.

These two situations are quite different, in nature and scale. Nevertheless it is not unreasonable to hope that some techniques that have been developed for dealing with the former class of phenomena might be adapted to deal with the latter one. In this sense, the most promising technique seems to be the use of

Residual-Free Bubbles. In the following sections, we are going to summarize the general idea behind this, trying to underlie its potential and its limitations. In particular we shall first present in Section 2 the basic principles of the strategy: *divide and conquer*, *static condensation* and *approximate solution*. In Section 3, as an example, we shall show how the strategy works on a simple model problem. In the final section we shall briefly discuss the possible extensions of the procedure to different problems.

2 Basic principles

At a very general (and generic) level, the procedure can be summarized as follows. We start with a given problem that, for simplicity, we assume to be linear, and in variational form:

$$\begin{cases} \text{find } u \in V \text{ such that :} \\ \mathcal{L}(u,v) = (f,v), \qquad \forall v \in V. \end{cases} \tag{2.1}$$

We assume that we are given a discretised problem:

$$\begin{cases} \text{find } u_h \in V_h \text{ such that :} \\ \mathcal{L}(u_h,v_h) = (f,v_h), \qquad \forall v_h \in V_h, \end{cases} \tag{2.2}$$

where $V_h \subset V$ is a finite element space, corresponding to a given decomposition $\mathcal{T}_h$ of the computational domain. We suppose, roughly speaking, that $\mathcal{T}_h$ is the finest grid we are ready to afford in the computation, in the sense that we are not ready to solve a final system having more unknowns than the dimension of V_h. Problem (2.2) is now, temporarily and artificially, augmented by considering a new subspace of V:

$$V_A := V_h + B_h, \tag{2.3}$$

where B_h is the (infinite dimensional) space of bubbles

$$B_h := \Pi_K \; B(K), \tag{2.4}$$

and, for every K in $\mathcal{T}_h$,

$$B(K) := \{v |\, v \in V, \; \text{supp}(v) \subset K\}. \tag{2.5}$$

As a typical example, if $V = H_0^1(\Omega)$, Ω being the computational domain, then $B(K) = H_0^1(K)$. The augmented problem now reads:

$$\begin{cases} \text{find } u_A \in V_A \text{ such that :} \\ \\ \mathcal{L}(u_A, v_A) = (f, v_A), \qquad \forall v_A \in V_A. \end{cases} \tag{2.6}$$

In principle, problem (2.6), although unsolvable, should be able to take into account all the small scales that *do not cross the boundaries of the elements* K. This is a severe limitation, but corresponds to a sort of **divide and conquer** principle that might, in the end, assure some feasibility to the whole procedure.

We now proceed to eliminate, at least formally, the bubble unknowns from problem (2.6). The technique that we are going to use is well known in the engineering practice, under the name of **static condensation**. However, here we apply it in a more general, infinite dimensional, case.

Assuming, for simplicity, that in (2.3) we have a direct sum of subspaces of V, we can write u_A and v_A, in a unique way, as:

$$u_A = u_h + u_B, \tag{2.7}$$

and

$$v_A = v_h + v_B, \tag{2.8}$$

respectively. Inserting (2.7) into (2.6), and taking $v = v_B$ we obtain the so-called **bubble equation**:

$$\begin{cases} \text{find } u_B \in B_h \text{ such that :} \\ \\ \mathcal{L}(u_B, v_B) = -\mathcal{L}(u_h, v_B) + (f, v_B), \qquad \forall v_B \in B_h. \end{cases} \tag{2.9}$$

The bubble equation (2.9) will play a fundamental role in the following discussion. We take advantage of the split nature of the space B_h. For every element K in $\mathcal{T}_h$ we define $u_{B,K}$ as the restriction of u_B to the element K. Then, for every $\varphi \in B(K)$ we have

$$\mathcal{L}(u_B, \varphi) = (f - Lu_h, \varphi), \tag{2.10}$$

where L indicates the operator associated with the bilinear form $\mathcal{L}$. Problem (2.10) can then be written in strong form as

$$Lu_{B,K} = f - Lu_h, \tag{2.11}$$

with the associated boundary conditions. In most cases, as we have seen, $B(K)$ will be equal to $H_0^1(K)$, and the boundary conditions will be simply

$$u_{B,K} = 0 \text{ on } \partial K. \tag{2.12}$$

More generally, the boundary conditions will be implicitly imposed by the two conditions $u_{B,K} \in V$ and $supp(u_{B,K}) \subset K$. We shall write the solution of (2.9) (or, in most examples, of the equivalent problem (2.11)-(2.12)) in compact form as

$$u_{B,K} = L_{B,K}^{-1}(f - Lu_h), \tag{2.13}$$

which implicitly defines the operator $L_{B,K}^{-1}$. With this notation we can write:

$$u_B = \sum_K L_{B,K}^{-1}(f - Lu_h). \tag{2.14}$$

Having made the dependence of u_B on u_h explicit in (2.14), we can go back to (2.6) and take $v_A = v_h$; inserting (2.7) and (2.14) we obtain:

$$\begin{cases} \text{find } u_h \in V_h \text{ such that :} \\ \mathcal{L}(u_h, v_h) + \sum_K \mathcal{L}(L_{B,K}^{-1}(f - Lu_h), v_h) = (f, v_h), \qquad \forall v_h \in V_h. \end{cases} \tag{2.15}$$

Note that (2.15) has the same form (and the same number of unknowns) of (2.2). However, the additional term

$$\sum_K \mathcal{L}(L_{B,K}^{-1}(f - Lu_h), v_h) \tag{2.16}$$

now takes into account the effect of *some* small scales (the ones that do not cross the inter-element boundaries) onto the scales that are visible on the computational grid.

The only, nonnegligible, difficulty is that, in general, (2.9) (or even the particular case of problem (2.11)-(2.12)) cannot be solved explicitly, so that the term (2.16) cannot be computed. We remark however that, in order to compute (2.16) in some approximate way, we do not need a very accurate solution of (2.9). Indeed, it is only *the effect of the small scales on the larger ones* that needs to be simulated, as comes out clearly by noting that in (2.16) the term $L_{B,K}^{-1}(f - Lu_h)$ is tested against v_h, which belongs to the coarse space. We can therefore hope that an approximate solution of (2.9) (or, more often, of problem (2.11)-(2.12)) can be sufficient to reproduce with reasonable accuracy the effect of the additional term (2.16). The most important aspect of the whole procedure is that such an element-by-element approximate solution can be performed in parallel, as a sort of preprocessing, and its results will then be used within the process of computing and assembling the final matrix corresponding to problem (2.15). In the next section we are going to see more practical aspects of the above discussion for a simple model problem.

3 A model problem

We consider here, as a model problem, the classical **toy problem** of advection-dominated linear equations. From the physical point of view, we may think to the problem of the passive transport of a scalar diffusive quantity in a fluid whose velocity is known. Let then Ω be, for instance, a convex polygon, ε a positive number (= diffusion coefficient), $\mathbf{c}$ a bounded mapping from Ω to $\mathbb{R}^2$ (= velocity field) and f, say, an element of $L^2(\Omega)$ (= source term). We consider then the problem of finding u in $H_0^1(\Omega)$ such that

$$-\varepsilon\Delta u + \mathbf{c}\cdot\nabla u = f \quad \textit{in } \Omega. \tag{3.1}$$

We can set $Lu := -\varepsilon\Delta u + \mathbf{c}\cdot\nabla u$, and

$$\mathcal{L}(u,v) := \varepsilon a(u,v) + c(u,v) \qquad \forall u,v \in H_0^1(\Omega), \tag{3.2}$$

where, in a natural way,

$$a(u,v) := \int_\Omega \nabla u\cdot\nabla v \,\mathrm{d}x, \quad \text{and} \quad c(u,v) := \int_\Omega \mathbf{c}\cdot\nabla u\, v \,\mathrm{d}x \quad \forall u,v \in H_0^1(\Omega). \tag{3.3}$$

Assume now that we are given a decomposition $\mathcal{T}_h$ of Ω into triangles, and assume moreover that $\mathbf{c}$ and f are piecewise constant on $\mathcal{T}_h$. We take then V_h to be the space of piecewise linear continuous functions vanishing on $\partial\Omega$, and B_h as in (2.4) with $B(K) = H_0^1(K)$ for each K. If we apply the theory of the previous section, the bubble equation (2.11) becomes, in each triangle K: find $u_{B,K}$ in $H_0^1(K)$ such that

$$-\varepsilon\Delta u_{B,K} + \mathbf{c}\cdot\nabla u_{B,K} = -(-\varepsilon\Delta u_h + \mathbf{c}\cdot\nabla u_h) + f \quad \textit{in } K. \tag{3.4}$$

As we already pointed out, equation (3.4) is unsolvable. As we shall see, there are ways to get around this difficulty in a more or less satisfactory way. Before discussing that, however, we want to point out how the solution in the model case can be used. In particular, it is not difficult to check that, in the present case, we have $a(u_B, v_h) = 0$ for every $u_B \in B_h$ and for every $v_h \in V_h$. Hence the additional term (2.16) arising in (2.15) becomes

$$\mathcal{L}(u_B, v_h) = c(u_B, v_h) = \int_\Omega \mathbf{c} \cdot \nabla u_B \, v_h \, \mathrm{d}x = -\int_\Omega u_B \mathbf{c} \cdot \nabla v_h \, \mathrm{d}x, \qquad (3.5)$$

with an obvious integration by parts. We also remark that the term $\mathbf{c} \cdot \nabla v_h$ is piecewise constant. Hence we see that **only the mean value** of u_B in each K will be used in the final system (2.15) for computing u_h. Moreover, still in our assumptions, we observe that the right-hand side of (3.4) is also constant in K, so that $u_{B,K}$, in each K, can be written as

$$u_{B,K} = b_K R_K, \qquad (3.6)$$

where

$$R_K := -(-\varepsilon \Delta u_h + \mathbf{c} \cdot \nabla u_h) + f \qquad (3.7)$$

is the *residual* in K (taking u_h as approximate solution) and the bubble b_K is the solution of the scaled problem:

$$\begin{cases} \text{find } b_K \in H_0^1(K) \text{ such that :} \\ -\varepsilon \Delta b_K + \mathbf{c} \cdot \nabla b_K = 1 \quad \textit{in } K. \end{cases} \qquad (3.8)$$

A simple computation shows that, inserting (3.6) in (3.5), the additional term (2.16) becomes

$$c(u_B, v_h) = \sum_K \frac{\int_K b_K \, \mathrm{d}x}{|K|} \int_K (\mathbf{c} \cdot \nabla u_h - f) \, \mathbf{c} \cdot \nabla v_h \, \mathrm{d}x, \qquad (3.9)$$

where b_K is still the solution of (3.8), which is still unsolvable. This, as already pointed out in [11] (see also [24], [5],) corresponds to the use of the well known SUPG (Streamline Upwind Petrov Galerkin) method (see [12], [14]) with the stabilising parameter chosen as

$$\tau_K = \frac{\int_K b_K \mathrm{d}x}{|K|}. \qquad (3.10)$$

We still have to tackle the problem of getting an approximate solution of (3.8). We shall consider several possibilities. As the present model problem is by far the most deeply studied among the various possible applications of the general technique, we shall obtain a reasonably complete overlook of the various options that can also be considered in different contexts.

The obvious, and most general, way of approximating (2.9) would consist in using a Galerkin approximation: we take, for each K, a subspace B_K^* of $H_0^1(K)$ and we look for $u_B^* = \sum_K u_{B,K}^*$ such that each $u_{B,K}^* \in B_K^*$ and satisfies

$$\mathcal{L}(u_{B,K}^*, v) = -\mathcal{L}(u_h, v) + (f, v) \quad \forall v \in B_K^*, \tag{3.11}$$

where, as usual, we identify a function in B_K^* with its extension by zero on the whole Ω. In our case, this amounts to solve in each K the following approximated version of the scaled equation (3.8):

$$\begin{cases} \text{find } b_K^* \in B_K^* \text{ such that :} \\ \\ \varepsilon a(b_K^*, v) + c(b_K^*, v) = \int_K v \, \mathrm{d}x, \quad \forall v \in B_K^*. \end{cases} \tag{3.12}$$

This would then give the solution $u_{B,K}^*$ of (3.11) as $u_{B,K}^* = b_K^* R_K$, with R_K always given by (3.7). For our model problem this will correspond, in the end, to use b_K^* instead of b_K in (3.9), obtaining an SUPG method with a stabilising parameter given by

$$\tau_K = \frac{\int_K b_K^* \mathrm{d}x}{|K|}. \tag{3.13}$$

In general B_K^* will correspond to a subgrid mesh. For our particular problem, the mesh might include a suitable refinement near the outflow boundary $\partial K \setminus \partial K^-$. For a similar approach, although on a different problem, see [17].

The use of a subgrid mesh is surely the most expensive and more general way. On the other hand, one can try to use a smart, cheaper choice, by using a one-dimensional subspace B_K^* of the form $B_K^* = span\{\varphi_K\}$, for a suitable choice of φ_K. In this case one can easily see that the solution of (3.12) is $b_K^* = \gamma_K \varphi_K$, with γ_K given by

$$\gamma_K = \frac{\int_K \varphi_K \, \mathrm{d}x}{\varepsilon a(\varphi_K, \varphi_K)} = \frac{\int_K \varphi_K \, \mathrm{d}x}{\varepsilon \int_K |\nabla \varphi_K|^2 \mathrm{d}x}. \tag{3.14}$$

Notice that this (as already pointed out in [3]) would again produce an SUPG formulation, but this time with a stabilising parameter

$$\tau_K = \frac{(\int_K \varphi_K \, \mathrm{d}x)^2}{\varepsilon |K| \int_K |\nabla \varphi_K|^2 \mathrm{d}x}, \tag{3.15}$$

corresponding (for small ε) to a huge value, unless φ_K is suitably chosen. In particular, an unrealistically blessed choice would be to take $\varphi_K = b_K$, solution of (3.8), which gives actually $u^*_{B,K} = u_{B,K}$. The need of a convenient shape for φ_K can also be traced back to the relationships between upwind methods and suitable versions of the Petrov-Galerkin method, pointed out a long time ago (e.g., [27]). In the present context (3.12), a realistic *ad hoc* choice for φ_K, which produces quite sensible results for all values of ε, is proposed in [9].

Another approach, proposed in [19], [20], is to choose φ_K in an arbitrary way (cubic bubble in K, or the pyramidal bubble with vertex in the barycenter of K) and add some artificial *subgrid viscosity* ε_A to (3.8), which becomes

$$-(\varepsilon + \varepsilon_A)\Delta b_K + \mathbf{c} \cdot \nabla b_K = 1 \quad \textit{in } K. \tag{3.16}$$

As pointed out in [7], if one approximates (3.16) with a one degree of freedom subspace $B^*_K = span\{\varphi_K\}$, the corresponding approximate solution has again the form $b^*_K = \gamma_K \varphi_K$, but now γ_K is given by

$$\gamma_K = \frac{\int_K \varphi_K \, \mathrm{d}x}{(\varepsilon + \varepsilon_A)\int_K |\nabla \varphi_K|^2 \mathrm{d}x}, \tag{3.17}$$

and the corresponding τ_K becomes

$$\tau_K = \frac{(\int_K \varphi_K \, \mathrm{d}x)^2}{(\varepsilon + \varepsilon_A)|K|\int_K |\nabla \varphi_K|^2 \mathrm{d}x}, \tag{3.18}$$

leaving us with the crucial problem of the choice of ε_A. For a heuristic attempt to get a sensible choice for ε_A, see [7]. It is interesting, and somehow surprising, that the stabilising effect of the procedure is minor for a big subgrid viscosity ε_A, and much bigger for a small subgrid viscosity, as shown by (3.18).

Another possibility (more specially tailored for the present model problem) to obtain a satisfactory approximate solution of (3.8) is to consider the associated *limit problem*: find $\tilde{b}_K$ in, say, $H^1(K)$, such that:

$$\begin{cases} \mathbf{c} \cdot \nabla \tilde{b}_K = 1 \quad \textit{in } K, \\ \\ \tilde{b}_K = 0 \quad \textit{on } \partial K^- = \{x \in \partial K \textit{ such that } \mathbf{c} \cdot \mathbf{n}_K < 0\}, \end{cases} \tag{3.19}$$

where $\mathbf{n}_K$ is the outward normal to K. It is easy to check that the difference between the integral of b_K and the integral of $\tilde{b}_K$ is $O(\varepsilon)$ for $\varepsilon \to 0$. On the other hand, the solution of (3.19) is elementary, and can be computed with paper and pencil. Hence, in practice, one substitutes $\tilde{b}_K$ in place of b_K in (3.9), obtaining an SUPG method with a stabilising parameter given by

$$\tau_K = \frac{\int_K \tilde{b}_K \mathrm{d}x}{|K|}. \tag{3.20}$$

This is essentially the approach proposed in [11], and gives quite reasonable answers even for ε only moderately small.

To conclude our discussion on the model problem (3.1), we summarize the possible choices to get an approximate solution of (3.4) (or of its scaled version (3.8).) We can use a Galerkin subgrid method, with a fine enough (or smart enough) grid. We can use a one-dimensional Galerkin method with a smart choice of the one-dimensional subspace, or we can use a plain one-dimensional subspace and add a smart artificial subgrid viscosity. Finally, we can solve a *reduced equation* (3.19) by hand. All of these approaches will produce in the end an SUPG method, with different values of the stabilising parameter.

We also note that similar procedures can also be adapted and applied, with minor additional complications, to the cases in which the original finite element space V_h is made, say, of piecewise polynomials of degree $k > 1$. The practical aspects of such an extension will become clearer in the next section. We just mention here that, for higher order polynomials, we obtain variants of SUPG that do not coincide with it anymore, and whose stabilising effect has still to be tested in practice.

Finally, we point out that, from the theoretical point of view, the convergence analysis developed for SUPG methods clearly applies to the case of our model problem (3.1), if V_h is made of piecewise linear functions (that is for $k = 1$.) However, recently, an independent analysis has been carried out starting directly from the formulation (2.15) (see [8]). This type of analysis can also be extended to the case of higher order polynomials. In particular, error estimates of usual type, for $k > 1$, have been obtained for $\varepsilon|u - u_h|_1^2$ and for $\mathbf{c} \cdot (u - u_h)$ in a weighted L^2-norm in [10]. By *usual type* we mean here error estimates that are half an order suboptimal: see [26], [21], [28]. More recently, global L^2-error estimates and local H^1-error estimates have been obtained, for the formulation (2.15), in [29].

4 Extensions to other problems

Having the example of Section 3 in mind, we now go back to the more general level of Section 2, for a brief discussion on possible applications to different problems. It is clear, from the above section, that all the viable strategies make use of the split nature of the bubble equation (2.9). This is the crucial point of the *divide and conquer* strategy, essentially contained in assumption (2.5). Splitting (2.9) among the subdomains, its *approximate solution* can be done in parallel. However, in order to perform the *static condensation*, one has to be able to substitute u_B (or, actually, its approximation) as a function of the original (and final) unknown u_h in (2.15). For this, we proceed in the following way: as a first step we identify, in each K, the smallest linear space that contains all possible residuals, namely

$$R_h^K := span\{f_{|K}, Lv_{h|K},\ v_h \in V_h\}. \tag{4.1}$$

In many cases, $f_{|K}$ can be approximated, without major loss of information, by means of elements of the space $(LV_h)_{|K}$, that can therefore be used in place of R_h^K. In the previous section, this was the space of constants on K.

The second step is then to choose, for each K, a basis $\{g_K^i\}_{\{i=1,..,N_K\}}$ of the space R_h^K. Clearly, N_K denotes the dimension of such space. Then, for each $i = 1,..,N_K$, we seek an approximate solution of the local problem:

$$\begin{cases} \text{find } \psi_K^i \in H_0^1(K) \text{ such that :} \\ \mathcal{L}(\psi_K^i, v) = (g_K^i, v) \quad \forall v \in H_0^1(K). \end{cases} \tag{4.2}$$

In the example of the previous section, there was just one function g_K^i, namely the constant 1 appearing in (3.8), and the corresponding function ψ_K^i was denoted by b_K.

The most general and widely applicable strategy in order to obtain an approximate solution of (4.2) consists, as we have seen, in the use of a Galerkin approximation, corresponding to a suitable choice of $B_K^* \subset H_0^1(K)$. Then one can solve N_K problems of the type:

$$\begin{cases} \text{find } \psi_K^{*,i} \in B_K^* \text{ such that :} \\ \mathcal{L}(\psi_K^{*,i}, v) = (g_K^i, v) \quad \forall v \in B_K^*. \end{cases} \tag{4.3}$$

The technique has been successfully applied to advection-dominated flows, in cases more complex than the one of the previous section, see, e.g., [17]. For an application of this technique to the Helmholtz equation see [15] and [16].

On more special classes of problems one might also think to extend some of the *tricks* of the previous section. For instance, the use of the limit problem (3.19) can surely be adapted to advection diffusion problems with a more general choice of V_h, e.g., [6].

On the other hand, for singularly perturbed problems where some artificial viscosity (or similar regularization) is usually employed, the idea of using only a kind of *subgrid viscosity* (or subgrid regularization,) as in (3.16), is surely appealing for its simplicity and rather wide range of applicability. However, as we have seen, the choice of the subgrid artificial viscosity ε_A appears to be crucial, and requires deeper investigations.

Another interesting area where these ideas can be applied is the solution of elliptic problems with rough coefficients. Consider for instance the problem of finding $u \in H_0^1(\Omega)$ such that

$$-\sum_{i,j=1,2} \frac{\partial}{\partial x_j}(a_{i,j}(\mathbf{x})\,\frac{\partial u}{\partial x_i}) = f \quad \text{in } \Omega, \tag{4.4}$$

where we assume that the matrix $a_{i,j}$ satisfies the usual uniform strong ellipticity conditions, but has jumps within Ω, on a scale that is only affordable when solving local problems in parallel. Problems of this type arise for instance in petroleum engineering, but the range of possible applications is clearly much wider. Notice that writing (4.4) in mixed form (e.g., [4]) the space V, in the notation of Section 2, becomes $H(\text{div};\Omega) \times L^2(\Omega)$. It is interesting to see that the present general strategy, in this case, gives back the upscaling method of [1], [2]. The mixed formulation, within this approach, seems to be particularly appealing. Indeed, for an element (σ, u) in $H(\text{div};\Omega) \times L^2(\Omega)$, the condition $supp((\sigma,u)) \subset K$ only requires $\sigma \cdot \mathbf{n}_K = 0$ on ∂K. Notice that a conventional variational formulation would use $V = H_0^1(\Omega)$, forcing the elements of the bubble space to satisfy $u = 0$ on ∂K, which looks as a more severe limitation. It would be interesting however to compare the two approaches on some practical problems. Similarly, the relations of these approaches with the one of Hou (e.g., [22], [23]) are surely worth investigating.

The application of the paradigm "divide and conquer/static condensation/approximate solution" to some nonlinear problems is currently under investigation. The obvious choice would be to apply it to the various linearized problems in an iterative procedure, but in particular cases the structure of the nonlinearity might suggest a better strategy.

References

[1] T. Arbogast, Numerical subgrid upscaling of two-phase flow in porous media. In preparation.

[2] T. Arbogast, S. E. Minkoff, and P. T. Keenan, An operator-based approach to upscaling the pressure equation. In: "Computational Methods in Water

Resources XII", v. 1, V.N. Burganos et al., eds. Computational Mechanics Publications, Southampton, U.K., 1998.

[3] F. Brezzi, M.-O. Bristeau, L. P. Franca, M. Mallet, and G. Rogé, A relationship between stabilized finite element methods and the Galerkin method with bubble functions, *Comput. Methods Appl. Mech. Engrg.* **96**, 117-129 (1992).

[4] F. Brezzi, M. Fortin, *Mixed and Hybrid Finite Element Methods.* Springer-Verlag, New York, Springer Series in Computational Mathematics **15**, 1991.

[5] F. Brezzi, L. P. Franca, T. J. R. Hughes, and A. Russo, $b = \int g$, *Comput. Methods Appl. Mech. Engrg.* **145**, 329-339 (1997).

[6] F. Brezzi, L. P. Franca, and A. Russo, Further considerations on residual free bubbles for advective-diffusive equations, *Comput. Methods Appl. Mech. Engrg.* **166**, 25-33 (1998).

[7] F. Brezzi, P. Houston, L. D. Marini, and E. Süli, Modeling subgrid viscosity for advection-diffusion problems. Submitted to *Comput. Methods Appl. Mech. Engrg.*

[8] F. Brezzi, T. J. R. Hughes, L. D. Marini, A. Russo, and E. Süli, A priori error analysis of a finite element method with residual-free bubbles for advection-dominated equations. To appear in *SIAM J. Num. Anal.*

[9] F. Brezzi, D. Marini, and A. Russo, Applications of pseudo residual-free bubbles to the stabilization of convection-diffusion problems, *Comput. Methods Appl. Mech. Engrg.* **166**, 51-63 (1998).

[10] F. Brezzi, D. Marini, and E. Süli, Residual-free bubbles for advection-diffusion problems: the general error analysis. To appear in *Numer. Math.*

[11] F. Brezzi, A. Russo, Choosing bubbles for advection-diffusion problems, *Math. Mod. and Meth. in Appl. Sci.* **4**, 571-587 (1994).

[12] A. N. Brooks, T. J. R. Hughes, Streamline Upwind/Petrov-Galerkin formulations for convection dominated flows with particular emphasis on the incompressible Navier-Stokes equations, *Comput. Methods Appl. Mech. Engrg.* **32**, 199-259 (1982).

[13] L. P. Franca, A. Russo, Deriving upwinding, mass lumping and selective reduced integration by residual-free bubbles, *Appl. Math. Lett.* **9**, 83-88 (1996).

[14] L. P. Franca, S. L. Frey, and T. J. R. Hughes, Stabilized finite element methods: I. Applications to advective-diffusive model, *Comput. Methods Appl. Mech. Engrg.* **95**, 253-276 (1992).

[15] L. P. Franca, C. Farhat, A. P. Macedo, and M. Lesoinne, Residual-free bubbles for the Helmholtz equation, *Int. J. Num. Meth. in Eng.* **40**, 4003-4009 (1997).

[16] L. P. Franca, A. P. Macedo, A two-level finite element method and its application to the Helmholtz equation, *Int. J. Num. Meth. in Eng.* **43**, 23-32 (1998).

[17] L. P. Franca, A. Nesliturk, and M. Stynes, On the stability of residual-free bubbles for convection-diffusion problems and their approximation by a two-level finite element method, *Comput. Methods Appl. Mech. Engrg.* **166**, 35-49 (1998).

[18] D. F. Griffiths, A. R. Mitchell, Spurious behaviour and nonlinear instability in discretised partial differential equations. In: "The Dynamics of Numerics and the Numerics of Dynamics". *Inst. Math. Appl. Conf. Ser., New Ser.* **34**, 215-242 (1992).

[19] J. L. Guermond, Stabilization of Galerkin approximations of transport equations by subgrid modeling. Submitted to *Math. Mod. Num. Anal.*

[20] J. L. Guermond, Stabilisation par viscosité de sous-maille pour l'approximation de Galerkin des opérateurs monotones, *C. R. Acad. Sci. Paris, Série I* **328**, 617-622 (1999).

[21] P. Hansbo, C. Johnson, Streamline diffusion finite element methods for fluid flow. *von Karman Institute Lectures,* 1995.

[22] T. Y. Hou, X. H. Wu, A multiscale finite element method for elliptic problems in composite materials and porous media, *J. Comput. Phys.* **134**, 169-189 (1997).

[23] T. Y. Hou, X. H. Wu, and Z. Cai, Convergence of a multiscale finite element method for elliptic problems with rapidly oscillating coefficients. To appear in *Math. Comp.*

[24] T. J. R. Hughes, Multiscale phenomena: Green's functions, the Dirichlet to Neumann formulation, subgrid scale models, bubbles and the origins of stabilised methods, *Comput. Methods Appl. Mech. Engrg.* **127**, 387-401 (1995).

[25] T. J. R. Hughes, G. Feijoo, L. Mazzei, and J.-B. Quincy, The variational multiscale method – A paradigm for computational mechanics, *Comput. Methods Appl. Mech. Engrg.* **166**, (1998).

[26] C. Johnson, U. Nävert, and J. Pitkäranta, Finite element methods for linear hyperbolic problems, *Comput. Methods Appl. Mech. Engrg.* **45**, 285-312 (1984).

[27] A. R. Mitchell, D. F. Griffiths, Generalised Galerkin methods for second order equations with significant first derivative terms. In: *Proc. Bienn. Conf., Dundee 1977, Lect. Notes Math.* **630**, 90-104 (1978).

[28] H.-G. Roos, M. Stynes, and L. Tobiska, *Numerical methods for singularly perturbed differential equations: convection diffusion and flow problems.* Springer-Verlag, Berlin, 1996.

[29] G. Sangalli, Global and local error analysis for the residual free bubble method applied to advection-dominated problems. In preparation.

Acknowledgement

This research was supported in part by grants of M.U.R.S.T. cofin 9801229483, and Consiglio Nazionale delle Ricerche (C.N.R.) under contracts 96.03847.PS01, 97.04704.PS01, 97.00892.01.

Franco Brezzi
Dipartimento di Matematica e Istituto di Analisi Numerica del C.N.R.
Università di Pavia
via Ferrata 1
I-27100 Pavia
Italy
brezzi@ian.pv.cnr.it
http://www.ian.pv.cnr.it/~brezzi

F. CHAITIN-CHATELIN

The computing power of Geometry

Abstract We compare, from the point of view of computation, the computing power of the four algebras of numbers, i.e., the real and the complex numbers, the quaternions and the octonions. We conjecture that computing with octonions is to biology what computation with quaternions is to physics.

1 Introduction

In the past few decades, there has been a number of distinguished scientists who played with the idea that the world is a giant computer. In particular there is a growing consensus that information is a more fundamental concept than energy, which may help explain the increase in complexity that everyone of us witnesses in real life.

Could it be possible that Nature does more than just arithmetic on real numbers : add, subtract, multiply and divide (except by 0)? A first hint came from the Italian "cossists" who were the algebraists of the 16th century. To solve 3rd and 4th degree equations, Cardano and Bombelli introduced a new variety of numbers that they called *imaginary* or *impossible*, in order to give meaning to $\sqrt{-1}$. More than two centuries passed before Gauss put complex numbers into the geometric setting of two-dimensional vectors over the reals. In the 19th century, the complex numbers became indispensable almost everywhere in physics. However, there has been a consensus amongst mathematicians to stop the axiomatic construction of numbers ($\mathbb{N} \to \mathbb{Q} \to \mathbb{R}$) with the set of complex numbers $\mathbb{C}$, viewed as a realization of $\mathbb{R}^2$.

We wish to review in this chapter some of the reasons why such a state of affairs is not necessary from a computational point of view. In particular we shall present some of the clues that Nature seems to give us about the role and the importance of computing with *hypercomplex* numbers.

2 Algebra within Geometry

How complex can a number be such that it is still possible to *compute* with it? The answer to such a question requires that we define what we mean by "computation". Here is the definition we take :

> *Two internal operations are defined on numbers, an addition + and a multiplication ×. The addition yields a structure of commutative group (0 as the special element). The multiplication is such that there exists an identity element $1 \neq 0$, and any non zero element has an inverse. The multiplication is not necessarily commutative nor associative. It is distributive with respect to +.*

Clearly $\mathbb{R}$ and $\mathbb{C} \sim \mathbb{R}^2$ qualify : they both are commutative fields over $\mathbb{R}$. How much further can this be extended into $\mathbb{R}^n$, that is to real vectors of dimension n, where the addition is the natural vector addition (addition of real components)?

Historically, two extensions are known corresponding to $n = 4$ and $n = 8$:

1. the non commutative field $\mathbb{H} \sim \mathbb{R}^4$ of the *quaternions* proposed by Hamilton (1843) [6], and
2. the non associative algebra $\mathbb{G} \sim \mathbb{R}^8$ of the *octonions* designed by Graves (1844) and later rediscovered by Cayley (1845) [6].

Is it possible to extend the geometric complexity of numbers to dimensions other than $n = 2, 4$ and 8? That the answer is NO came as a big surprise. This will be discussed in Section 4. Before, in Section 3, we present the four algebras $\mathbb{R}$, $\mathbb{C}$, $\mathbb{H}$ and $\mathbb{G}$, which display a remarkable unity of design.

3 The four real division algebras A_k of dimension $n = 2^k$, $k = 0, 1, 2, 3$

3.1 The multiplication ×

The three algebras $A_1 = \mathbb{C} \sim \mathbb{R} \times \mathbb{R}$, $A_2 = \mathbb{H} \sim \mathbb{C} \times \mathbb{C}$ and $A_3 = \mathbb{G} \sim \mathbb{H} \times \mathbb{H}$ are defined iteratively from $A_0 = \mathbb{R}$ by the law of multiplication in A_k defined on ordered pairs of vectors in A_{k-1} according to the formula

$$\begin{aligned} x \times y &= xy = (x_1, x_2)(y_1, y_2) \\ &= (x_1 y_1 - \overline{y}_2 x_2, x_2 \overline{y}_1 + y_2 x_1) \end{aligned} \tag{3.1}$$

where the overbar denotes the conjugacy operator:

$$x = (x_1, x_2) \longrightarrow \overline{x} = (\overline{x}_1, -x_2). \tag{3.2}$$

Two remarks are in order about (3.1) :

1. for $A_1 = \mathbb{C}$, the overbars are superfluous since y_1 and y_2 are reals in $A_0 = \mathbb{R}$,

2. for $A_3 = \mathbb{G}$, the order in the product of quaternions is essential.

3.2 The norm $|\cdot|$

Formulae (3.1) and (3.2) imply that the product $x\overline{x} = (x_1\overline{x}_1 + \overline{x}_2 x_2, 0)$ is a real positive number for $x \neq 0$. This allows to define the norm or absolute value : $|x| = \sqrt{x\overline{x}}$.

Therefore $x\overline{x} = |x|^2$ and $x^{-1} = \frac{\overline{x}}{|x|^2}$ for $x \neq 0$. The norm satisfies the product rule,

$$|xy| = |x||y|, \tag{3.3}$$

that is : multiplication is isometric.

3.3 Real and imaginary parts of A_k, $k = 1, 2, 3$.

When there is no ambiguity, the subscript k is omitted.

One identifies the real part ReA of the algebra A with $\mathbb{R}$, its real subalgebra. The imaginary part ImA is then defined as

$$ImA = \{x \in A; x^2 \in \mathbb{R}, x \notin \mathbb{R} - \{0\}\}.$$

ImA is isomorphic to the vector spaces $\mathbb{R}, \mathbb{R}^3$ and $\mathbb{R}^7$ for $k = 1, 2, 3$, respectively. For this reason, it is also called the vector or pure part of A.

3.4 Scalar and vector products in A_k, $k = 1, 2, 3$

The scalar product $\langle \cdot, \cdot \rangle$ of x and y in A is defined as

$$\langle x, y \rangle = Re(x\overline{y}) = Re(\overline{x}y) \tag{3.4}$$

Therefore x and y are orthogonal if and only if

$$x\overline{y} \in ImA.$$

ImA is an algebra with vector product $\wedge$ defined by

$$u \wedge v = \frac{1}{2}(uv - vu) = uv + \langle u, v \rangle, \quad \forall\, u, v \in ImA \tag{3.5}$$

For $k = 1$ the vector product in $Im\mathbb{C}$ is always zero ($Im\mathbb{C}$ is the zero algebra in $\mathbb{R}$). For $k = 2$, one gets the classical vector product in $\mathbb{R}^3$, which satisfies the Jacobi identity :

$$u \wedge (v \wedge w) + v \wedge (w \wedge u) + w \wedge (u \wedge v) = 0.$$

For $k = 3$, one gets a vector product in $\mathbb{R}^7$, which does not satisfy a Jacobi identity.

The following assertions are equivalent in ImA :

1. $|u| = |v| = 1$ and $\langle u, v\rangle = 0 \Rightarrow |u \wedge v| = 1$
2. $|u \wedge v|^2 + \langle u, v\rangle^2 = |uv|^2 = |u|^2|v|^2$
3. $u \wedge (u \wedge v) = \langle u, v\rangle u - |u|^2 v$

4 Only four real division algebras $A_k, k = 0, 1, 2, 3$.

The characterization of all real division algebras of finite dimensions turned out to be a much more formidable task than anticipated in the days of Hamilton and Graves. Some of these algebraic properties have required sophisticated topological proofs. The main results can be grouped into three categories :

1. dimension of real division algebras : The dimension should be a power of 2 (Hopf, 1940) [9]. The only powers of 2 are 1,2,4 and 8 (Milnor, 1958) [12].
2. uniqueness of $\mathbb{R}$, $\mathbb{C}$, $\mathbb{H}$ and $\mathbb{G}$:

   Up to an isomorphism, $\mathbb{R}$, $\mathbb{C}$ and $\mathbb{H}$ are the only associative division algebras, with finite dimension on $\mathbb{R}$ (Frobenius, 1877) [7]. There is a similar uniqueness result for $\mathbb{G}$ without associativity (Zorn, 1933) [13].
3. dimension of algebras of vector product :

   The only possible dimensions are 1, 3 and 7. Such algebras are isometrically isomorphic to $\mathbb{R}$, $\mathbb{R}^3$ and $\mathbb{R}^7$, the imaginary parts of $\mathbb{C}$, $\mathbb{H}$ and $\mathbb{G}$, respectively.

There is yet another set of related results of importance for number theory, the celebrated squares theorems : for $n = 1, 2, 4$ and 8 there exist n bilinear forms with integer coefficients

$$c_i = \sum_{j,k}^{n} \alpha_{jk}^{(i)} a_j b_k, \; i = 1, \ldots, n$$

such that for any real a_j, b_k, then

$$\sum_{i=1}^{n} c_i^2 = \left(\sum_{j=1}^{n} a_j^2\right)\left(\sum_{k=1}^{n} b_k^2\right) \tag{4.1}$$

For $n = 2$, the theorem of *two* squares was known to Diophantus [6]. For $n = 4$, the theorem of *four* squares was stated by Fermat (1659) and proved by Lagrange [6]. For $n = 8$, the theorem of *eight* squares was proposed by Graves with his octonions (1844), but it had been discovered first by Degen (1818) [6]. Legendre (1830) [6] showed that the result (4.1) was impossible for $n = 3$. Hurwitz (1898) [10] proved that (4.1) could be satisfied only for the known values $n = 1, 2, 4$ and 8.

This is truly an impressive set of results showing that one cannot extend the geometric complexity of computation beyond the dimension 8. But this also indicates powerfully that there is no good mathematical reason to stop the construction of hypercomplex numbers at $n = 2$, other than the fact that we, as human beings, have a very limited ability to picture vectors of dimension 4 and 8 !

The fact that Nature does compute with hypercomplex numbers can be inferred from the fact that :

1. quaternions are useful in physics, and more and more heavily used in computer graphics [8] to compute 3D-geometry.
2. octonions play a role in particle physics [11].

5 Powers of hypercomplex numbers

Let p in A_k, $k = 1, 2, 3$ be written as $p = x + X$ where x is the real part of p and $X \in ImA_k$ is the vector (or imaginary, or pure) part. An easy calculation shows that, for $m \geq 1$,

$$p^m = x_m + y_m X \text{ with } x_1 = x,\ y_1 = 1 \text{ and}$$

$$\begin{cases} x_{m+1} = x x_m - y_m |X|^2, \\ y_{m+1} = x y_m + x_m. \end{cases}$$

For $m = 2$, $p^2 = x^2 - |X|^2 + 2xX$, and $p^3 = x(x^2 - 3|X|^2) + (3x^2 - |X|^2)X$ for $m = 3$.

What is the role of complex numbers in algebra? They allow any polynomial equation of degree n to have exactly n roots in the complex plane, when it may have none on the real line. The fundamental theorem of algebra (d'Alembert-Gauss) expresses an identity between the degree of the polynomial and the number of roots expressed as points in a plane. From this vantage point, the set of complex numbers seems perfectly adequate. At this point, the reason to go to hypercomplex numbers is essentially curiosity. We are like the infant who is tired of exploring the floor and begins to discover the joys of walking erect.

5.1 $p^2 + 1 = 0$

Let us take a closer look at the quadratic equation

$$p^2 + 1 = 0, \tag{5.1}$$

which was the historical reason for introducing the complex numbers.

Set of numbers	$\mathbb{R}$	$\mathbb{C}$	$\mathbb{H}$	$\mathbb{G}$
Geometric dimension	1	2	4	8
Number of solutions to (5.1)	0	2	∞	∞
Solutions to (5.1)	–	$\pm i$	S^2	S^6

Table 5.1: Equation $p^2 + 1 = 0$: its geometric solutions

In Table 5.1, S^2 (resp. S^6) denotes the unit sphere in $\mathbb{R}^3$ (resp. $\mathbb{R}^7$) that is the set of vectors of unit Euclidean norm. Instead of just two complex solutions, one gets the infinity of points on the unit sphere with 2 or 6 degrees of freedom, respectively.

5.2 $p^3 + 1 = 0$

We turn to the 3rd degree equation

$$p^3 + 1 = 0, \tag{5.2}$$

which has a unique real solution $x = -1$. What are the additional solutions provided by hypercomplex numbers? They are listed in Table 5.2.

Set of numbers	$\mathbb{R}$	$\mathbb{C}$	$\mathbb{H}$	$\mathbb{G}$
Geometric dimension	1	2	4	8
Number of solutions to (5.2)	1	3	∞	∞
Solutions to (5.2)	-1	-1 and $\frac{1}{2} \pm i\frac{\sqrt{3}}{2}$	-1 and $\frac{1}{2} + \frac{\sqrt{3}}{2}S^2$	-1 and $\frac{1}{2} + \frac{\sqrt{3}}{2}S^6$

Table 5.2: Equation $p^3 + 1 = 0$: its geometric solutions

5.3 An example of quadratic iteration: the logistic

The logistic iteration :

$$p_0,\ p_{j+1} = rp_j(1 - p_j),\ j \geq 0 \tag{5.3}$$

has been extensively studied over $\mathbb{R}$ and $\mathbb{C}$. See for example [3] *pp. 33-37* and *pp. 188-189*. The iteration (5.3) depends on two parameters p_0 and r. The convergence properties of the logistic iteration can be analyzed by means of two types of sets :

1. a Julia set $J(r)$ for r fixed :

$$J(r) = \left\{ p_0;\ \lim_{j\to\infty} p_j \not\to \infty \right\},$$

2. a Mandelbrot set $\mathcal{M}(p_o)$ for p_0 fixed :

$$\mathcal{M}(p_o) = \left\{ r;\ \lim_{j\to\infty} p_j \not\to \infty \right\}.$$

In his original papers, Mandelbrot chooses p_0 to be the critical value p_c for which the derivative $r - 2rp$ of $rp - rp^2$, with respect to p, vanishes, that is $p_c = \frac{1}{2}$. The comparison between real and complex Julia (resp. Mandelbrot) sets is well known [4].

What are the new features when the computation of $J(r)$ or $\mathcal{M}(p_0)$ is performed in the quaternions $\mathbb{H}$ or in the octonions $\mathbb{G}$? Because the corresponding sets live in $\mathbb{R}^4$ or $\mathbb{R}^8$, they cannot be displayed directly. In $\mathbb{H} \sim \mathbb{R}^4$, a 3D-rendering of a 3D-section is often used in computer graphics [8]. No such option exists in $\mathbb{G} \sim \mathbb{R}^8$. We have therefore chosen to analyze all possible 2D-sections.

To the best of our knowledge, the systematic study of the information conveyed by carrying the iteration (5.3) in $\mathbb{H}$ or $\mathbb{G}$ has not been done. We report below some preliminary results obtained for the Julia set for the values of r being :

$$\begin{aligned} r &= (-0.6 \ -0.85) \text{ in } \mathbb{C} \\ &= (-0.6 \ -0.85\ 0\ 0) \text{ in } \mathbb{H} \\ &= (-0.6 \ -0.85\ 0\ 0\ 0\ 0\ 0\ 0) \text{ in } \mathbb{G}. \end{aligned}$$

We look at the 2D-sections of the Julia set obtained by forcing to zero all but two components of p_j. We list the summary of our findings :

1. going from $\mathbb{R}$ to $\mathbb{C}$ we add to the Cantor set on a line, the classical complex Julia set in the plane,
2. going from $\mathbb{C}$ to $\mathbb{H}$ and $\mathbb{G}$, we add to the classical complex Julia set three new types of pictures : two of them are fractals and the third type reveals a cylindrical symmetry.

The most interesting feature is that *visually* there is no difference whether a 2D-section is obtained by computation with quaternions or with octonions. The classical Julia set picture itself can be interpreted as the graphical display of a computation in $\mathbb{C}$, $\mathbb{H}$ or $\mathbb{G}$.

6 A philosophical conclusion

If, following Hamilton, one is willing to bet on the idea that Nature does compute with geometric numbers in the four algebras $\mathbb{R}$, $\mathbb{C}$, $\mathbb{H}$ and $\mathbb{G}$, the natural question is, of course, why? What can be done with multiplication in $\mathbb{H}$ or $\mathbb{G}$ which cannot be expressed as easily in $\mathbb{R}$ or $\mathbb{C}$?

The answer is clearly to do geometry in three or seven dimensions. In the usual space $\mathbb{R}^3$, Hamilton [6] showed that most of the formulae of vector calculus (including *div* and $\overrightarrow{rot}$) could be elegantly written by means of scalar or vector products with the vector of partial derivative $\overrightarrow{\nabla}$. Most of the equations of the classical physics are amenable to that format.

One can therefore propose the conjecture that computing with octonions is to biology what computation with quaternions is to physics.

It is conceivable that geometry plays a much more fundamental role in Nature's computation (or Nature's laws) than is currently realized. An indication that this may indeed be the case is obtained by contrasting two fundamental results by Gödel and Tarski in the domain of the foundations of mathematics [5] :

1. the incompleteness result of Gödel (1931) shows that $\mathbb{N}$ is too scarce to always imply decidability in arithmetic,

2. the decidability of elementary geometry on $\mathbb{R}$ proved by Tarski (1929) shows that $\mathbb{R}$ is rich enough as a basis field.

As a psychological curiosity, it is interesting to note that Gödel's theorem, which closes a door, is by far more famous than Tarski's theorem, which opens the door to the theory of elimination and computer algebra. The limit imposed by the discreteness of the arithmetic seems intolerable [2], to the point of overshadowing the power granted by the continuity of geometry.

References

[1] Blanchard A. *Les Corps Non Commutatifs.* PUF, Paris, 1972.

[2] Chaitin G. *The Unknowable.* Springer-Verlag, Berlin, 1999.

[3] Chaitin-Chatelin F., Frayssé V. *Lectures on Finite Precision Computations.* SIAM Publ., Philadelphia, 1996.

[4] Devaney R. L. *An Introduction to Chaotic Dynamical Systems.* 2nd edition, Addison Wesley, New York, 1989.

[5] Dickmann M. A. Le continu du 1e ordre in *Le Labyrinthe du continu.* Salanskis J.-M., Sinaceur H., eds., Colloque de Cerisy. Springer-Verlag, Paris, 1992. *pp. 117-133.*

[6] Ebbinghaus H.-D., Hermes H. and Hirzebruch F. *Les Nombres.* Vuibert, Paris, 1998.

[7] Frobenius F. G. Über linear Substitutionen und bilineare Formen. *Ges. Abhandl.,* **1**, *pp. 343-405*, 1877.

[8] Hart J. C., Sandin D. J. and Kauffman L. H. Ray tracing deterministic 3-D fractals. *Comp. Graphics,* **23**, *pp. 289-296*, 1989.

[9] Hopf H. Ein topologischer Beitrag zur reellen Algebra. *Comm. Math. Helvetici,* **13**, *pp. 219-239*, 1940/41.

[10] Hurwitz A. Über die Komposition der quadratischen Formen von beliebig vielen Variablen. *Nachr. der k. Gesell. Wiss. Göttingen, pp. 309-318,* 1898.

[11] Michel L., Radicati L. A. The geometry of the octet. *Ann. Inst. Henri Poincaré,* **18**, *pp. 185-214*, 1973.

[12] Milnor J. Some consequences of a theorem of Bott. *Ann. of Math.,* **68**, *pp. 444-449*, 1958.

[13] Zorn M. Alternativkörper und quadratische Systeme. *Abh. Math. Sem. Hamburg,* **9**, *pp. 395-402,* 1933.

Acknowledgement

The author is greatly indebted to Dr. Valérie Fraysśe for performing the quaternion and octonion computations and graphics.

Françoise Chaitin-Chatelin
Université de Toulouse 1 et
CERFACS, 42 av. G. Coriolis
31057 Toulouse Cedex 1
`chatelin@cerfacs.fr`
`http://www.cerfacs.fr/~chatelin.html`

C. M. ELLIOTT AND V. M. STYLES

Numerical approximation of vortex density evolution in a superconductor

Abstract A finite volume/element approximation of a mean field model of superconducting vortices in one and two space dimensions is presented. The model incorporates positive and negative vorticity with time dependent applied magnetic field, flux pinning and boundary nucleation. Stability bounds on the solutions of these approximations are proved. Some computed results are presented.

1 Introduction

We consider numerical approximations of a two-dimensional version of the mean field model of superconducting vortices considered in [3, 4]. We derive a finite volume/element discretization of the model and prove estimates on the solution. For Ω a given bounded domain in $\mathbb{R}^2$, being the cross section of an infinitely long cylindrical superconductor, the equations are:

$$w_t + \nabla \cdot (w\mathbf{v}) = 0 \qquad \text{in } \Omega_{\mathrm{T}}, \tag{1.1}$$

$$-\lambda^2 \Delta u + u = w \qquad \text{in } \Omega_{\mathrm{T}}, \tag{1.2}$$

$$\mathbf{v} = -\mathrm{sign}(w)\frac{[|\nabla u| - J_p]_+ \nabla u}{|\nabla u|}, \tag{1.3}$$

$$u = u_{\mathrm{b}}(t) \qquad \text{on } \partial\Omega_{\mathrm{T}}, \tag{1.4}$$

$$-\mathbf{n}\cdot\mathbf{v}w = \alpha(u_{\mathrm{b}}(t))[|\mathbf{j}| - J_n]_+ \qquad \text{on } \partial\Omega_{\mathrm{T}}^-. \tag{1.5}$$

Here w and u, respectively, denote the 'vorticity' and the average magnetic field of the superconducting sample Ω, aligned parallel to the axis of the cylinder. The electric current is denoted by $\mathbf{j} = (\partial_2 u, -\partial_1 u, 0) = \mathrm{curl}\ (0,0,u)$. The constant λ is a material parameter related to the so-called penetration depth of the magnetic field, u_{b} denotes the externally applied magnetic field, α depends on the material and $\alpha(r)r \geq 0$, $\partial\Omega_{\mathrm{T}}^{-} = \{(\mathbf{x},t) \in \partial\Omega_{\mathrm{T}} : \mathbf{v}(\mathbf{x},t)\cdot\mathbf{n} < 0\}$, and $\mathbf{n}$ is the outward unit normal to $\partial\Omega$. Equation (1.1) is a conservation law with $\mathbf{v}$ denoting the velocity field for the vorticity. The magnetic field satisfies the averaged version of London's equation (1.2). The form of the velocity law (1.3) models the effect of pinning sites in the superconductor, which hinders the motion of vortices. The current J_p is the magnitude of the pinning current that the current must exceed for the vorticity to move. The boundary condition (1.5), suggested by Chapman [3], allows the nucleation of vorticity at the boundary provided the magnitude of the current is larger than the nucleation current J_n. In the following we consider simplified versions of the above model by setting $J_p = J_n$ and $w = w_{\mathrm{b}}(t) := \alpha(u_{\mathrm{b}}(t))$ on all inflow boundary sections so that (1.4)–(1.5) become

$$u = u_{\mathrm{b}}(t) \qquad \text{on } \partial\Omega_{\mathrm{T}}, \quad w = w_{\mathrm{b}}(t) \qquad \text{on } \partial\Omega_{\mathrm{T}}^{-}, \tag{1.6}$$

and we rewrite (1.3) as

$$\mathbf{v} = -\mathrm{sign}(w)F(|\nabla u|)\nabla u, \tag{1.7}$$

where we assume F is a given Lipschitz continuous function such that

$$0 \leq F(r) \leq 1 \quad r \geq 0; \quad \text{and} \quad 0 \leq f(r) - f(s) \leq r - s \quad \forall\, r \geq s, \tag{1.8}$$

where $f(r) = rF(|r|)$. We also assume that $u_{\mathrm{b}}(t)$ and $w_{\mathrm{b}}(t) \in C^1(0,T)$.

In [12] the existence of a weak solution $\{w,u\}$ of (1.1)-(1.7) with $w \geq 0$, $\mathbf{v} = -\nabla u$, $\alpha \equiv 0$ and constant $u_{\mathrm{b}}(t) = u_{\mathrm{b}} > 0$, was proved which was shown to satisfy

$$0 \leq w, u \leq \max(||w_0(x)||_{L^\infty(\Omega)}, u_{\mathrm{b}}).$$

Furthermore the weak solution in this one-dimensional model was shown to be unique.

In [8] a finite volume/element discretization of (1.1)-(1.7) with $w \geq 0$, $\mathbf{v} = -\nabla u$, $\alpha \equiv 0$ and $u_{\mathrm{b}}(t) =$ constant was derived. Various estimates on the numerical solution were obtained and the long-time behaviour and steady-state were studied. In the case of one space dimension it was shown that in the limit, as the mesh size and the time step tend to zero, the approximate solution converges to the unique weak solution of the continuous model. In [6] the spatial discretization proposed in [8] was shown to converge weakly to a weak solution in two space dimensions. Uniqueness of solutions in two space dimensions is still an open question. The case of flux pinning and boundary nucleation of vorticity was considered in [9] for positive vorticity. Existence and uniqueness of a solution in one space dimension was proved together with convergence of the numerical approximation. In this article we consider the general case of positive and negative vorticity. We note that the vortex density model for an infinite cylinder in a transverse magnetic field leads to an elliptic equation coupled with a Hamilton-Jacobi equation, which has been studied analytically and numerically in [2, 7].

In Section 2 we adapt the techniques used in [8, 9] to derive a numerical discretization of (1.1)-(1.6) in a polygonal domain and in a bounded interval in one space dimension, using a fixed spatial mesh and a uniform time step. We obtain stability estimates in Section 3. We conclude with Section 4 in which we display some computed results.

2 Finite volume/element approximation of the model

In this section we derive a combined finite volume/element approximation of (1.1)-(1.6), for a polygonal domain Ω. In order to do so we triangulate the domain Ω and create a dual mesh by perpendicularly bisecting the edges of the triangulation T_h, cf. [11]. Using this dual mesh a finite volume discretization is derived. We denote the edge of T_h connecting the ith node $\mathbf{x}_i$, to the jth node by σ_{ij} and its length by h_{ij}. Similarly we denote the co-edge, that is the perpendicular bisector of σ_{ij}, by σ'_{ij} and its length by h'_{ij}, see Figure 2.1. We define C_j to be the set of nodes k adjacent to the node j such that $h'_{jk} \neq 0$ and V_j to be the cell of the dual mesh enclosed by the edges σ'_{jk} for all $k \in C_j$. We let I denote the set of nodes of T_h and $\{\mathbf{x}_i\}_{i \in I}$ the coordinates of the vertices of this triangulation. We divide I into I_B, the boundary nodes, i.e., $I_B = \{i \in I : \mathbf{x}_i \in \partial\Omega\}$ and $I_I = I \backslash I_B$, the interior nodes. We assume that the triangulation has no angle greater than $\pi/2$ and that $\min h_{ij}/h > \rho$, where $h = \max h_{ij}$. We now introduce some notation that will be useful in defining our approximations. We set Ω to be a polygon and we define S_h, S_h^0, W_h, W_h^0 and $a_h(u, v)$ by

$$S_h = \{\xi \in C(\overline{\Omega}) : \xi|_T \text{ is linear } \forall\, T \in T_h\},$$

$$S_h^0 = \{\xi \in S_h : \xi_i = 0 \;\forall\, i \in I_B\},$$

$$W_h = \{\xi \in L^\infty(\Omega) : \xi|_{V_j} := \xi_j \in \mathbb{R} \;\forall\, j \in I\},$$

$$W_h^0 = \{\xi \in W_h : \xi|_{V_j} = 0 \;\forall\, j \in I_B\},$$

$$a_h(\eta, \xi) \equiv \lambda^2 \int_\Omega \nabla\eta\nabla\xi d\mathbf{x} + (\eta, \xi)_h \quad \forall\, \eta, \xi \in H^1(\Omega) \cap C(\overline{\Omega}),$$

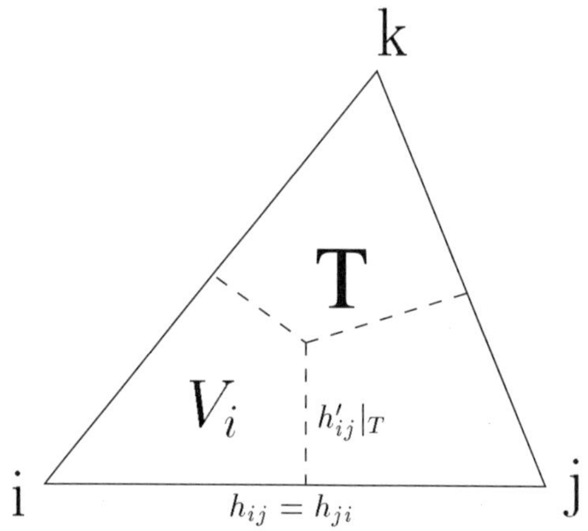

Figure 2.1: Dual mesh.

where $(\cdot,\cdot)_h$ is the discrete replacement of the L^2 inner product over Ω, $(\cdot,\cdot)$, defined by the following 'vertex' integration rule

$$(\eta,\xi)_h = \sum_{j\in I} m_j\eta_j\xi_j = \underline{\eta}^T M\underline{\xi} \quad \forall\ \eta,\xi \in C(\overline{\Omega}), \tag{2.1}$$

where $\eta_j = \eta(\mathbf{x}_j)$ are the nodal values of η_h, $\underline{\eta} = (\eta_j)$, m_j is one third of the sum of the areas of elements attached to the node j, and the positive diagonal matrix M is called the lumped mass matrix. It is well known that for all $\eta,\xi \in S_h^0$, we have

$$|(\eta,\xi) - (\eta,\xi)_h| \le Ch^2|\eta|_1|\xi|_1 \le Ch|\eta|_0|\xi|_1, \tag{2.2}$$

and constants C_1 and C_2 exist independent of h such that

$$C_1|\xi|_0 \le |\xi|_h \le C_2|\xi|_0 \quad \forall\ \xi \in S_h, \tag{2.3}$$

where

$$|\eta|_0 = \left(\int_\Omega \eta^2 d\mathbf{x}\right)^{\frac{1}{2}}, \quad |\eta|_1 = \left(\int_\Omega |\nabla\eta|^2 d\mathbf{x}\right)^{\frac{1}{2}}.$$

We denote by χ_j the piecewise linear basis function satisfying

$$\chi_j(\mathbf{x}_k) = \delta_{jk} \quad \forall\ j,k \in I. \tag{2.4}$$

We associate with any function $\eta \in W_h$ the nodal values η_j where

$$\eta_h(\mathbf{x}) = \eta_j \quad \text{for all } \mathbf{x} \in V_j,\ j \in I,$$

and we note that any $\xi \in S_h$ may be written as

$$\xi = \sum_{j \in I} \xi_j \chi_j(\mathbf{x}).$$

By virtue of the assumptions on the mesh we have the existence of a K independent of h such that

$$\max_{j \in I} \left(\sum_{k \neq j, k \in \in I} \frac{|a_{jk}|}{m_j} \right) h^2 \leq K \tag{2.5}$$

where

$$a_{jk} = \int_\Omega \nabla \chi_j(\mathbf{x}) \nabla \chi_k(\mathbf{x}) d\mathbf{x}$$

and, since $\sum_{k \in I} a_{jk} = 0 \quad \forall\ j \in I_I$, we have that

$$|\xi|_1^2 = -\frac{1}{2} \sum_{j \in I} \sum_{k \in I} a_{jk}(\xi_j - \xi_k)^2 \quad \forall\ \xi \in S_h^0. \tag{2.6}$$

We now define a numerical approximation of (1.1)-(1.6). We begin by approximating the initial data $w_0(\mathbf{x})$ by $w_h^0(\mathbf{x}) \subset S_h^0$ and $\hat{w}_h^0(\mathbf{x}) \in W_h^0$, in the following way

$$w_j^0 = \frac{1}{|V_j|} \int_{V_j} w_0(\mathbf{x}) d\mathbf{x} \quad \forall\ j \in I_I,$$

$$w_j^0 = w_b(0) \quad \forall\ j \in I_B,$$

$$w_h^0(\mathbf{x}) = \sum_{j \in I} w_j^0 \chi_j(\mathbf{x}) \in S_h, \tag{2.7}$$

$$\hat{w}_h^0(\mathbf{x})|_{V_j} := w_j^0, \quad \forall j \in I. \tag{2.8}$$

We now discretize (1.2) and (1.7) using a standard finite element approximation,

$$a_h(u_h^n, \chi) = (w_h^n, \chi)_h \qquad \forall\, \chi \in S_h^0,\ \forall\, n \geq 0, \tag{2.9}$$

$$u_h^n - u_b^n \in S_h^0,\ \forall\, n \geq 0, \tag{2.10}$$

where $w_h^n = \sum_{j\in I} w_j^n \chi_j(\mathbf{x}) \in S_h$ and $u_h^n = \sum_{j\in I} u_j^n \chi_j(\mathbf{x}) \in S_h$ approximate, respectively, w and u at time $n\Delta t$ where Δt denotes the time step.

Noting that $\lambda^2 \sum_{k\in I} a_{jk} = 0$ for all $j \in I_I$, we see that (2.9) and (2.10) can be written explicitly as

$$\lambda^2 \sum_{k\in I} a_{jk}(u_k^n - u_j^n) = m_j(w_j^n - u_j^n) \quad \forall\, j \in I_I,\ \forall\, n \geq 0, \tag{2.11}$$

$$u_j^n = u_b^n \quad \forall\, j \in I_B,\ \forall\, n \geq 0. \tag{2.12}$$

In order to approximate (1.1) we use an equivalent upwinding finite volume scheme to the ones derived in [8, 9, 10]. First we formally integrate (1.1) over each interior dual mesh cell V_j to obtain

$$\int_{V_j} w_t d\mathbf{x} = \int_{V_j} \nabla(|w|F(|\nabla u|)\nabla u) d\mathbf{x} = \int_{\partial V_j} |w|F(|\nabla u|)\frac{\partial u}{\partial \mathbf{n}} dS. \tag{2.13}$$

Noting that $\mathbf{v} = -\mathrm{sign}(w)F(|\nabla u|)\nabla u$ and that $|w| = [w]_+ - [w]_-$ (where we set $[r]_+ := \max(r,0)$ and $[r]_- := \min(r,0)$, $r \in \mathbb{R}$) we use upwinding to approximate the right-hand side of (2.13) bearing in mind that, for $w_h \in W_h$, on the interior of an edge of ∂V_j, w_h has a possibly different value on either side of the edge. This leads to the following approximations of the right-hand side of (2.13) for $u \in S_h$ and $w \in W_h$:

$$\begin{aligned} &\sum_{k\in C_j} \frac{h'_{jk}}{h_{jk}} F_{jk}(u)\left\{[w_j]_+[u_k - u_j]_- + [w_k]_+[u_k - u_j]_+\right\} \\ - &\sum_{k\in C_j} \frac{h'_{jk}}{h_{jk}} F_{jk}(u)\left\{[w_j]_-[u_k - u_j]_+ + [w_k]_-[u_k - u_j]_-\right\}. \end{aligned}$$

where
$$F_{jk}(u) := \left(F(|\nabla u|)|_{T_+} h'_{jk,T_+} + F(|\nabla u|)|_{T_-} h'_{jk,T_-} \right) / h'_{jk},$$

T_+ and T_- are the triangles with common edge σ_{jk} and the lengths of the portions of σ'_{jk} in the corresponding triangles are h'_{jk,T_+} and h'_{jk,T_-}. Note that in this derivation we have used the fact that σ_{jk} and σ'_{jk} are orthogonal in order to compute $\partial u/\partial \mathbf{n}$ on σ'_{jk}. We use an explicit method in time and approximate the left-hand side of (2.13) by $m_j(w_j^{n+1} - w_j^n)/\Delta t$ yielding the scheme,

$$\frac{m_j(w_j^{n+1} - w_j^n)}{\Delta t} = - \sum_{k \in I} a_{jk} F_{jk}(u_h^n) \left([w_j^n]_+ - [w_k^n]_- \right) [u_k^n - u_j^n]_-$$

$$- \sum_{k \in I} a_{jk} F_{jk}(u_h^n) \left([w_k^n]_+ - [w_j^n]_- \right) [u_k^n - u_j^n]_+ \quad \forall\, j \in I_I, \tag{2.14}$$

$$w_j^{n+1} = w_b^{n+1} \quad \forall\, j \in I_B. \tag{2.15}$$

Here we have used the fact that $a_{jk} = -h'_{jk}/h_{jk}$ for $k \in C_j$ and $a_{jk} = 0$ $j \neq k$, otherwise. Note that the interior values of w_h^{n+1} are only affected by the boundary values w_b^{n+1} on the inflow boundary points:

$$k \in I_B \cap C_j \;\; : - \quad \text{sign } w_b^{n+1} \text{sign}(u_k^n - u_j^n) > 0.$$

Also we note that for h sufficiently small $|V_j| \simeq m_j$; in fact for some triangulations, particularly ones in which all triangles are equilateral, or uniform right-angled triangulations, we have $|V_j| = m_j$.

It follows from (2.14) that for all $\eta \in S_h^0$ we have

$$(w_h^{n+1} - w_h^n, \eta)_h = -\Delta t b_h(w_h^n; u_h^n, \eta) \tag{2.16}$$

where the triple form $b_h(\cdot;\cdot,\cdot)$ is defined for $(\nu, \xi, \eta) \in S_h \times S_h \times S_h$ by

$$\begin{aligned} b_h(\nu; \xi, \eta) \;\; := \;\; & \sum_{j,k \in I} a_{jk} F_{jk}(\xi) \left([\nu_j]_+ - [\nu_k]_- \right) [\xi_k - \xi_j]_- \eta_j \\ & + \sum_{j,k \in I} a_{jk} F_{jk}(\xi) \left([\nu_k]_+ - [\nu_j]_- \right) [\xi_k - \xi_j]_+ \eta_j. \end{aligned} \tag{2.17}$$

The numerical scheme is then to construct $\{w_h^n, u_h^n\} \in S_h \times S_h$ such that $w_h^n - w_b^n \in S_h^0$, $u_h^n - u_b^h \in S_h^0$, $\forall\, n \geq 0$, the initial condition (2.7) holds and the equations (2.9) and (2.16) hold for $n \geq 0$. Here we identify $\hat{w}_h^n \in W_h$ and $w_h^n \in S_h$ through their nodal values.

It is convenient to note here some properties of b_h. First observe that it is linear in the third variable. By rearranging the sums we have equivalent formulae for b_h:

$$\begin{aligned} b_h(\nu;\xi,\eta) &:= \frac{1}{2}\sum_{j,k\in I} a_{jk}F_{jk}(\xi)\left([\nu_j]_+ - [\nu_k]_-\right)[\xi_k - \xi_j]_-(\eta_j - \eta_k) \\ &\quad + \frac{1}{2}\sum_{j,k\in I} a_{jk}F_{jk}(\xi)\left([\nu_k]_+ - [\nu_j]_-\right)[\xi_k - \xi_j]_+(\eta_j - \eta_k) \qquad (2.18) \\ &= \sum_{j,k\in I} a_{jk}F_{jk}(\xi)\left([\nu_j]_+ - [\nu_k]_-\right)[\xi_k - \xi_j]_-(\eta_j - \eta_k). \end{aligned}$$

We recognize that $b_h(\nu;\xi,\eta)$ is a discrete form of

$$\int_\Omega |\nu| F(|\nabla\xi|)\nabla\xi\nabla\eta,$$

and that $b_n(\nu;\xi - c_1, \eta - c_2) = b_h(\nu;\xi,\eta)$ for any constants c_1 and c_2. Using the Cauchy-Schwarz inequality we have

$$\begin{aligned} |b_h(\nu;\xi,\eta)| \;\leq\; & \overbrace{\left(-\sum_{j,k\in I} a_{jk}F_{jk}(\xi)\left([\nu_j]_+ - [\nu_k]_-\right)[\xi_k - \xi_j]_-^2\right)^{1/2}}^{b_h(\nu;\xi,\xi)^{1/2}} \\ & \left(-\sum_{j,k\in I} a_{jk}F_{jk}(\xi)\left([\nu_j]_+ - [\nu_k]_-\right)(\eta_j - \eta_k)^2\right)^{1/2} . \qquad (2.19) \end{aligned}$$

Also using (2.6) we have

$$|b_h(\nu;\xi,\eta)| \leq 2\|\nu\|_\infty |\xi|_1 |\eta|_1 \quad \forall\, \xi,\eta \in S_h^0. \qquad (2.20)$$

In one-space dimension, for $\Omega = (0, L)$, we use a uniform mesh $T_h = \{[(j-1)h, jh)]; j = 1, \ldots, J\}$, $h = L/J$ with I being the set of nodes, I_B being the two boundary nodes and I_I the set of interior nodes. It follows that $C_j = \{j-1, j+1\}$ and $V_j = ((j-1/2)h, (j+1/2)h))$ for $j \in I_I$ with $V_0 = [0, h/2)$

and $V_J = ((J - 1/2)h, Jh)]$. Furthermore $m_j = h$, $a_{jj+1} = a_{jj-1} = -1/h$ and we can use the same notation for $a_h(\cdot,\cdot)$, and $b_h(\cdot;\cdot,\cdot)$, W_h, S_h, etc. as in two-space dimensions. Node-wise the approximations are, for $j \in I_I$,

$$-\lambda^2(u_{j+1}^n - 2u_j^n + u_{j-1}^n) = h(w_j^n - u_j^n) \tag{2.21}$$

$$\begin{aligned}\frac{h(w_j^{n+1} - w_j^n)}{\Delta t} &= \frac{1}{h}F_{jj+1}(u_h^n)\left\{([w_{j+1}^n]_+ - [w_j^n]_-)[u_{j+1}^n - u_j^n]_+ \right. \\ &\qquad \left. + ([w_j^n]_+ - [w_{j+1}^n]_-)[u_{j+1}^n - u_j^n]_-\right\} \\ &+ \frac{1}{h}F_{jj-1}(u_h^n)\left\{([w_{j-1}^n]_+ - [w_j^n]_-)[u_{j-1}^n - u_j^n]_+ + ([w_j^n]_+ \right. \\ &\qquad \left. - [w_{j-1}^n]_-)[u_{j-1}^n - u_j^n]_-\right\},\end{aligned} \tag{2.22}$$

where $F_{jj+1}(u_h^n) = F\left(\frac{(u_{j+1}^n - u_j^n)}{h}\right)$ and $F_{jj-1}(u_h^n) = F\left(\frac{(u_j^n - u_{j-1}^n)}{h}\right)$. This one-dimensional scheme is related to the upwind scheme for the one-dimensional, non-local, Hamilton-Jacobi equation formulation of the model which is one of the schemes analyzed in [2].

In the following section we set Ω to be a convex polygonal domain in two space dimensions. However the results can be extended to the case where Ω is a polygonal approximation of a smooth domain.

3 Stability bounds

Lemma 3.1 *For each $n \geq 0$ and w_h^n there exists a unique solution u_h^n of (2.9)-(2.10) and the following bounds hold*

$$\begin{aligned}||u_h^n||_\infty &\leq \max(||w_h^n||_\infty, |u_b^n|) \\ ||\nabla u_h^n||_\infty &\leq C^* \max(||w_h^n||_\infty, |u_b^n|)\end{aligned}$$

where C^* is a constant dependent on Ω and independent of h.

Proof: Existence and uniqueness are standard. To check the L^∞ bound on u_h^n we need only consider interior nodes. Suppose $j_0 \in I_I$ such that $u_{j_0}^n =$

$\max_{j\in I} u_j^n$. Since $a_{jk} \leq 0$ for all $j \neq k$ it follows from (2.11) that

$$m_{j_0}\left(u_{j_0}^n - w_{j_0}^n\right) = -\lambda^2 \sum_{k\in I} a_{jk}(u_k^n - u_{j_0}^n) \leq 0,$$

which implies $u_{j_0}^n \leq w_{j_0}^n$. The lower bound follows in the same way.

Turning to the L^∞ bound on the gradient we introduce u^* and u_h^* being the unique solutions of: $u^* - u_b^n \in H_0^1(\Omega)$, $u_h^* - u_b^n \in S_h^0$

$$a(u^*, v) = (w_h^n, v) \quad \forall\, v \in H_0^1(\Omega), \quad a(u_h^*, \chi) = (w_h^n, \chi) \quad \forall\, \chi \in S_h^0.$$

It follows from [5] and elliptic regularity theory that

$$\|u_h^*\|_{W^{1,\infty}(\Omega)} \leq \tilde{C}\, \|u^*\|_{W^{1,\infty}(\Omega)} \leq C \max\left(\|w_h^n\|_\infty, |u_b^n|\right).$$

Since

$$\begin{aligned} a(u_h^* - u_h^n, u_h^* - u_h^n) &= (w_h^n - u_h^n, u_h^n - u_h^*)_h - (w_h^n - u_h^n, u_h^n - u_h^*) \\ &\leq Ch|w_h^n - u_h^n|_0 |u_h^n - u_h^*|_1, \end{aligned}$$

by the numerical integration estimate (2.2) and from [1], for space dimension d,

$$\|\nabla\chi\|_{L^\infty(\Omega)} \leq Ch^{-d/2}\|\nabla\chi\|_{L^2(\Omega)} \quad \forall\, \chi \in S_h$$

we conclude the lemma. □

Lemma 3.2 *Set* $\Lambda = \max\left\{\left\|w_h^0\right\|_\infty, \|u_b\|_{L^\infty(0,T)}, \|w_b\|_{L^\infty(0,T)}\right\}$. *Let* h *and* Δt *satisfy*

$$\frac{\Delta t}{h} \leq \frac{1-\delta}{K\mathcal{L}}, \quad \Delta t \leq \frac{\lambda^2\delta}{\Lambda}. \tag{3.1}$$

Then there exists a unique sequence $\{w_h^n, u_h^n\}$ *for each* $0 \leq n\Delta t \leq T$ *satisfying*

$$\|w_h^n\|_\infty \leq \Lambda, \quad \|u_h^n\|_\infty \leq \Lambda, \quad \|\nabla u_h^n\|_\infty \leq \mathcal{L} = \mathcal{L}(\Lambda) \quad \forall\, n \geq 0, \tag{3.2}$$

provided in two-space dimensions that $f(r) \equiv 1$.

Proof: We prove by induction. Suppose $\{w_h^n, u_h^n\}$ satisfy the bounds. Clearly by Lemma 3.1 we need only check the bound on w_h^{n+1}. Rewriting (2.14) for $F \equiv 1$, we have that for $j \in I_I$,

$$w_j^{n+1} = [w_j^n]_+ \left(1 - \frac{\Delta t}{m_j} \sum_{k \in I} a_{jk}[u_k^n - u_j^n]_-\right) - \sum_{k \in I} a_{jk}[w_k^n]_+[u_k^n - u_j^n]_+$$
$$+ [w_j^n]_- \left(1 + \frac{\Delta t}{m_j} \sum_{k \in I} a_{jj}[u_k^n - u_j^n]_+\right) + \sum_{k \in I} a_{jk}[w_k^n]_-[u_k^n - u_j^n]_-. \quad (3.3)$$

By the assumption on the mesh ratio $\Delta t/h$ and the bound on ∇u_h^n in Lemma 3.1, it follows that

$$1 - \frac{\Delta t}{m_j} \sum_{k \in I} a_{jk}[u_k^n - u_j^n]_- \text{ and } 1 + \frac{\Delta t}{m_j} \sum_{k \in I} a_{jj}[u_k^n - u_j^n]_+$$

are positive. Hence we have

$$w_j^{n+1} \le [w_j^n]_+ \left(1 - \frac{\Delta t}{m_j} \sum_{k \in I} a_{jk}[u_k^n - u_j^n]_-\right) - \sum_{k \in I} a_{jk}[w_k^n]_+[u_k^n - u_j^n]_+.$$

If $w_j^n \le 0$ then

$$w_j^{n+1} \le -\Lambda \frac{\Delta t}{m_j} \sum_{k \in I} a_{jk}[u_k^n - u_j^n]_+ \le \Lambda.$$

We now prove the bound for $w_j^n > 0$. For this case we have

$$\begin{aligned} w_j^{n+1} &\le w_j^n \Big(1 - \frac{\Delta t}{m_j} \sum_{k \in I} a_{jk}[u_k^n - u_j^n]_-\Big) - \sum_{k \in I} a_{jk}[w_k^n]_+[u_k^n - u_j^n]_+ \\ &\le \Lambda\Big(1 - \frac{\Delta t}{m_j} \sum_{k \in I_I} a_{jk}(u_k^n - u_j^n)\Big) \\ &= \Lambda(1 + \Delta t \lambda^{-2}(u_j^n - w_j^n)), \end{aligned} \quad (3.4)$$

from which we see that if $w_j^n \ge u_j^n$ then (3.2) holds. Thus it remains to prove that $w_j^{n+1} \le \Lambda$ if $0 < w_j^n < u_j^n$. Noting from (3.1) that Δt is such that

$\Delta t\Lambda/\lambda^2 < \delta$, from (3.3) we have

$$\begin{aligned}
w_j^{n+1} &\le w_j^n\Big(1 - \frac{\Delta t}{m_j}\sum_{k\in I} a_{jk}[u_k^n - u_j^n]_-\Big) - \frac{\Delta t}{m_j}\sum_{k\in I} a_{jk}[w_k^n]_+[u_k^n - u_j^n]_+ \\
&\le w_j^n\Big(1 - \frac{\Delta t}{m_j}\sum_{k\in I} a_{jk}[u_k^n - u_j^n]_-\Big) - \Lambda\frac{\Delta t}{m_j}\sum_{k\in I} a_{jk}[u_k^n - u_j^n]_+ \\
&= w_j^n\Big(1 - \frac{\Delta t}{m_j}\sum_{k\in I} a_{jk}[u_k^n - u_j^n]_-\Big) + \Delta t\Lambda\lambda^{-2}(u_j^n - w_j^n) \\
&\qquad + \Lambda\frac{\Delta t}{m_j}\sum_{k\in I} a_{jk}[u_k^n - u_j^n]_- \\
&< w_j^n\Big(1 - \delta - \frac{\Delta t}{m_j}\sum_{k\in I} a_{jk}[u_k^n - u_j^n]_-\Big) + \delta u_j^n + \Lambda\frac{\Delta t}{m_j}\sum_{k\in I} a_{jk}[u_k^n - u_j^n]_-.
\end{aligned}$$

Since Δt satisfies (3.1) we have that $\Big(1 - \delta - \frac{\Delta t}{m_j}\sum_{k\in I} a_{jk}[u_k^n - u_j^n]_-\Big) > 0$ and hence

$$\begin{aligned}
w_j^{n+1} &< \Lambda\Big(1 - \delta - \frac{\Delta t}{m_j}\sum_{k\in I} a_{jk}[u_k^n - u_j^n]_-\Big) + \delta\Lambda + \Lambda\frac{\Delta t}{m_j}\sum_{k\in I} a_{jk}[u_k^n - u_j^n]_- \\
&= \Lambda. \qquad (3.5)
\end{aligned}$$

A similar argument yields the corresponding lower bound $w_j^{n+1} \le -\Lambda$. Thus (3.2) holds for $n = n+1$ and by induction it holds for all $n \ge 0$.

In the case of one-space dimension we can treat the case of non-constant F. It is convenient to rewrite (2.22) as

$$\begin{aligned}
\frac{h(w_j^{n+1} - w_j^n)}{\Delta t} &= ([w_{j+1}^n]_+ - [w_j^n]_-)\left[f\left(\frac{u_{j+1}^n - u_j^n}{h}\right)\right]_+ \\
&\quad + ([w_j^n]_+ - [w_{j+1}^n]_-)\left[f\left(\frac{u_{j+1}^n - u_j^n}{h}\right)\right]_- \\
&\quad + ([w_{j-1}^n]_+ - [w_j^n]_-)\left[f\left(\frac{u_{j-1}^n - u_j^n}{h}\right)\right]_+ \\
&\quad + ([w_j^n]_+ - [w_{j-1}^n]_-)\left[f\left(\frac{u_{j-1}^n - u_j^n}{h}\right)\right]_- .
\end{aligned}$$

Again it is sufficient to consider the upper bound for w_j^{n+1}. We have that

$$w_j^{n+1} \le [w_j^n]_+ \left(1 + \frac{\Delta t}{h}\left\{\left[f\left(\frac{u_{j+1}^n - u_j^n}{h}\right)\right]_- + \left[f\left(\frac{u_{j-1}^n - u_j^n}{h}\right)\right]_-\right\}\right)$$
$$+ \frac{\Delta t}{h}\left([w_{j-1}^n]_+ \left[f\left(\frac{u_{j-1}^n - u_j^n}{h}\right)\right]_+ + [w_{j+1}^n]_+ \left[f\left(\frac{u_{j+1}^n - u_j^n}{h}\right)\right]_+\right).$$

If $w_j^n \le 0$ then, using the properties of f,

$$w_j^{n+1} \le \frac{2\Lambda\mathcal{L}\Delta t}{h}$$

and since $K = 2$ in one-space dimension, we have the bound. We now consider $w_j^n > 0$. Since $f(s) = -f(-s)$, we have

$$\begin{aligned} w_j^{n+1} \le\ & [w_j^n]_+ \left(1 + \frac{\Delta t}{h}\left\{\left[f\left(\frac{u_{j+1}^n - u_j^n}{h}\right)\right]_- + \left[f\left(\frac{u_{j-1}^n - u_j^n}{h}\right)\right]_-\right\}\right) \\ & + \frac{\Delta t\Lambda}{h}\left(f\Big(\frac{u_{j+1}^n - u_j^n}{h}\Big) - f\Big(\frac{u_j^n - u_{j-1}^n}{h}\Big)\right) \\ & - \frac{\Delta t\Lambda}{h}\left(\left[f\left(\frac{u_{j+1}^n - u_j^n}{h}\right)\right]_- + \left[f\left(\frac{u_{j-1}^n - u_j^n}{h}\right)\right]_-\right). \end{aligned}$$

Either the middle term of the above inequality is non-positive and the bound easily follows or it is positive and

$$0 \le f\Big(\frac{u_{j+1}^n - u_j^n}{h}\Big) - f\Big(\frac{u_j^n - u_{j-1}^n}{h}\Big) \le \frac{u_{j+1}^n - 2u_j^n + u_{j-1}^n}{h} = \frac{h}{\lambda^2}(u_j^n - w_j^n).$$

In which case

$$\begin{aligned} w_j^{n+1} \le\ & w_j^n \left(1 - \delta + \frac{\Delta t}{h}\left\{\left[f\left(\frac{u_{j+1}^n - u_j^n}{h}\right)\right]_- + \left[f\left(\frac{u_{j-1}^n - u_j^n}{h}\right)\right]_-\right\}\right) \\ & + \frac{\Delta t\Lambda u_j^n}{\lambda^2} + \left(\delta - \frac{\Delta t\Lambda}{\lambda^2}\right) w_j^n \\ & - \frac{\Delta t\Lambda}{h}\left(\left[f\left(\frac{u_{j+1}^n - u_j^n}{h}\right)\right]_- + \left[f\left(\frac{u_{j-1}^n - u_j^n}{h}\right)\right]_-\right). \end{aligned}$$

From which the bound follows using the constraints on Δt. □

It is convenient to introduce the notation

$$\partial\eta^n := (\eta^{n+1} - \eta^n)/\Delta t, \quad \bar{u}_h^n := u_h^n - u_\flat^n. \tag{3.6}$$

Lemma 3.3 *If the bounds (3.2) hold then for each $n \geq 0$,*

$$|\partial u_h^n|_1 + |\partial u_h^n|_h \leq C(\Lambda, u_\flat). \tag{3.7}$$

Proof: From (2.9) we have

$$a_h(\partial u_h^n, \chi) = (\partial w_h^n, \chi)_h \quad \forall\, \chi \in S_h^0,$$

and taking $\chi = \partial\bar{u}_h^n$ we find using (2.16),

$$\begin{aligned} a_h(\partial\bar{u}_h^n, \partial\bar{u}_h^n) &= (\partial w_h^n, \partial\bar{u}_h^n)_h - a_h(\partial u_\flat^n, \partial\bar{u}_h^n) \\ &= -b_h(w_h^n; u_h^n, \partial\bar{u}_h^n) - (\partial u_h^\flat, \partial\bar{u}_\flat^n)_h \\ &\leq 2\Lambda|u_h^n|_1|\partial\bar{u}_h^n|_1 + |\partial u_\flat^n|_h|\partial\bar{u}_h^n|_h. \end{aligned}$$

The conclusion easily follows. □

We now introduce the discrete energy functional

$$E_h(\chi) := \frac{1}{2}a_h(\chi, \chi) \quad \forall\, \chi \in S_h^0.$$

Lemma 3.4 *If the bounds (3.2) hold and $\Delta t \leq \frac{\varepsilon\lambda^2}{2\Lambda}$ then for each $n \geq 0$*

$$E_h(\bar{u}_h^{n+1}) - E_h(\bar{u}_h^n) + (1-\varepsilon)\Delta t b_n(w_h^n; u_h^n, u_h^n) \leq -\Delta t(\partial u_\flat^n, \bar{u}_h^{n+1})_h. \tag{3.8}$$

Proof: A calculation shows that

$$\begin{aligned} E_h(\bar{u}_h^{n+1}) \;-\; & E_h(\bar{u}_h^n) + \frac{\Delta t^2}{2} a_h(\partial\bar{u}_h^n, \partial\bar{u}_h^{n+1}) \\ &= \Delta t a_h(\partial\bar{u}_h^n, \bar{u}_h^{n+1}) \text{ and using (2.9) and (2.16) we have} \\ &= \Delta t((\partial w_h^n, \bar{u}_h^{n+1})_h - a_h(\partial u_\flat^n, \bar{u}_h^{n+1})) \\ &= -\Delta t b_h(w_h^n; u_h^n; \bar{u}_h^n) - (\Delta t)^2 b_h(w_h^n; u_h^n, \partial\bar{u}_h^n) - \Delta t(\partial u_\flat^n, \partial\bar{u}_h^n)_h. \end{aligned}$$

Using (2.19) we have

$$\begin{aligned}\Delta t b_h(w_h^n; u_h^n, \partial\bar{u}_h^n) &\le \left(4\Lambda(\Delta t)^2|\partial\bar{u}_h^n|_1^2\right)^{1/2} b_h(w_h^n; u_h^n, u_h^n)^{1/2}\\ &\le (\Delta t)^2\Lambda/\varepsilon|\partial\bar{u}_h^n|_1^2 + \varepsilon b_h(w_h^n; u_h^n, u_h^n).\end{aligned}$$

Hence

$$\begin{aligned}&E_h(\bar{u}_h^{n+1}) - E_h(\bar{u}_h^n) + \frac{\Delta t^2}{2} a_h(\partial\bar{u}_h^n, \partial\bar{u}_h^n)\\ &\quad\le -(1-\varepsilon)\Delta t b_h(w_h^n; u_h^n, u_h^n) + \frac{(\Delta t)^3\Lambda}{\varepsilon}|\partial\bar{u}_h^n|_1^2 - \Delta t(\partial u_{\mathrm{b}}^n, \bar{u}_h^{n+1})_h.\end{aligned}$$

This concludes the proof. □

Using Lemma 3.4 we can deduce the asymptotic behaviour of the numerical solution as $n \to \infty$ in the case of constant boundary conditions. The functional $E_h(u_h^n - u_{\mathrm{b}})$ serves as a Lyapunov functional. By summing (3.8) over n we see that any limit $\{u_h^\infty, w_h^\infty\}$ of $\{u_h^n, w_h^n\}$ satisfies

$$a_h(u_h^\infty, \eta) = (w_h^\infty, \eta)_h \quad \forall\ \eta \in S_h^0,$$

$$u_h^\infty - u_{\mathrm{b}} \in S_h^0,$$

$$b_h(w_h^\infty; u_h^\infty, u_h^\infty) = 0.$$

These are discrete versions of

$$-\lambda^2\Delta u^\infty + u^\infty = w^\infty \quad \text{in } \Omega,$$

$$u^\infty = u_{\mathrm{b}} \quad \text{in } \partial\Omega,$$

$$f(|\nabla u^\infty|)|w^\infty|\nabla u^\infty = 0 \quad \text{in } \Omega.$$

4 Numerical results

In this section we display some one- and two-dimensional computations obtained using the numerical schemes of Section 2. We take $f(r) := \mathrm{sgn}\,(r)[|r| - J_p]_+$. For the one-dimensional results we set $\Omega = (0, L)$ for $L = 1$ and $L = 10$ with $h = 0.01$ and $h = 0.1$, respectively, while for the two-dimensional results we set $\Omega = (0, 10) \times (0, 10)$ and $h = 0.1$. In each case we have Δt satisfying (3.1).

4.1 Numerical computations

The one-dimensional results are displayed in Figures 4.1–4.4, in which $w_h(t)$ (bold line) and $u_h(t)$ (dashed line) are displayed together in one plot, while for the two-dimensional results, Figures 4.5 and 4.6, we just display the vorticity $w_h(t)$. Figures 4.1 and 4.2 show the evolution of w_h and u_h from some given initial data (top left-hand subplots) to their steady-states (bottom right-hand subplots). Figure 4.1 displays the evolution of two blocks of vorticity (one positive, one negative) under the influence of a positive constant magnetic field with no flux pinning and no boundary nucleation of vorticity.

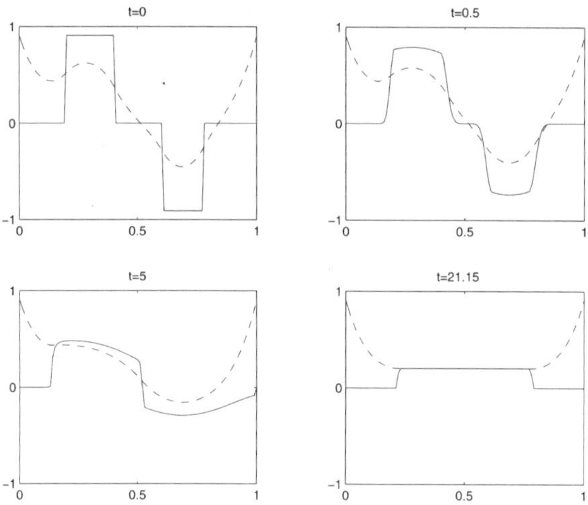

Figure 4.1: Evolution of vorticity $J_p = 0$, $u_\flat = 1$, $w_\flat = 0$, $\lambda = 0.01$.

We see that as time evolves some of the positive and negative vorticity cancel each other out. However no vorticity leaves the domain and a block of positive vorticity remains in the steady-state and the total amount of vorticity is conserved. Figure 4.2 displays the evolution of vorticity, under the influence of a time-dependent applied magnetic field, $u_\flat(t) = \sin(0.01t)$, with $w = u_\flat(t)$ on the boundary and $J_p = 0.05$. We begin the calculation with no vorticity in the sample and we see that as time evolves from $t = 0$ to $t = 200$ the applied magnetic field increases from 0 to 1 and positive vorticity enters the domain. Then as t increases from $t = 200$ to $t = 500$ the applied magnetic field decreases from 1 to -1, the positive vorticity leaves the sample and negative vorticity enters. Similarly as t increases from $t = 500$ to $t = 800$ the applied magnetic field increases from -1 to 1, the negative vorticity leaves the sample

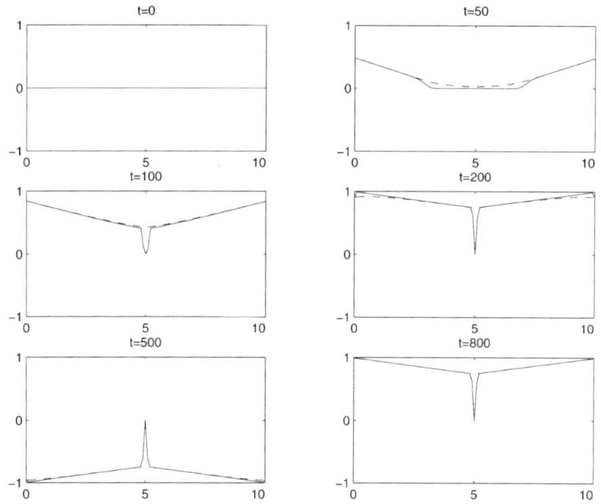

Figure 4.2: Evolution of vorticity $J_p = 0.05 \; u_b(t) = w_b(t) = \sin(0.01t)$, $\lambda = 1$.

and positive vorticity enters. This leads to a time periodic evolution. The spike in the vorticity is a consequence of small current in the centre of the domain and thus no vorticity can move into this region.

In Figures 4.3 and 4.4 we display steady-state solutions obtained using different values of J_p and the same initial data, displayed in the top left-hand subplot of each figure. Figure 4.3 displays the steady-state solutions obtained taking two blocks of vorticity (one positive, one negative) as initial data, under the influence of a positive constant magnetic field with flux pinning and no boundary nucleation of vorticity.

From this figure we see that the minimum value of the magnitude of the current $|\mathbf{j}| := |u_{\mathrm{x}}|$ occurs at the centre of the initial blocks of vorticity and as a result the vorticity remains pinned to these sites throughout the computation. While the vorticity situated away from the centre of the block spreads out either side of the block. Clearly the greater the value of the critical pinning current J_p, the less the vorticity is free to move and the less spread out the blocks become.

Figure 4.4 displays the steady-state solutions obtained taking one block of negative vorticity as initial data, under the influence of a positive constant magnetic field with flux pinning and boundary nucleation of vorticity. From this figure we see that positive vorticity enters the sample at the boundary, and the negative vorticity situated away from the centre of the block spreads out, while the vorticity at the centre of the block remains pinned to its original

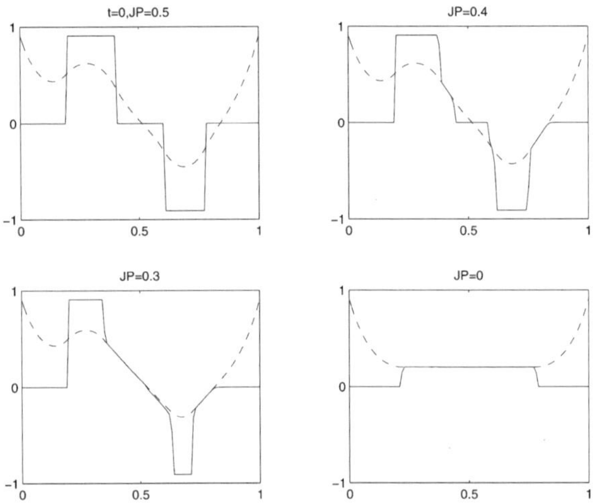

Figure 4.3: Steady-state solutions with varying J_p, $u_b = 1$, $w_b = 0$, $\lambda = 0.01$.

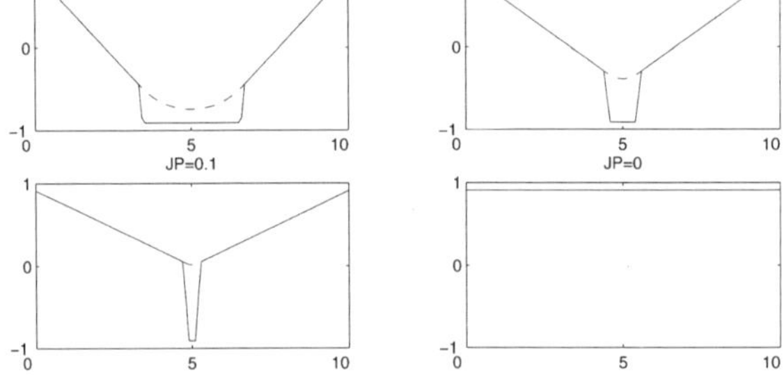

Figure 4.4: Steady-state solutions with varying J_p, $u_b = w_b = 1$, $\lambda = 1$.

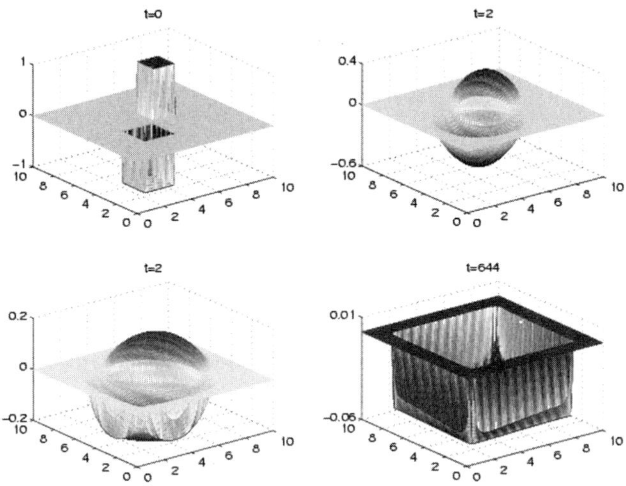

Figure 4.5: Evolution with inflow $J_p = 0$, $\lambda = 1$.

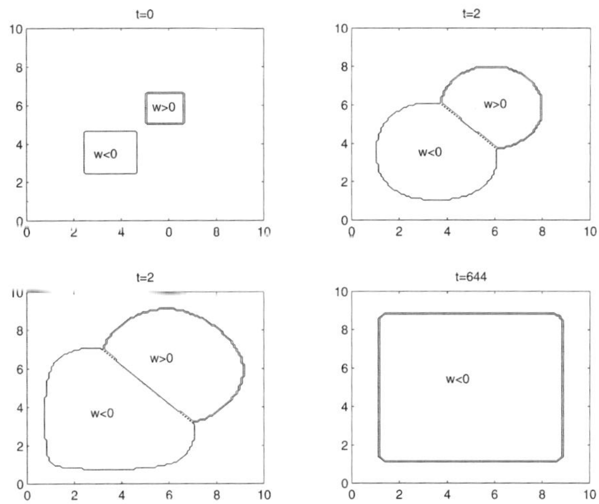

Figure 4.6: Contour plot of Figure 4.5.

position, apart from when $J_p = 0$. It can be seen in these pictures there are regions in which the current takes its critical pinning values $\pm J_p$. In these regions the magnetic field is linear and from London's equation the vorticity is equal to the magnetic field. For $J_p \neq 0$ the structure of the set of steady states is much richer than for $J_p = 0$.

In Figure 4.5 we display the evolution of two blocks of vorticity (one positive, one negative) in two-space dimensions, evolving under the influence of a constant negative magnetic field $u_b = -1$, with no flux pinning and no boundary nucleation of vorticity. As in the equivalent one-dimensional figure, Figure 4.1, we see that as time evolves some of the positive and negative vorticity cancel each other out. However in this case a block of negative vorticity remains in the steady state. Figure 4.6 displays the same results as Figure 4.5, with contour plots of the vorticity w_h. The steady-state problem in this situation can be formulated as an obstacle problem [12].

References

[1] Brenner, S.C., Ridgway-Scott, L. (1994) The Mathematical Theory of Finite Element Methods. Springer-Verlag, Berlin.

[2] Briggs, A., Claisse, J., Elliott, C.M. (1999) Finite Difference Approximation of a Non-Local Hamilton-Jacobi Equation Arising in Superconductivity. CMAIA Research Report 99/09, University of Sussex, Brighton, England.

[3] Chapman, S.J. (1995) A Mean-field Model of Superconducting Vortices in Three Dimensions. SIAM J. Appl. Math. **55**, 1252-1274.

[4] Chapman, S.J., Rubinstein, J., Schatzman, M. (1995) A Mean-field Model of Superconducting Vortices. Euro. J. Appl. Math. **7**, 97-111.

[5] Ciarlet, P.G. (1978) The Finite Element Method for Elliptic Problems. North Holland, Amsterdam.

[6] Du, Q. Convergence analysis of a Numerical Method for a Mean Field Model of Superconducting Vortices. (Preprint).

[7] Elliott, C. M., Schätzle, R., Stoth, B. E. E. (1998) Viscosity Solutions of a Degenerate Parabolic-Elliptic System Arising in the Mean-Field Theory of Superconductivity. Arch. Rat. Mech. Anal. **145**, 99-127.

[8] Elliott, C.M., Styles, V. Numerical Analysis of a Mean Field Model of Superconducting Vortices. IMA J. Num. Anal. (to appear).

[9] Elliott, C.M., Styles, V. Flux Pinning and Boundary Nucleation of a Mean Field Model of Superconducting Vortices. Interface and Free Boundaries. (to appear).

[10] Eymard, R., Ghilani, M. (1994) Convergence d'un schéma implicite de type éléments finis-volumes finis pour un système formé d'une équation ellipitique et d'une équation hyperbolique. C. R. Acad. Sci. Paris. **319**, 1095-1100.

[11] Godlewski, E., Raviart, P. (1996) Numerical Approximation of Hyperbolic Systems of Conservation Laws. Springer-Verlag, Berlin.

[12] Schätzle, R., Styles, V. (1999) Analysis of a Mean Field Model of Superconducting Vortices. EJAM **10**, 319-352.

Charles M. Elliott
Centre for Mathematical Analysis and its Applications
School of Mathematical Sciences
University of Sussex
Brighton, BN1 9QH
England
`C.M.Elliott@sussex.ac.uk`

Vanessa M. Styles
School of Computing and Mathematical Sciences
Oxford Brookes University
Headington, Oxford, OX3 0BP
England
`vmstyles@brookes.ac.uk`

A. C. Faul and M. J. D. Powell

Krylov subspace methods for radial basis function interpolation

Abstract Radial basis function methods for interpolation to values of a function of several variables are particularly useful when the data points are in general positions, but hardly any sparsity occurs in the matrix of the linear system of interpolation equations. Therefore an iterative procedure for solving the system is studied. The k-th iteration calculates the element in a k-dimensional linear subspace of radial functions that is closest to the required interpolant, the subspaces being generated by a Krylov construction that employs a self-adjoint operator A. Distances between functions are measured by the semi-norm that is induced by the well-known conditional positive or negative definite properties of the matrix of the interpolation problem, and conditions on A are found that guarantee successful termination of the iterations in exact arithmetic. A particular choice of A is recommended and is tried in numerical experiments. Fortunately, the number of iterations to achieve high accuracy is much less than the number that provides termination in theory, at most ten iterations being usual in practice. Thus the given procedure solves the interpolation equations very efficiently when the number of data is large.

1 Introduction

In many applications, some values of an unknown function from $\mathcal{R}^d$ to $\mathcal{R}$ are measured or are given, and one has to choose a function, s say, that fits these data exactly. Therefore the conditions on s are the equations

$$s(\underline{x}_i) = f_i, \qquad i=1,2,\ldots,n, \tag{1.1}$$

where the points $\underline{x}_i \in \mathcal{R}^d$, $i = 1,2,\ldots,n$, and the right hand sides $f_i \in \mathcal{R}$, $i=1,2,\ldots,n$, are the data. We allow the data points to be in general positions in $\mathcal{R}^d$, provided that they are all different, and they may have to satisfy a polynomial unisolvency condition that will be specified later. We apply a usual way of constructing s, which is to choose an n-dimensional linear space, $\mathcal{S}$ say, of functions from $\mathcal{R}^d$ to $\mathcal{R}$. Then one seeks $s\in\mathcal{S}$ that satisfies the interpolation conditions (1.1). It is elementary that this procedure defines the required s

uniquely if and only if $\mathcal{S}$ has the property that $s \in \mathcal{S}$ and $s(\underline{x}_i)=0$, $i=1,2,\ldots,n$, imply that s is identically zero.

It is known that this property can fail in the multivariable case $d \geq 2$ if the choice of $\mathcal{S}$ is independent of the positions of the data points. On the other hand, in the simplest form of the radial basis function method the general element of $\mathcal{S}$ has the form

$$s(\underline{x}) = \sum_{j=1}^{n} \lambda_j \, \phi(\|\underline{x}-\underline{x}_j\|_2), \qquad \underline{x} \in \mathcal{R}^d, \tag{1.2}$$

where ϕ is a prescribed function from $\mathcal{R}^+$ to $\mathcal{R}$, and where each λ_j is a real coefficient. It follows that the required s exists and is unique if and only if the $n \times n$ symmetric matrix Φ with the elements

$$\Phi_{ij} = \phi(\|\underline{x}_i-\underline{x}_j\|_2), \qquad 1 \leq i,j \leq n, \tag{1.3}$$

is nonsingular. The choice $\phi(r) = e^{-\rho r^2}$, $r \geq 0$, or $\phi(r) = (r^2+\rho^2)^{-1/2}$, $r \geq 0$, where ρ is any positive constant, causes Φ to be not only nonsingular but also positive definite (see Powell, 1992, for instance). Furthermore, when $n \geq 2$, the choice $\phi(r)=r$, $r \geq 0$, provides Φ with one positive and n–1 negative eigenvalues, which also implies nonsingularity.

This chapter studies a useful generalization of the form (1.2), that depends on a small integer m that satisfies $m \geq -1$. The value $m=-1$ is reserved for the case (1.2) of the previous paragraph, but otherwise we let Π_m be the linear space of polynomials from $\mathcal{R}^d$ to $\mathcal{R}$ of total degree at most m, and we let p_j, $j=1,2,\ldots,\hat{m}$, be a basis of Π_m. Then the general element of $\mathcal{S}$ has the form

$$s(\underline{x}) = \sum_{j=1}^{n} \lambda_j \, \phi(\|\underline{x}-\underline{x}_j\|_2) + \sum_{j=1}^{\hat{m}} \gamma_j \, p_j(\underline{x}), \qquad \underline{x} \in \mathcal{R}^d, \tag{1.4}$$

where the coefficients λ_j, $j=1,2,\ldots,n$, have to satisfy the constraints

$$\sum_{i=1}^{n} \lambda_i \, p_j(\underline{x}_i) = 0, \qquad j=1,2,\ldots,\hat{m}. \tag{1.5}$$

These constraints and the interpolation conditions (1.1) provide the square system of linear equations

$$\left(\begin{array}{c|c} \Phi & P \\ \hline P^T & 0 \end{array}\right) \left(\begin{array}{c} \underline{\lambda} \\ \hline \underline{\gamma} \end{array}\right) = \left(\begin{array}{c} \underline{f} \\ \hline 0 \end{array}\right), \tag{1.6}$$

Radial function	$\phi(r)$	Condition on m	Value of σ
Gaussian	$e^{-\rho r^2}$	$m\geq -1$	$\sigma=1$
Inverse multiquadric	$(r^2+\rho^2)^{-1/2}$	$m\geq -1$	$\sigma=1$
Linear	r	$m\geq 0$	$\sigma=-1$
Multiquadric	$(r^2+\rho^2)^{1/2}$	$m\geq 0$	$\sigma=-1$
Thin plate spline	$r^2\log r$	$m\geq 1$	$\sigma=1$
Cubic	r^3	$m\geq 1$	$\sigma=1$

Table 1: Values of σ for several radial basis functions

where Φ has been defined already, where P is the $n\times\hat{m}$ matrix with the elements $P_{ij} = p_j(\underline{x}_i)$, $1 \leq i \leq n$, $1 \leq j \leq \hat{m}$, where the components of $\underline{\lambda}$ and $\underline{\gamma}$ are the coefficients of the function (1.4), and where $\underline{f}$ has the components f_i, $i=1,2,\ldots,n$.

The $(n+\hat{m})\times(n+\hat{m})$ matrix of the system (1.6) is nonsingular only if the columns of P are linearly independent. Now, for any $\underline{\gamma}\in\mathcal{R}^{\hat{m}}$, the i-th component of the product $P\underline{\gamma}$ is $p(\underline{x}_i)$, where p is the polynomial $\sum_{j=1}^{\hat{m}}\gamma_j\,p_j$. Thus the rank of P is $\hat{m}$ if and only if $p\in\Pi_m$ and $p(\underline{x}_i)=0$, $i=1,2,\ldots,n$, imply $p\equiv 0$. We assume that m is small enough to give this property for the positions of the interpolation points, which is the polynomial unisolvency condition that has been mentioned.

Now the usual choices of ϕ in the radial basis function method are such that, if m is large enough, and if $\underline{\lambda}$ is a nonzero vector in $\mathcal{R}^n$ that satisfies $P^T\underline{\lambda}=0$, then $\underline{\lambda}^T\Phi\underline{\lambda}$ is nonzero. Further, by continuity, the sign of $\underline{\lambda}^T\Phi\underline{\lambda}$ is independent of $\underline{\lambda}$. We define $\sigma=\pm 1$ by the equation

$$|\underline{\lambda}^T\Phi\underline{\lambda}| = \sigma\,\underline{\lambda}^T\Phi\underline{\lambda}. \tag{1.7}$$

This condition on m and the value of σ for several useful radial basis functions are stated in Table 1, where ρ is still a positive constant. The $m\geq -1$ entries occur when Φ is positive definite, because then the polynomial term of expression (1.4) is optional. Proofs of the assertions in the table are given by Micchelli (1986).

From now on, we assume not only the full rank of P (polynomial unisolvency) but also the condition on m that is stated in the previous paragraph. It follows that the system (1.6) is nonsingular. We prove this assertion by deducing that, if $\underline{f}$ were zero, then $\underline{\lambda}$ and $\underline{\gamma}$ would be zero too. Indeed, by pre-multiplying equation (1.6) by $(\underline{\lambda}^T\ \underline{\gamma}^T)$ and by using $P^T\underline{\lambda}=0$, we find the identity

$$\underline{\lambda}^T\Phi\underline{\lambda}+\underline{\lambda}^T P\underline{\gamma}+\underline{\gamma}^T P^T\underline{\lambda}=\underline{\lambda}^T\Phi\underline{\lambda}=\underline{\lambda}^T\underline{f}. \tag{1.8}$$

Therefore $\underline{f}=0$ implies $\underline{\lambda}^T\Phi\underline{\lambda}=0$, and then the previous paragraph gives $\underline{\lambda}=0$. Moreover, $\Phi\underline{\lambda}+P\underline{\gamma}=\underline{f}$ with $\underline{\lambda}=0$ and $\underline{f}=0$ give $\underline{\gamma}=0$ by polynomial unisolvency, which completes the proof. Thus the interpolation conditions (1.1) are satisfied uniquely by a function of the form (1.4) subject to the constraints (1.5). Further, the set of functions s that can occur for different right hand side vectors $\underline{f}\in\mathcal{R}^n$ is an n-dimensional linear space, namely the space $\mathcal{S}$ of functions of the form (1.4) whose coefficients are subject to the constraints (1.5).

These assumptions allow the system (1.6) to be solved by a direct method, that usually gives excellent accuracy even when Φ is highly ill-conditioned, and that was proposed originally for thin plate spline radial functions by Sibson and Stone (1991). Specifically, we let $\mathcal{Z}$ be the $(n-\hat{m})$-dimensional linear subspace of $\mathcal{R}^n$ such that $\underline{\lambda}$ is in $\mathcal{Z}$ if and only if its components satisfy the conditions (1.5), and we let Z be an $n\times(n-\hat{m})$ real matrix whose columns are a basis of $\mathcal{Z}$. Then the $\underline{\lambda}$ that is defined by the system (1.6) can be written as $\underline{\lambda}=Z\underline{\mu}$ for some $\underline{\mu}$ in $\mathcal{R}^{n-\hat{m}}$. Further, because Z^TP is zero, the first part of the system (1.6) provides the relation

$$(\sigma\, Z^T\Phi Z)\,\underline{\mu}=\sigma\, Z^T\underline{f}. \tag{1.9}$$

One chooses Z so that it is not expensive to calculate all the elements of the symmetric matrix $\sigma\, Z^T\Phi Z=W$, say. It follows from the paragraph that includes equation (1.7) that W is positive definite. Therefore the direct method solves the system (1.9) for $\underline{\mu}$ by calculating the Cholesky factorization of W. Then the formula $\underline{\lambda}=Z\underline{\mu}$ gives the radial basis function coefficients of the required interpolant (1.4). Finally, the polynomial coefficients of s can be derived conveniently from the following construction. By multiplying the first part of expression (1.6) by P^T, we obtain the square system of normal equations

$$(P^TP)\,\underline{\gamma}=P^T(\underline{f}-\Phi\underline{\lambda}), \tag{1.10}$$

which defines the coefficients γ_j, $j=1,2,\ldots,\hat{m}$. Ill-conditioning can be avoided in the matrix P^TP by a careful choice of the basis p_j, $j=1,2,\ldots,\hat{m}$.

The amount of work of this procedure for solving the system (1.6) is $\mathcal{O}(n^3)$ operations, which restricts the number of data points to about one thousand. Therefore, because of the very wide range of applications of the radial basis function method (see Hardy, 1990, for instance), there has been much research on iterative techniques for calculating the coefficients of the interpolant (1.4).

In particular, the procedure of Dyn, Levin and Rippa (1986) is analogous to the solution of the system (1.9) by the conjugate gradient algorithm, after picking Z in a clever way that assists the conditioning of $Z^T \Phi Z$. Another approach is suggested by Beatson and Powell (1994). It is based on the Lagrange form

$$s(\underline{x}) = \sum_{j=1}^{n} f_j \, \ell_j(\underline{x}), \qquad \underline{x} \in \mathcal{R}^d, \tag{1.11}$$

of the solution of the interpolation equations (1.1), where ℓ_j is the s that would occur if the right hand sides had the values $f_i = \delta_{ij}$, $i=1,2,\ldots,n$. Each ℓ_j in formula (1.11) is replaced by an estimate of the Lagrange function that employs only a small subset of the interpolation points $\underline{x}_i$, $i=1,2,\ldots,n$, so the formula provides an approximation to s. The error of the approximation is corrected by iterative refinement.

Huge improvements to these methods have been developed during the last six years by Rick Beatson, George Goodsell and the authors. It is usual for the new algorithms to solve the interpolation equations (1.1) to eight decimal places of accuracy in fewer than ten iterations for general positions of the data points, even when n is large. The main task of each iteration is the calculation of the values $s(\underline{x}_i)$, $i=1,2,\ldots,n$, for a function s of the form (1.4). Therefore fast multipole methods are highly useful when they can be applied (see Beatson and Newsam, 1992, for instance). The research in Cambridge provided a procedure of this kind about three years ago (Powell, 1997), s being the current estimate of the interpolant, because the residuals $f_i - s(\underline{x}_i)$, $i=1,2,\ldots,n$, are required. Further, many intermediate residuals were calculated. Then George Goodsell found experimentally that in most cases much work can be saved by replacing the intermediate residuals by the ones that occur at the beginning of the iteration. Occasionally that method fails to converge, so a Krylov subspace technique has been added to it. Further, we now regard the Krylov construction as the main part of the method, so the most successful development of the previous five years reduces to a particular choice of a self-adjoint operator, namely A. Thus the description of the recommended algorithm becomes easier. Therefore we are going to address the algorithm and its convergence properties from this point of view.

The Krylov subspace method requires a norm or a semi-norm for the n-dimensional linear space $\mathcal{S}$, which is the subject of Section 2. We will find there that the native semi-norm of Schaback (1993) is very suitable. Then Section 3 considers the Krylov technique when the operator A is general, including conditions on A that are sufficient for convergence, and that allow an implementation that is analogous to the conjugate gradient method. The particular A and some other details of the recommended algorithm are specified in Section 4. Finally, in Section 5, the efficiency of the given procedure is investigated numerically on some test problems, and some plans for future work are mentioned.

2 The semi-norm of $\mathcal{S}$

Let the functions (1.4) and

$$t(\underline{x}) = \sum_{j=1}^{n} \mu_j \, \phi(\|\underline{x}-\underline{x}_j\|_2) + \sum_{j=1}^{\hat{m}} \delta_j \, p_j(\underline{x}), \qquad \underline{x}\in\mathcal{R}^d, \tag{2.1}$$

be elements of the n-dimensional linear space $\mathcal{S}$ that has been defined already. Therefore the radial basis function coefficients satisfy the constraints

$$\sum_{i=1}^{n} \lambda_i \, p_j(\underline{x}_i) = \sum_{i=1}^{n} \mu_i \, p_j(\underline{x}_i) = 0, \qquad j=1,2,\ldots,\hat{m}. \tag{2.2}$$

We introduce the bilinear form

$$\langle s,t\rangle = \sigma \, \underline{\lambda}^T \Phi \underline{\mu}, \tag{2.3}$$

where Φ is still the matrix (1.3). The symmetry of Φ implies the condition $\langle t,s\rangle = \langle s,t\rangle$. Moreover, the property (1.7) shows that $\langle s,s\rangle$ is nonnegative for every s in $\mathcal{S}$. Therefore, by defining expression (2.3) to be the scalar product between s and t, and by defining the semi-norm

$$\|s\| = \langle s,s\rangle^{1/2}, \qquad s\in\mathcal{S}, \tag{2.4}$$

the space $\mathcal{S}$ becomes a semi-Hilbert space. We will use the notation (2.3) and (2.4) extensively.

It follows from the paragraph that includes equation (1.7) that, if the function (1.4) is an element of $\mathcal{S}$, then $\|s\|$ is zero if and only if all the coefficients λ_j, $j=1,2,\ldots,n$, are zero. In other words, the semi-norm of $s\in\mathcal{S}$ is zero if and only if s is in the space Π_m of polynomials of degree at most m. Here we take the view that Π_{-1} contains just the zero function, because the semi-Hilbert space becomes a Hilbert space in the case $m=-1$.

Our algorithm requires a reformulation of expression (2.3). Specifically, equations (2.3), (1.3), (1.5) and (2.1) imply the identities

$$\begin{aligned}
\langle s,t\rangle &= \sigma \sum_{i=1}^{n} \lambda_i \Big\{ \sum_{j=1}^{n} \phi(\|\underline{x}_i-\underline{x}_j\|_2)\,\mu_j \Big\} \\
&= \sigma \sum_{i=1}^{n} \lambda_i \Big\{ \sum_{j=1}^{n} \mu_j\,\phi(\|\underline{x}_i-\underline{x}_j\|_2) + \sum_{j=1}^{\hat{m}} \delta_j\,p_j(\underline{x}_i) \Big\} \\
&= \sigma \sum_{i=1}^{n} \lambda_i\,t(\underline{x}_i) = \sigma \sum_{i=1}^{n} \mu_i\,s(\underline{x}_i), \qquad (2.5)
\end{aligned}$$

the last assertion being valid due to symmetry. One advantage of this remark is that the matrix Φ will not occur in the description of the algorithm. Another advantage is that, if the function (1.4) is a general element of $\mathcal{S}$ and if $s^\star$ is the required interpolant, then the interpolation conditions provide the formula

$$\langle s, s^\star\rangle = \sigma \sum_{i=1}^{n} \lambda_i\,s^\star(\underline{x}_i) = \sigma \sum_{i=1}^{n} \lambda_i\,f_i, \qquad s\in\mathcal{S}. \qquad (2.6)$$

Therefore the value of $\langle s, s^\star\rangle$ can be calculated before the coefficients of $s^\star$ are known, which is highly useful.

The semi-norm has a fundamental property that provides excellent corroboration for the claim that it is very suitable for the Krylov subspace method. The property is taken from Schaback (1993), and will not be employed by our algorithm, because it applies to a linear space of functions $\mathcal{S}$ that is larger than the one that has been defined already. Indeed, just for the remainder of this section, we let $\hat{s}$ be in $\mathcal{S}$ if it has the form

$$\hat{s}(\underline{x}) = \sum_{j=1}^{\hat{n}} \hat{\lambda}_j\,\phi(\|\underline{x}-\hat{\underline{x}}_j\|_2) + \sum_{j=1}^{\hat{m}} \hat{\gamma}_j\,p_j(\underline{x}), \qquad \underline{x}\in\mathcal{R}^d, \qquad (2.7)$$

the parameters being subject to the usual constraints

$$\sum_{i=1}^{\hat{n}} \hat{\lambda}_i\,p_j(\hat{\underline{x}}_i) = 0, \qquad j=1,2,\ldots,\hat{m}. \qquad (2.8)$$

Here ϕ and p_j, $j=1,2,\ldots,\hat{m}$, are as before, but $\hat{n}$ can be much larger than n. The only restriction on the set $\{\hat{\underline{x}}_i : i=1,2,\ldots,\hat{n}\}$ is that it has to include all the original data points $\underline{x}_i$, $i=1,2,\ldots,n$. It follows that the previous $\mathcal{S}$ is an n-dimensional linear subspace of the new $\mathcal{S}$, that the polynomial unisolvency condition is preserved, and that at least one element of the new $\mathcal{S}$ interpolates the data. Further, $\hat{s} \in \mathcal{S}$ is an interpolant if and only if it can be written in the form $\hat{s} = s^\star + \hat{t}$, where $s^\star$ is still the required interpolant, and where $\hat{t} \in \mathcal{S}$ satisfies $\hat{t}(\underline{x}_i) = 0$, $i = 1, 2, \ldots, n$.

Now the analogue of expression (2.3) for the new space $\mathcal{S}$ is the bilinear form

$$\langle \hat{s}, \hat{t} \rangle = \sigma \sum_{i=1}^{\hat{n}} \sum_{j=1}^{\hat{n}} \hat{\lambda}_i \, \phi(\|\hat{\underline{x}}_i - \hat{\underline{x}}_j\|_2) \, \hat{\mu}_j, \tag{2.9}$$

where $\hat{s}$ and $\hat{t}$ are the functions (2.7) and

$$\hat{t}(\underline{x}) = \sum_{j=1}^{\hat{n}} \hat{\mu}_j \, \phi(\|\underline{x} - \hat{\underline{x}}_j\|_2) + \sum_{j=1}^{\hat{m}} \hat{\delta}_j \, p_j(\underline{x}), \qquad \underline{x} \in \mathcal{R}^d, \tag{2.10}$$

respectively. Thus there is no change to the previous value of $\langle \hat{s}, \hat{t} \rangle$ if $\hat{s}$ and $\hat{t}$ happen to be in the original space $\mathcal{S}$. Moreover, the definition (2.4) of the semi-norm is valid for the new space $\mathcal{S}$.

The fundamental property that supports the choice (2.4) is that, if $\hat{s}$ is any function in the new space $\mathcal{S}$ that satisfies the interpolation equations (1.1), then $\langle \hat{s}, \hat{s} \rangle$ is never less than $\langle s^\star, s^\star \rangle$. We prove this assertion by writing $\hat{s} = s^\star + \hat{t}$ as before, and by considering the scalar product

$$\langle s^\star, \hat{t} \rangle = \sigma \sum_{i=1}^{\hat{n}} \hat{\lambda}_i^\star \, \hat{t}(\hat{\underline{x}}_i), \tag{2.11}$$

where $\hat{\lambda}_i^\star$, $i = 1, 2, \ldots, \hat{n}$, are the radial basis function coefficients of $s^\star$, and where the right hand side is taken from the identity (2.5). The coefficient $\hat{\lambda}_i^\star$ is nonzero only if $\hat{\underline{x}}_i$ is one of the given interpolation points, and then $\hat{t}(\hat{\underline{x}}_i)$ is zero. Thus expression (2.11) is zero, which gives the required bound

$$\langle \hat{s}, \hat{s} \rangle = \langle s^\star + \hat{t}, s^\star + \hat{t} \rangle = \langle s^\star, s^\star \rangle + \langle \hat{t}, \hat{t} \rangle \geq \langle s^\star, s^\star \rangle. \tag{2.12}$$

This bound is a generalization of the well-known minimum second derivative property of thin plate spline interpolation in two dimensions (Duchon, 1977), if one restricts attention to functions in the new space $\mathcal{S}$. From now on, however, we revert to the original space $\mathcal{S}$ that has dimension n.

3 The Krylov subspace method

The Krylov subspace method is iterative and depends on a linear operator A from $\mathcal{S}$ to $\mathcal{S}$ and on the semi-norm (2.4). It is similar to the GMRES algorithm of Saad and Schultz (1986), except that A need not depend on the interpolation equations and the semi-norm need not be a norm. We let k be the iteration number of the method, and we let $\mathcal{S}_k$ be the linear subspace of $\mathcal{S}$ that is spanned by the functions $A^j s^\star$, $j=1,2,\ldots,k$, where $s^\star$ is still the required interpolant. If the k-th iteration is not the final one, then its main task is to calculate an element of $\mathcal{S}_k$, say s_{k+1}, that satisfies the condition

$$||s^\star - s_{k+1}|| \leq ||s^\star - s||, \qquad s \in \mathcal{S}_k, \tag{3.1}$$

for the semi-norm (2.4).

An implementation of this method will be described that has the following useful properties for the choice of A that is given in Section 4. If there are no computer rounding errors, and if a tolerance parameter is set to zero, then the required interpolant $s^\star$ is obtained. The sequence of semi-norms $||s^\star - s_j||$, $j=1,2,\ldots,k^\star$, decreases strictly monotonically, where s_1 is the zero function and where $k^\star$ is the number of the final iteration. Condition (3.1) is achieved for every integer k in $[1, k^\star\!-\!1]$. The coefficients of the functions $A^j s^\star$, $j=1,2,\ldots,k$, are not found explicitly, because these functions tend to provide a basis of $\mathcal{S}_k$ that is highly ill-conditioned. Instead, for $k=1,2,\ldots,k^\star$, the function s_{k+1} is calculated from s_k by applying the formula

$$s_{k+1} = s_k + \alpha_k \, d_k \tag{3.2}$$

in a way that is analogous to an iteration of the conjugate gradient method. Specifically, d_k is the search direction

$$d_k = A\,(s^\star - s_k) + \beta_k \, d_{k-1}, \tag{3.3}$$

the multiplier $\beta_k \in \mathcal{R}$ being defined by the orthogonality condition $\langle d_k, d_{k-1}\rangle = 0$, except that the $\beta_k \, d_{k-1}$ term is deleted when $k=1$, so $s_1=0$ implies $d_1 = As^\star$. The step-length $\alpha_k \in \mathcal{R}$ of formula (3.2) is derived from inequality (3.1). The calculation ends when the function values $s_{k+1}(\underline{x}_i)$, $i=1,2,\ldots,n$, are sufficiently close to the data f_i, $i=1,2,\ldots,n$, the termination condition being the requirement

$$|s_{k+1}(\underline{x}_i) - f_i| \leq \varepsilon, \qquad i=1,2,\ldots,n, \tag{3.4}$$

where $\varepsilon \geq 0$ is the tolerance parameter that has been mentioned. Therefore every iteration, before employing the step-length that provides the property (3.1), tries to choose α_k in equation (3.2) to satisfy inequality (3.4), the calculation being completed if and only if the attempt is successful. It is necessary for A to be such that the attempt gives termination in the case $\|s^\star - s_k\| = 0$, because then the usual reduction $\|s^\star - s_{k+1}\| < \|s^\star - s_k\|$ cannot be attained. The construction of the search directions provides the orthogonality conditions

$$\langle d_k, d_j \rangle = 0, \qquad j = 1, 2, \ldots, k-1, \tag{3.5}$$

for every integer k in the interval $[2, k^\star - 1]$.

Instead of showing that the remarks of the previous paragraph are valid for the particular A that is specified in Section 4, we are going to establish that it is sufficient if A has the properties

$$\left.\begin{array}{llll} s \in \mathcal{S} \text{ and } s \neq 0 & \Rightarrow & As \neq 0 & \text{(nonsingularity)} \\ s \in \Pi_m & \Rightarrow & As = s & \text{(polynomial reproduction)} \\ s \in \mathcal{S} \text{ and } s \notin \Pi_m & \Rightarrow & \langle s, As \rangle > 0 & \text{(ellipticity)} \\ s \in \mathcal{S} \text{ and } t \in \mathcal{S} & \Rightarrow & \langle t, As \rangle = \langle At, s \rangle & \text{(self-adjointness)} \end{array}\right\}. \tag{3.6}$$

Then only these properties will have to be verified in Section 4.

If the required interpolant $s^\star$ is in the space Π_m, then polynomial reproduction and $s_1 = 0$ imply that the first iteration generates the direction $d_1 = As^\star = s^\star$. Thus the choice $\alpha_1 = 1$ in formula (3.2) gives $s_2 = s^\star$. It follows that the termination condition (3.4) is achieved on the first iteration, even if ε is zero.

Alternatively, in the case $s^\star \notin \Pi_m$, termination occurs on the first iteration if ε is sufficiently large, but usually s_2 is calculated by satisfying condition (3.1) for $k = 1$. Specifically, s_2 is the function $\alpha_1 d_1$, where d_1 is the direction $As^\star$, and where α_1 minimizes the scalar product

$$\langle s^\star - \alpha_1 d_1, s^\star - \alpha_1 d_1 \rangle = \|s^\star - s_2\|^2, \qquad \alpha_1 \in \mathcal{R}. \tag{3.7}$$

Now $s^\star \notin \Pi_m$ implies $d_1 = As^\star \notin \Pi_m$, because otherwise A would map not only Π_m but also $s^\star$ into Π_m, which would contradict nonsingularity. Therefore $\langle d_1, d_1 \rangle$ is positive, as stated in the second paragraph of Section 2. It follows that the minimization of expression (3.7) defines $\alpha_1 = \langle s^\star, d_1 \rangle / \langle d_1, d_1 \rangle$ uniquely. Further, α_1 is positive, because $s^\star \notin \Pi_m$, $d_1 = As^\star$ and the ellipticity condition (3.6) provide $\langle s^\star, d_1 \rangle > 0$. Hence, if $k^\star \geq 2$, we deduce the strict inequality

$$\|s^\star - s_2\| < \|s^\star\| = \|s^\star - s_1\|. \tag{3.8}$$

For $k \geq 2$, we are going to prove the assertions in the second paragraph of this section by induction. Therefore we assume that they hold if k is replaced by $k-1$. Thus s_k and d_{k-1} are in the space $\mathcal{S}_{k-1}$. It follows from the definition of $\mathcal{S}_k$ and from equations (3.3) and (3.2) that, for any choice of β_k and α_k, both d_k and s_{k+1} are in the space $\mathcal{S}_k$. Further, because the inductive hypothesis includes the relations $s_k = s_{k-1} + \alpha_{k-1} d_{k-1}$ and $\|s^\star - s_k\| < \|s^\star - s_{k-1}\|$, the value of $\langle d_{k-1}, d_{k-1}\rangle$ is positive. Hence equation (3.3) and the orthogonality condition $\langle d_k, d_{k-1}\rangle = 0$ define the parameter β_k uniquely.

The possibility $\|s^\star - s_k\| = 0$ can occur. Then $s^\star - s_k$ is an element of Π_m, which is nonzero because the calculation was not terminated by the previous iteration. Moreover, the polynomial reproduction condition (3.6) shows that the search direction (3.3) is the function

$$d_k = s^\star - s_k + \beta_k \, d_{k-1}. \tag{3.9}$$

Now the Cauchy–Schwarz inequality

$$|\langle s^\star - s_k, d_{k-1}\rangle| \le \|s^\star - s_k\| \, \|d_{k-1}\| \tag{3.10}$$

and $\|s^\star - s_k\| = 0$ provide $\langle s^\star - s_k, d_{k-1}\rangle = 0$, while the definition of β_k provides $\langle d_k, d_{k-1}\rangle = 0$. It follows from expression (3.9) that β_k is zero. Hence equation (3.2) takes the form $s_{k+1} = s_k + \alpha_k \, (s^\star - s_k)$. Thus the termination condition (3.4) is achieved by the choice $\alpha_k = 1$. It is also achieved when $\|s^\star - s_k\|$ is nonzero if the tolerance parameter ε is sufficiently large. Our analysis of the case $k = k^\star$ is complete, so the conditions $\|s^\star - s_k\| > 0$ and $2 \le k < k^\star$ are assumed for the remainder of the inductive argument.

We recall that the step-length α_k of formula (3.2) is chosen to minimize the function

$$\|s^\star - s_{k+1}\|^2 = \langle s^\star - s_k - \alpha_k \, d_k, s^\star - s_k - \alpha_k \, d_k\rangle, \qquad \alpha_k \in \mathcal{R}. \tag{3.11}$$

We are going to deduce from the assumption $\|s^\star - s_k\| > 0$ that α_k is positive. We see that, if α_k is well defined, then it has the value $\langle s^\star - s_k, d_k\rangle / \langle d_k, d_k\rangle$. Further, equation (3.3) gives the identity

$$\langle s^\star - s_k, d_k\rangle = \langle s^\star - s_k, A(s^\star - s_k)\rangle + \beta_k \, \langle s^\star - s_k, d_{k-1}\rangle. \tag{3.12}$$

Now $\|s^\star - s_k\| > 0$ is equivalent to $(s^\star - s_k) \notin \Pi_m$, so the ellipticity condition of expression (3.6) shows that $\langle s^\star - s_k, A(s^\star - s_k)\rangle$ is positive. Moreover, $d_{k-1} \in \mathcal{S}_{k-1}$ and the inductive hypothesis

$$\|s^\star - s_k\| \le \|s^\star - s\|, \qquad s \in \mathcal{S}_{k-1}, \tag{3.13}$$

imply that the least value of $\|s^\star - s_k - \theta \, d_{k-1}\|$, $\theta \in \mathcal{R}$, occurs at $\theta = 0$, so the scalar product $\langle s^\star - s_k, d_{k-1}\rangle$ is zero. Thus the identity (3.12) implies $\langle s^\star - s_k, d_k\rangle > 0$, and then the Cauchy–Schwarz inequality provides $\langle d_k, d_k\rangle > 0$. It follows that the algorithm generates a positive step-length as required. The property $\alpha_k \neq 0$ with the construction of α_k yield the strict inequality

$$||s^\star - s_{k+1}|| < ||s^\star - s_k||, \tag{3.14}$$

which establishes another of the assertions that are being addressed.

Equation (3.5) is satisfied when $j=k-1$, because the choice of β_k provides $\langle d_k, d_{k-1}\rangle = 0$. Therefore, if $k \geq 3$, we have to show that $\langle d_k, d_j\rangle$ is zero for $1 \leq j \leq k-2$. Definition (3.3) gives the identity

$$\langle d_k, d_j\rangle = \langle A(s^\star - s_k), d_j\rangle + \beta_k \langle d_{k-1}, d_j\rangle, \tag{3.15}$$

and the inductive hypothesis includes expression (3.5) when k is replaced by $k-1$. Thus, using the self-adjointness condition (3.6), we deduce the equation

$$\langle d_k, d_j\rangle = \langle s^\star - s_k, Ad_j\rangle, \qquad j=1,2,\ldots,k-2. \tag{3.16}$$

Now $d_j \in \mathcal{S}_j$ implies $Ad_j \in \mathcal{S}_{j+1}$, so Ad_j is an element of $\mathcal{S}_{k-1}$ for $j \leq k-2$. It follows from the hypothesis (3.13) that the least value of $||s^\star - s_k - \theta A d_j||$, $\theta \in \mathcal{R}$, occurs at $\theta = 0$. Therefore the scalar product (3.16) is zero, which completes the proof of assertion (3.5).

We now know that the search directions $d_j \in \mathcal{S}_k$, $j = 1,2,\ldots,k$, are mutually orthogonal with respect to the scalar product of the semi-Hilbert space $\mathcal{S}$. Further, because the Krylov subspace definition of $\mathcal{S}_k$ implies $\dim(\mathcal{S}_k) \leq k$, these directions are a basis of $\mathcal{S}_k$. Therefore the least value of the right hand side of expression (3.1) is achieved when s is the function $s = \sum_{j=1}^{k} \theta_j\, d_j$, where the coefficients $\theta_j \in \mathcal{R}$, $j=1,2,\ldots,k$, minimize the quadratic form

$$\Big\| s^\star - \sum_{j=1}^{k} \theta_j\, d_j \Big\|^2 = \langle s^\star, s^\star\rangle - 2 \sum_{j=1}^{k} \theta_j \langle s^\star, d_j\rangle + \sum_{j=1}^{k} \theta_j^2 \langle d_j, d_j\rangle. \tag{3.17}$$

Hence the coefficients of the optimal s take the unique values

$$\theta_j^\star = \langle s^\star, d_j\rangle / \langle d_j, d_j\rangle, \qquad j=1,2,\ldots,k. \tag{3.18}$$

Similarly, the inductive hypothesis (3.13) implies that s_k is the function $\sum_{j=1}^{k-1} \theta_j^\star\, d_j$. It follows that $||s^\star - s_k - \theta_k^\star\, d_k||$ is the least value of $||s^\star - s||$, $s \in \mathcal{S}_k$. Now the given method sets $s_{k+1} = s_k + \alpha_k\, d_k$, where α_k minimizes $||s^\star - s_k - \alpha_k\, d_k||$, $\alpha_k \in \mathcal{R}$. Thus $s_{k+1} = s_k + \theta_k^\star\, d_k$ must occur, which provides the required property (3.1).

We also have to prove that the given method terminates in exact arithmetic. Therefore, for any iteration number $k < k^\star$, we consider the search directions

d_j, $j=1,2,\ldots,k$, and the basis p_j, $j=1,2,\ldots,\hat{m}$, of Π_m. We pick coefficients θ_j, $j=1,2,\ldots,k$, and γ_j, $j=1,2,\ldots,\hat{m}$, such that $s=\sum_{j=1}^{k}\theta_j\,d_j+\sum_{j=1}^{\hat{m}}\gamma_j\,p_j$ is the zero element of $\mathcal{S}$. The mutual orthogonality of the search directions and the form (2.5) of the scalar product provide the equation

$$\langle s,s\rangle = \sum_{j=1}^{k}\theta_j^2\,\langle d_j,d_j\rangle = 0, \tag{3.19}$$

for general values of γ_j, $j=1,2,\ldots,\hat{m}$. Because our analysis gives $\langle d_j,d_j\rangle>0$, $j=1,2,\ldots,k$, it follows that every θ_j is zero. Then we deduce $\gamma_j=0$, $j=1,2,\ldots,\hat{m}$, from $s\equiv 0$, as the polynomials p_j, $j=1,2,\ldots,\hat{m}$, are a basis of Π_m. Therefore we have found $k+\hat{m}$ linearly independent elements of $\mathcal{S}$, which implies $k+\hat{m}\le\dim(\mathcal{S})=n$ for each $k<k^\star$. In other words, assuming exact arithmetic, termination occurs on an iteration whose number $k^\star$ is at most $n-\hat{m}+1$. The justification of the remarks in the second paragraph of this section is complete.

The nonsingularity condition (3.6) on A is important, because the functions that can be calculated by the Krylov subspace method are restricted to the set $\{As : s\in\mathcal{S}\}$. Indeed, if A were singular, then the unique solution $s^\star$ of the interpolation equations would not be in this set for most values of the right hand sides f_i, $i=1,2,\ldots,n$. The polynomial reproduction condition (3.6) is also important, because it provides a way of generating the correct polynomial part of $s^\star$. This task is not done by the techniques of the method that are designed to achieve inequality (3.1), because the definition (2.4) of the semi-norm ignores any polynomial terms of s. The purpose of the ellipticity condition (3.6) is to give the method the property $s_{k+1}\ne s_k$, where s_{k+1} is defined by condition (3.1). If $s_{k+1}=s_k$ occurred, however, then formula (3.3) would be unsuitable for generating the search direction d_{k+1}, because the relation

$$A(s^\star-s_{k+1}) = A(s^\star-s_k) = d_k-\beta_k\,d_{k-1} \tag{3.20}$$

would put d_{k+1} in the linear space that is spanned by the first k search directions. The self-adjointness condition (3.6) implies by induction that all the orthogonality properties (3.5) are satisfied when d_k has the simple form (3.3). Only one change has to be made to the given method if this condition fails, namely that the definition (3.3) is replaced by the equation

$$d_k = A(s^\star-s_k)-\sum_{j=1}^{k-1}\langle d_j, A(s^\star-s_k)\rangle\,\langle d_j,d_j\rangle^{-1}\,d_j, \tag{3.21}$$

the sum being deleted when $k=1$. The authors have studied the use of this formula in a Krylov subspace method, because each iteration of the algorithm in

Powell (1997) can be regarded as an operator A from $\mathcal{S}$ to $\mathcal{S}$, and this operator possesses only the first three of the properties (3.6).

The polynomial reproduction and ellipticity conditions imply nonsingularity. Specifically, given any nonzero $s \in \mathcal{S}$, we ask if it is in Π_m. If the answer is affirmative, then polynomial reproduction provides $As = s \neq 0$. Otherwise, ellipticity gives $\langle s, As \rangle > 0$, which shows $As \neq 0$ as required.

4 The recommended algorithm

The proof of termination in the previous section is misleading, because it suggests that the number of iterations of the Krylov subspace method may be close to n. Instead, as mentioned in Section 1, fewer than ten iterations are usual when the recommended algorithm is applied in practice, even when n becomes large. Therefore we seem to have made an excellent choice of the operator A, which will be addressed in this section. We begin by deriving this A when there are no polynomial terms ($m = -1$), so the general element of $\mathcal{S}$ has the form (1.2). This derivation will make it easy to allow $m \geq -1$ later.

When $m = -1$, we consider the choice

$$As = \sum_{j=1}^{n} \langle \psi_j, s \rangle \, \langle \psi_j, \psi_j \rangle^{-1} \, \psi_j, \qquad s \in \mathcal{S}, \tag{4.1}$$

where ψ_j, $j = 1, 2, \ldots, n$, is a basis of $\mathcal{S}$. The self-adjointness of A is shown by the symmetry of the right hand side of the equation

$$\langle t, As \rangle = \sum_{j=1}^{n} \langle \psi_j, s \rangle \, \langle \psi_j, \psi_j \rangle^{-1} \, \langle \psi_j, t \rangle. \tag{4.2}$$

There are no polynomials to be reproduced except for the trivial case of the zero function, and the following argument establishes ellipticity. It is sufficient to prove that $\langle s, As \rangle = 0$ implies $s = 0$. We make use of the remark that expression (2.4) is a norm in the case $m = -1$. We let $s = \sum_{j=1}^{n} \theta_j \, \psi_j$ be any element of $\mathcal{S}$ such that the scalar product

$$\langle s, As \rangle = \sum_{j=1}^{n} \langle \psi_j, s \rangle^2 / \, \langle \psi_j, \psi_j \rangle \tag{4.3}$$

is zero. It follows that $\langle \psi_j, s \rangle$ is zero for every j. Therefore $\|s\|^2$ has the value

$$\langle s, s \rangle = \langle \textstyle\sum_{j=1}^{n} \theta_j \, \psi_j, s \rangle = \displaystyle\sum_{j=1}^{n} \theta_j \, \langle \psi_j, s \rangle = 0, \tag{4.4}$$

which provides $s=0$ as required. Further, we find at the end of Section 3 that the ellipticity condition implies that A is nonsingular.

We achieve good efficiency in the recommended algorithm by trying to make A close to the identity operator. Therefore it is relevant that equation (4.1) provides $A=I$, where I is the identity, if the basis functions ψ_j, $j=1,2,\ldots,n$, are mutually orthogonal with respect to the scalar product (2.3). Indeed, in this case the definition (4.1) gives $A\psi_\ell=\psi_\ell$, $\ell=1,2,\ldots,n$, which implies $As=s$, $s\in\mathcal{S}$, because ψ_j, $j=1,2,\ldots,n$, is a basis of $\mathcal{S}$. The advantage of $A=I$ in the Krylov subspace method is that, on the first iteration, equation (3.2) takes the form $s_2=\alpha_1 s^\star$. Thus, because the value $\alpha_1=1$ is allowed, termination occurs immediately. Therefore we require a convenient way of choosing orthogonal, or nearly orthogonal, basis functions.

A method is suggested by the form (2.5) of the scalar product $\langle s,t\rangle$, where s and t are the functions (1.4) and (2.1). We see that t is orthogonal to s if all of the products $\mu_i\, s(\underline{x}_i)$, $i=1,2,\ldots,n$, are zero. Therefore, if $m=-1$ and the matrix (1.3) is positive definite, then, for every integer j in $[1,n]$, we define ψ_j to be the function of the form

$$\psi_j(\underline{x}) = \sum_{\ell=j}^{n} \lambda_{j\ell}\,\phi(\|\underline{x}-\underline{x}_\ell\|_2), \qquad \underline{x}\in\mathcal{R}^d, \tag{4.5}$$

that satisfies the Lagrange equations

$$\psi_j(\underline{x}_\ell) = \delta_{\ell j}, \qquad \ell=j,j+1,\ldots,n. \tag{4.6}$$

Thus $\langle\psi_i,\psi_j\rangle$ is zero for $j>i$, because the least index of the sum (4.5) is $\ell=j$, and all of the values $\psi_i(\underline{x}_\ell)$, $\ell=j,j+1,\ldots,n$, are zero. A development of this construction and a reformulation of equation (4.1) will provide the recommended choice of A.

A development is needed, because the calculation of the coefficients $\lambda_{j\ell}$, $j\le\ell\le n$, of expression (4.5) is comparable to the solution of the original interpolation problem when there are $n-j+1$ data, and there are many values of j. Therefore we seek a modification of the given functions ψ_j, $j=1,2,\ldots,n$, that reduces the number of coefficients greatly by allowing mild departures from the orthogonality properties $\langle\psi_i,\psi_j\rangle=0$, $i\neq j$. Our research has shown that it is highly suitable to employ Lagrange function estimates of the kind that are proposed by Beatson and Powell (1994). Specifically, we pick a positive integer q, the value $q=30$ being typical, and then, for each j, we let ψ_j have the form

$$\psi_j(\underline{x}) = \sum_{\ell\in\mathcal{L}_j} \lambda_{j\ell}\,\phi(\|\underline{x}-\underline{x}_\ell\|_2), \qquad \underline{x}\in\mathcal{R}^d, \tag{4.7}$$

instead of expression (4.5), where the set $\mathcal{L}_j$ contains j and at most $q-1$ of the integers $j+1, j+2, \ldots, n$. Further, the coefficients of the new ψ_j are defined by the Lagrange conditions

$$\psi_j(\underline{x}_\ell) = \delta_{\ell j}, \qquad \ell \in \mathcal{L}_j. \tag{4.8}$$

We let $\mathcal{L}_j$ be the set $\{j, j+1, \ldots, n\}$ if j satisfies $n-j+1 \leq q$, but otherwise we prefer to include in $\mathcal{L}_j$ the q different integers from $[j, n]$ that minimize the Euclidean distances $\|\underline{x}_\ell - \underline{x}_j\|_2$, $\ell \in \mathcal{L}_j$. Then equations (4.1), (4.7) and (4.8) define the recommended operator A in the case $m=-1$.

In order to prove that this A has the required properties (3.6), we ask whether the new functions ψ_j, $j=1,2,\ldots,n$, are linearly independent. Let Γ be the $n \times n$ matrix with the elements $\Gamma_{j\ell}$, $1 \leq j, \ell \leq n$, where $\Gamma_{j\ell}$ is 0 or $\lambda_{j\ell}$ if $\ell \notin \mathcal{L}_j$ or $\ell \in \mathcal{L}_j$, respectively. It follows from equation (4.7) and the nonsingularity of Φ that linear independence occurs if and only if Γ is nonsingular. The diagonal elements of Γ are nonzero, because equations (2.5), (4.7) and (4.8) imply the strict inequality

$$0 < \langle \psi_j, \psi_j \rangle = \sigma \sum_{\ell \in \mathcal{L}_j} \lambda_{j\ell} \, \psi_j(\underline{x}_\ell) = \sigma \, \lambda_{jj}, \qquad j=1,2,\ldots,n. \tag{4.9}$$

Moreover, Γ is upper triangular due to the choice of the sets $\mathcal{L}_j$, $j=1,2,\ldots,n$. It follows that Γ is nonsingular. Therefore linear independence is achieved, so the remarks of the second paragraph of this section hold, which establish the conditions (3.6).

In the reformulation of the recommended A when $m=-1$, we preserve the sets $\mathcal{L}_j$, $j=1,2,\ldots,n$, and equations (4.7) and (4.8), but we write the definition (4.1) as the expression

$$As = \sum_{j=1}^{n-q} \langle \psi_j, s \rangle \, \langle \psi_j, \psi_j \rangle^{-1} \, \psi_j + Hs, \qquad s \in \mathcal{S}, \tag{4.10}$$

where H is the operator

$$Hs = \sum_{j=n-q+1}^{n} \langle \psi_j, s \rangle \, \langle \psi_j, \psi_j \rangle^{-1} \, \psi_j, \qquad s \in \mathcal{S}. \tag{4.11}$$

Thus H is a linear mapping from $\mathcal{S}$ to $\mathcal{T}$, where $\mathcal{T}$ is the q-dimensional linear subspace of $\mathcal{S}$ that contains functions of the form

$$t(\underline{x}) = \sum_{j=n-q+1}^{n} \lambda_j \, \phi(\|\underline{x} - \underline{x}_j\|_2), \qquad \underline{x} \in \mathcal{R}^d. \tag{4.12}$$

The reason for giving special attention to H is that it has the property

$$Ht = t, \qquad t \in \mathcal{T}. \tag{4.13}$$

We begin the proof of this assertion by noting that, for every j in $[n-q+1, n]$, the choice of $\mathcal{L}_j$ makes the functions (4.5) and (4.7) identical. Therefore the mutual orthogonality of the functions ψ_j, $j=n-q+1, n-q+2, \ldots, n$, is preserved. Hence the definition (4.11) implies $H\psi_\ell = \psi_\ell$, $\ell = n-q+1, n-q+2, \ldots, n$, so the required condition (4.13) follows from the remark that these functions ψ_ℓ are a basis of $\mathcal{T}$.

The property (4.13) provides a view of H that is highly useful. Indeed, we deduce from equation (4.11) and the scalar product (2.5) that the dependence of Hs on s is through the values $s(\underline{x}_i)$, $i=n-q+1, n-q+2, \ldots, n$. Therefore $Hs = Ht$ occurs, where t is the unique element of $\mathcal{T}$ that satisfies the equations

$$t(\underline{x}_i) = s(\underline{x}_i), \qquad i=n-q+1, n-q+2, \ldots, n. \tag{4.14}$$

Further, the identity (4.13) shows $Ht = t$. Thus the term Hs of expression (4.10) is the function of the form (4.12) whose coefficients are defined by the interpolation conditions (4.14).

This remark is important for $m \geq -1$, because the operator (4.11) is degenerate in the cases with $m \geq 0$. Indeed, the denominator $\langle \psi_n, \psi_n \rangle$ becomes zero, because, when $j=n$ and $\mathcal{L}_j = \{n\}$, equation (4.8) is satisfied by the constant polynomial $\psi_n(\underline{x}) = 1$, $\underline{x} \in \mathcal{R}^d$. Therefore, for the remainder of this section, we let H be the mapping from $\mathcal{S}$ to $\mathcal{T}$ such that Hs, $s \in \mathcal{S}$, is the element of $\mathcal{T}$ of the form

$$t(\underline{x}) = \sum_{j-n-q+1}^{n} \lambda_j \, \phi(\|\underline{x} - \underline{x}_j\|_2) + \sum_{j=1}^{\hat{m}} \gamma_j \, p_j(\underline{x}), \qquad \underline{x} \in \mathcal{R}^d, \tag{4.15}$$

that is defined by the interpolation conditions (4.14). Further, for $j=1, 2, \ldots, n-q$, we let ψ_j be the element of $\mathcal{S}$ that has the form

$$\psi_j(\underline{x}) = \sum_{\ell \in \mathcal{L}_j} \lambda_{j\ell} \, \phi(\|\underline{x} - \underline{x}_\ell\|_2) + \sum_{\ell=1}^{\hat{m}} \gamma_{j\ell} \, p_\ell(\underline{x}), \qquad \underline{x} \in \mathcal{R}^d, \tag{4.16}$$

and that satisfies the Lagrange conditions (4.8). Then expression (4.10) is the recommended choice of A for $m \geq -1$. The sets $\mathcal{L}_j$, $j = 1, 2, \ldots, n-q$, are as before, except for a new constraint on the points $\underline{x}_\ell$, $\ell \in \mathcal{L}_j$, to ensure that the denominator $\langle \psi_j, \psi_j \rangle$ of expression (4.10) is positive. In other words, the solution (4.16) of the equations (4.8) must not be an element of Π_m. Therefore the new constraint is that the polynomial unisolvency condition must hold on the set $\{\underline{x}_\ell : \ell \in \mathcal{L}_j \backslash \{j\}\}$ for every integer j in $[1, n-q]$. Thus the case $j = n-q$ ensures that H is well-defined. Our description of the recommended A is complete, but some further remarks on the choice of $\mathcal{L}_j$, $j=1,2,\ldots,n-q$, and on the ordering of the interpolation points, will be made in Section 5.

We have to show that the conditions (3.6) hold for this A. If s is any element of Π_m, then the scalar products $\langle \psi_j, s\rangle$, $j=1,2,\ldots,n-q$, are zero, so the definition (4.10) gives $As = Hs$. Moreover, $s \in \Pi_m$ implies $s \in \mathcal{T}$, so the unique solution of the equations (4.14) is $t=Hs=s$. Therefore the polynomial reproduction condition $As=s$, $s\in\Pi_m$, is achieved.

Our proof of the other conditions involves the scalar product $\langle Hs, H\hat{s}\rangle$, where s and $\hat{s}$ are general elements of $\mathcal{S}$. Because Hs is in $\mathcal{T}$, it has the form

$$(Hs)(\underline{x}) = \sum_{j=n-q+1}^{n} \nu_j\, \phi(\|\underline{x}-\underline{x}_j\|_2) + \text{element of } \Pi_m, \qquad \underline{x}\in\mathcal{R}^d. \tag{4.17}$$

Moreover, the definition of H provides $(H\hat{s})(\underline{x}_i)=\hat{s}(\underline{x}_i)$, $i=n-q+1, n-q+2, \ldots, n$. Therefore, by applying expression (2.5) twice, we deduce the identity

$$\langle Hs, H\hat{s}\rangle = \sigma \sum_{i=n-q+1}^{n} \nu_i \{(H\hat{s})(\underline{x}_i)\} = \sigma \sum_{i=n-q+1}^{n} \nu_i\, \hat{s}(\underline{x}_i) = \langle Hs, \hat{s}\rangle. \tag{4.18}$$

It follows from the symmetry of $\langle Hs, H\hat{s}\rangle$ that H has the self-adjoint property $\langle Hs, \hat{s}\rangle = \langle s, H\hat{s}\rangle$. Further, remembering our use of equation (4.2), we find that the operator (4.10) is self-adjoint, as required.

Next we let s be any element of $\mathcal{S}$ that is not in Π_m, and we consider the value of $\langle s, As\rangle$. Equations (4.10) and (4.18) provide the formula

$$\langle s, As\rangle = \sum_{j=1}^{n-q} \langle \psi_j, \psi_j\rangle^{-1} \langle \psi_j, s\rangle^2 + \|Hs\|^2, \tag{4.19}$$

and we wish to show that $\langle s, As\rangle$ is positive. We see that the ellipticity condition $\langle s, As\rangle > 0$ holds if $\|Hs\| \neq 0$, so we turn our attention to the alternative case when Hs is in Π_m. The function $\hat{s} = s - Hs$ is nonzero due to $s \notin \Pi_m$ and $Hs\in\Pi_m$. Hence there exists a greatest integer j in $[1, n]$ such that $\hat{s}(\underline{x}_j)\neq 0$. It satisfies $j \leq n-q$, because Hs interpolates s at the last q data points. Therefore $\langle \psi_j, \psi_j\rangle^{-1}\langle \psi_j, s\rangle^2$ is one of the terms in the sum of expression (4.19). Further, $Hs \in \Pi_m$ implies $\langle \psi_j, \hat{s}\rangle = \langle \psi_j, s\rangle$. It follows that $\langle s, As\rangle > 0$ is achieved if

$\langle\psi_j, \hat{s}\rangle$ is nonzero. Now $\langle\psi_j, \psi_j\rangle$ is positive, and the derivation of inequality (4.9) provides $\langle\psi_j, \psi_j\rangle = \sigma\,\lambda_{jj}$ when ψ_j is the function (4.16). Thus, because the choice of j gives $\hat{s}(\underline{x}_\ell) = 0$, $\ell = j+1, j+2, \ldots, n$, equations (2.5) and (4.16) with the form of $\mathcal{L}_j$ imply the required property

$$\langle\psi_j, \hat{s}\rangle = \sigma \sum_{\ell\in\mathcal{L}_j} \lambda_{j\ell}\, \hat{s}(\underline{x}_\ell) = \sigma\,\lambda_{jj}\, \hat{s}(\underline{x}_j) = \langle\psi_j, \psi_j\rangle\, \hat{s}(\underline{x}_j) \neq 0, \tag{4.20}$$

which completes the proof of the ellipticity condition (3.6). The remaining condition, namely nonsingularity, is deduced at the end of Section 3.

The preliminary work of the algorithm is the choice of $\mathcal{L}_j$, and the calculation of the coefficients of the function (4.16), for every integer j in $[1, n-q]$. Further, because the interpolation equations (4.14) are solved by the direct procedure of the paragraph that includes expressions (1.9) and (1.10), the $(q-\hat{m})\times(q-\hat{m})$ matrix W of this procedure is formed and factorized. These operations are completed before the iterative method of Section 3 is begun. The time that they require in our numerical experiments is only of order n, because we have restricted attention to $d=2$, and then Dirichlet tessellations provide a convenient way of generating the sets $\mathcal{L}_j$ (Goodsell, 1997). Here we are assuming that a small value of q, say $q=30$, remains adequate if n becomes large.

The iterative part of the algorithm calculates the coefficients of several functions in $\mathcal{S}$. Therefore we let $\underline{\lambda}(s) \in \mathcal{R}^n$ and $\underline{\gamma}(s) \in \mathcal{R}^{\hat{m}}$ be the vectors whose components are the coefficients of the general element (1.4). At the beginning of the k-th iteration of the Krylov subspace method, the vectors $\underline{\lambda}(s_k)$ and $\underline{\gamma}(s_k)$, which define s_k, and the residuals

$$r_i^{(k)} = f_i - s_k(\underline{x}_i), \qquad i=1,2,\ldots,n, \tag{4.21}$$

are available, their values being obvious when $k=1$, because s_1 is the zero function. Further, $\underline{\lambda}(d_{k-1})$ and $\underline{\gamma}(d_{k-1})$, and the values $d_{k-1}(\underline{x}_i)$, $i=1,2,\ldots,n$, are also available when $k\geq 2$. The first task of the k-th iteration is to generate the coefficients of the term $A(s^\star - s_k)$ of the search direction (3.3), which requires only $\mathcal{O}(nq)$ operations. Indeed, we put $s = s^\star - s_k$ in the definition (4.10) of A, we write each scalar product $\langle\psi_j, s^\star - s_k\rangle$ in the form $\sigma\sum_{\ell\in\mathcal{L}_j} \lambda_{j\ell}\, r_\ell^{(k)}$, and we make use of the property $|\mathcal{L}_j| = q$, $j=1,2,\ldots,n-q$. When $k=1$, the required components of $\underline{\lambda}(d_k)$ and $\underline{\gamma}(d_k)$ are just the coefficients of $A(s^\star - s_k)$, because the $\beta_k\, d_{k-1}$ term of the search direction (3.3) is suppressed on the first iteration. Alternatively, when $k\geq 2$, the orthogonality condition $\langle d_k, d_{k-1}\rangle = 0$ gives the value

$$\beta_k = -\frac{\langle d_{k-1}, z_k\rangle}{\langle d_{k-1}, d_{k-1}\rangle} = -\frac{\sum_{i=1}^n \lambda(z_k)_i\, d_{k-1}(\underline{x}_i)}{\sum_{i=1}^n \lambda(d_{k-1})_i\, d_{k-1}(\underline{x}_i)}, \tag{4.22}$$

where $z_k = A(s^\star - s_k)$, and where $\lambda(s)_i$ denotes the i-th component of $\underline{\lambda}(s)$, $s \in \mathcal{S}$. Thus β_k is found in $\mathcal{O}(n)$ operations, and the algorithm applies the formulae

$$\underline{\lambda}(d_k) = \underline{\lambda}(z_k) + \beta_k\, \underline{\lambda}(d_{k-1}) \quad \text{and} \quad \underline{\gamma}(d_k) = \underline{\gamma}(z_k) + \beta_k\, \underline{\gamma}(d_{k-1}). \tag{4.23}$$

When n is large, the most expensive part of an iteration is the calculation of all the values $s(\underline{x}_i)$, $i = 1, 2, \ldots, n$, for a particular s in $\mathcal{S}$. Two attractive alternatives are $s = d_k$ and $s = s_k + d_k$. We prefer $s = d_k$, because it avoids complications if rounding errors cause substantial inconsistencies between the residuals (4.21) and the coefficients of s_k. Therefore the numbers

$$d_k(\underline{x}_i) = \sum_{j=1}^{n} \lambda(d_k)_j\, \phi(\|\underline{x}_i - \underline{x}_j\|_2) + \sum_{j=1}^{\hat{m}} \gamma(d_k)_j\, p_j(\underline{x}_i), \qquad i = 1, 2, \ldots, n, \tag{4.24}$$

are computed. One of their uses is the evaluation of new residuals, because equations (3.2) and (4.21) provide the formula

$$r_i^{(k+1)} = f_i - s_{k+1}(\underline{x}_i) = r_i^{(k)} - \alpha_k\, d_k(\underline{x}_i), \qquad i = 1, 2, \ldots, n. \tag{4.25}$$

Thus the algorithm seeks $\alpha_k \in \mathcal{R}$ that satisfies the termination condition (3.4). If termination does not occur, however, then we recall from Section 3 that the property (3.1) implies the step-length

$$\alpha_k = \langle s^\star - s_k, d_k \rangle \,/\, \langle d_k, d_k \rangle = \sum_{i=1}^{n} \lambda(d_k)_i\, r_i^{(k)} \Big/ \sum_{i=1}^{n} \lambda(d_k)_i\, d_k(\underline{x}_i). \tag{4.26}$$

Expressions (4.25) and (4.26) give the residuals for the next iteration if $k < k^\star$. Finally, the procedure observes equation (3.2) by setting the coefficient vectors $\underline{\lambda}(s_{k+1})$ and $\underline{\gamma}(s_{k+1})$ to $\underline{\lambda}(s_k) + \alpha_k\, \underline{\lambda}(d_k)$ and $\underline{\gamma}(s_k) + \alpha_k\, \underline{\gamma}(d_k)$, respectively. Then k is increased by one if another iteration is required.

This version of the algorithm has the property that the coefficients of s_k contribute only to the coefficients of s_{k+1}. Thus any errors in these coefficients are inherited by the next iteration. Therefore, when high accuracy is required, it is advisable to generate the values $s_{k^\star}(\underline{x}_i)$, $i = 1, 2, \ldots, n$, using the coefficients that occur at termination. If the differences $f_i - s_{k^\star}(\underline{x}_i)$, $i = 1, 2, \ldots, n$, are unacceptably large, then the estimate of $s^\star$ can be improved by iterative refinement, using the Krylov subspace method again to compute a correction to $s_{k^\star}$.

5 Numerical results and further developments

Four test problems will be considered in this section, the value of d being two in each case. The positions of the interpolation points are shown schematically in Figure 1. Specifically, in Problem 1 the points $\underline{x}_i$, $i=1,2,\ldots,n$, are equally spaced on a circle, in Problem 2 the points are the vertices of a square grid, in Problem 3 they are generated randomly on the unit disc, and in Problem 4 they are equally spaced on two concentric circular arcs, where the ratio of the distance between the arcs to the smaller radius is only 10^{-5} in practice, but a greater distance is shown in the figure. In all of the calculations, the right hand sides f_i, $i=1,2,\ldots,n$, are random numbers from the uniform distribution on $[-1,1]$.

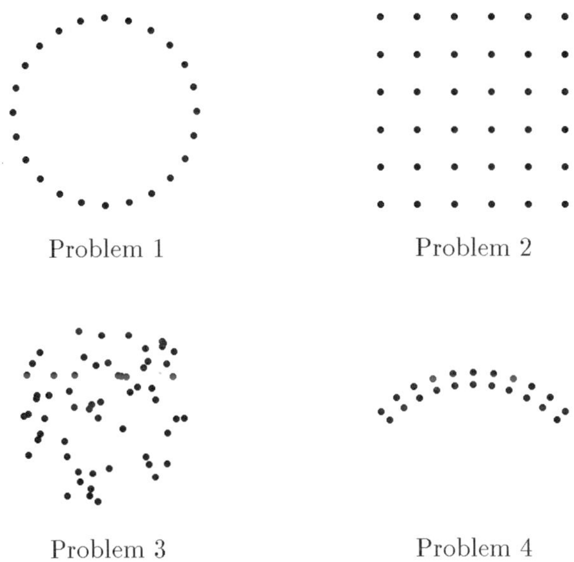

Figure 1: Positions of the data points

These test problems were highly useful in the development of the given algorithm. Indeed, the perfect symmetry of Problem 1 makes it easy for expression (4.16) to be a good approximation to a Lagrange function when $|\mathcal{L}_j|$ is much less than n, which provided much encouragement during the early research of Beatson and Powell (1994). Problem 2 was frustrating, however, when we were seeking estimates of Lagrange functions that have small errors at every data point, because such estimates may not exist for large n and a moderate value of $|\mathcal{L}_j|$, in the case when the true Lagrange function is one at a corner of the grid. The successful solution to this difficulty was the discovery of an algorithm that does not require small errors at the data points, provided that the scalar products $\langle \psi_i, \psi_j \rangle$, $i \neq j$, are sufficiently close to zero, but we did not know about the relevance of the scalar product of Section 2 when the new algorithm was found. That algorithm was developed for the square grid and could be applied for general positions of the data points. Then numerical tests with the random data points of Problem 3 gave the brilliant result that no further modifications were needed to achieve good efficiency. The thin plate spline radial function was used throughout that work, and we looked for an example of divergence or some other unacceptable behaviour of the iterative procedure. Thus the data of Problem 4 became of interest, because they caused the rate of convergence to be very slow. They also caused the failures, mentioned in Section 1, that were corrected by the Krylov subspace technique.

Tables 2 and 3 provide some numerical results for the linear radial function $\phi(r) = r$ with $m=0$ and the thin plate spline radial function $\phi(r) = r^2 \log r$ with $m=1$. The main entries of the tables are the numbers of iterations that occur when the tolerance parameter of the termination condition (3.4) is given the value $\varepsilon = 10^{-8}$. The results show overwhelmingly that the proof of termination in at most $n - \hat{m}$ iterations is irrelevant to practical computation. Further, they support the claim in Section 1 that it is usual to achieve eight places of accuracy in fewer than ten iterations, because $q=30$ should be preferred to $q=10$. On the other hand, our experimental work so far has addressed only interpolation problems in two variables.

If the values $\alpha_k = 1$ and $\beta_k = 0$ are set on every iteration, then expressions (3.2) and (3.3) imply the formula

$$s_{k+1} = s_k + A\,(s^\star - s_k), \qquad k = 1, 2, 3, \ldots . \tag{5.1}$$

The test problems of Figure 1 were used to construct algorithms that satisfy this equation, and then the Krylov subspace technique was added less than a year ago. Because equation (5.1) gives the relation

$$s_{k+1} - s^\star = (I - A)\,(s_k - s^\star), \qquad k = 1, 2, 3, \ldots , \tag{5.2}$$

those algorithms have linear rates of convergence that depend on the spectral radius of $I - A$. The addition of the Krylov subspace technique was found to be highly successful when the positions of the interpolation points $\underline{x}_i$, $i = 1, 2, \ldots, n$, cause the spectral radius of $I - A$ to exceed 0.5, but in other cases there is usually little change to the number of iterations, whatever the value of the tolerance parameter ε. For example, for the first three test problems of

n	q	Problem 1	Problem 2	Problem 3	Problem 4
	10	5	15	13	5
400	30	4	8	6	5
	50	4	7	5	5
	10	5	15	14	5
900	30	4	7	7	5
	50	4	7	6	5

Table 2: Numbers of iterations when $\phi(r) = r$

n	q	Problem 1	Problem 2	Problem 3	Problem 4
	10	7	27	33	70
400	30	4	10	8	42
	50	4	8	7	29
	10	7	28	42	87
900	30	4	10	10	61
	50	4	8	8	51

Table 3: Numbers of iterations when $\phi(r) = r^2 \log r$

Table 3 with $q \geq 30$, the technique reduced the number of iterations by at most one third. Therefore, when the method of Section 3 requires few iterations in practice, we expect the simple procedure (5.1) to be efficient too. Further, if the right hand sides f_i, $i = 1, 2, \ldots, n$, are general, then the spectral radius of $I - A$ may provide a good estimate of the number of iterations of the Krylov subspace algorithm, especially when the operator A is self-adjoint.

These remarks suggest correctly that there is little advantage in including the Krylov subspace technique when solving most radial basis function interpolation problems. On the other hand, the technique provides not only a guarantee of convergence in theory but also large gains in efficiency in difficult cases. Further, one may not be able to tell in advance if a case is difficult, and the extra work per iteration of the technique is trivial. In order to assess the extra work, we take the view that, instead of setting $\beta_k = 0$ and $\alpha_k = 1$ in equations (3.3) and (3.2), the technique prefers the values (4.22) and (4.26), which have to be calculated. Now both procedures require the coefficients of the function $z_k = A(s^\star - s_k)$ for the search direction (3.3), which takes $\mathcal{O}(nq)$ operations, but we deduce from equations (4.22), (3.3), (4.26) and (3.2) that the extra work of the Krylov subspace technique is only $\mathcal{O}(n)$ operations. Further, after generating the coefficients $\lambda(d_k)_j$, $j=1,2,\ldots,n$, and $\gamma(d_k)_j$, $j=1,2,\ldots,\hat{m}$, of d_k, both procedures apply formula (4.24), which is an $\mathcal{O}(n^2)$ calculation, except that in some cases the fast multipole method replaces formula (4.24) by an $\mathcal{O}(n\log n)$ procedure. Thus the additional computation of the Krylov subspace technique is negligible. Therefore there are strong reasons for adding it to algorithms that apply the simple formula (5.1), if the required interpolating function is in a linear space $\mathcal{S}$ that has the semi-norm of Section 2, and if A satisfies the conditions (3.6).

The relatively large number of iterations in the last column of Table 3 is a topic for future research. That inefficiency was found before the results of Table 2 were computed, so the excellent performance of the linear radial function $\phi(r)=r$ was unexpected. A major difference between the calculations of the two tables occurs in the sets $\mathcal{L}_j$, $j=1,2,\ldots,n-q$. When $\phi(r)=r$, each $\mathcal{L}_j$ contains the q different integers from $[j,n]$ that minimize the distances $\|\underline{x}_\ell-\underline{x}_j\|_2$, $\ell\in\mathcal{L}_j$, as suggested in Section 4. When $\phi(r)=r^2\log r$, however, that choice of $\mathcal{L}_j$ may be unacceptable, because the polynomial unisolvency constraint states that, for $d=2$ and $m=1$, the points x_ℓ, $\ell\in\mathcal{L}_j\backslash\{j\}$, must not be collinear. We responded to this requirement by picking three data points initially that are far from collinear, by reordering the data so that these points are $\underline{x}_{n-2}$, $\underline{x}_{n-1}$ and $\underline{x}_n$, and by forcing the integers $n-2$, $n-1$ and n to be in all of the sets $\mathcal{L}_j$, $j=1,2,\ldots,n-q$. The remaining elements of each $\mathcal{L}_j$ are selected as before by minimizing $\|\underline{x}_\ell-\underline{x}_j\|_2$, $\ell\in\mathcal{L}_j$. If future investigations show that a different choice of sets can reduce greatly the iteration counts in Table 3 for Problem 4, then we will seek a form of the new choice that can be applied for general positions of the interpolation points.

Another question for future research is the dependence of the number of iterations on changes to the ordering of the data $\underline{x}_i$, $i=1,2,\ldots,n$. A decision on the ordering was taken for the first algorithm with approximations to Lagrange functions that was tested on Problem 3, and that ordering has been retained. It is based on the assumption that, if points are removed in sequence from the beginning of the set $\underline{x}_i$, $i=1,2,\ldots,n$, then the surviving points should provide good coverage of the original set in $\mathcal{R}^d$. Therefore, in a preliminary calculation, the integer j runs through the values $1,2,\ldots,n-q$, and, for each j, $\underline{x}_j$ is exchanged with one of the points $\underline{x}_{j+1},\underline{x}_{j+2},\ldots,\underline{x}_{\hat{n}}$, if necessary, so that the nearest neighbour distance $\min\{\|\underline{x}_\ell-\underline{x}_j\|_2 : \ell=j+1,j+2,\ldots,n\}$ becomes as small as possible, where $\hat{n}=n$ is usual, but the technique of the last paragraph for thin plate splines requires $\hat{n}=n-3$. This device is included in the calculations of Tables 2 and 3. Recently, however, we tried some experiments with random orderings of the data points, using the linear radial function and the test problems of Table 2. We found no appreciable changes to the iteration counts. Therefore we intend to explore the idea of preferring an ordering that is helpful to the task of generating the sets $\mathcal{L}_j$, $j=1,2,\ldots,n-q$, automatically. This task can be done conveniently by Dirichlet tessellations in two dimensions, but some assistance from the ordering of the data may be important to efficiency for $d\geq 3$.

Most of the coding of the computer program that generated the results of Tables 2 and 3 is for the construction of Dirichlet tessellations, because it is easy to implement the iterative part of the algorithm of Section 4, if the numbers (4.24) are calculated directly instead of by the fast multipole method. The computer software will be available from the authors after attention has been given to the proposed future work that is sketched in the previous two paragraphs.

The software will be able to solve the interpolation equations using linear and multiquadric radial functions without a polynomial term, although these $m=-1$ cases are absent from Table 1. Indeed, one calculates $s^\star$ in the way that has been described with $m=0$, and also one calculates $\hat{s}^\star$, say, in this way by interpolation to the right hand sides $\phi(\|\underline{x}_i-\underline{x}_\ell\|_2)$, $i=1,2,\ldots,n$, where ℓ is now a prescribed integer from $[1,n]$. Then the function

$$\breve{s}^\star(\underline{x}) = s^\star(\underline{x}) + \hat{c}\,[\hat{s}^\star(\underline{x})-\phi(\|\underline{x}-\underline{x}_\ell\|_2)], \qquad \underline{x}\in\mathcal{R}^d, \tag{5.3}$$

interpolates the data for every choice of $\hat{c}\in\mathcal{R}$, because the expression in square brackets vanishes at all the interpolation points. Further, the constant term of this expression must be nonzero, because the matrix (1.3) is nonsingular. Thus a unique value of $\hat{c}$ annihilates the constant term of $\breve{s}^\star$, which forces $\breve{s}^\star$ to be the required interpolant.

The low iteration counts of Tables 2 and 3 suggest that the given Krylov subspace technique will become highly useful for radial basis function interpolation, when n is so large that the $\mathcal{O}(n^3)$ cost of direct methods is prohibitive. Thus the technique may increase greatly the range of applications that can take advantage of the smoothness and accuracy properties of radial functions. Franke (1982) states that these properties compare very favourably with those of interpolating functions from other linear spaces. Further, the radial basis function method is particularly useful for interpolation on nonlinear mainfolds in $\mathcal{R}^d$, because the nonsingularity of the system (1.6) imposes hardly any (or no) restrictions on the positions of the data points $\underline{x}_i$, $i=1,2,\ldots,n$.

References

R.K. Beatson and G.N. Newsam (1992), "Fast evaluation of radial basis functions: I", *Comput. Math. Applic.*, Vol. 24, pp. 7–19.

R.K. Beatson and M.J.D. Powell (1994), "An iterative method for thin plate spline interpolation that employs approximations to Lagrange functions", in *Numerical Analysis 1993*, eds. D.F. Griffiths and G.A. Watson, Longman Scientific & Technical (Burnt Mill), pp. 17–39.

J. Duchon (1977), "Splines minimizing rotation - invariant seminorms in Sobolev spaces", in *Constructive Theory of Functions of Several Variables, Lecture Notes in Mathematics 571*, eds. W. Schempp and K. Zeller, Springer-Verlag (Berlin), pp. 85–100.

N. Dyn, D. Levin and S. Rippa (1986), "Numerical procedures for surface fitting of scattered data by radial functions", *SIAM J. Sci. Statist. Comput.*, Vol. 7, pp. 639–659.

R. Franke (1982), "Scattered data interpolation: tests of some methods", *Math. Comp.*, Vol. 38, pp. 181–200.

G. Goodsell (1997), "On finding p-th nearest neighbours of scattered points in two dimensions for small p", Report No. DAMTP 1997/NA01, University of Cambridge, England (to be published in *Computer Aided Geometric Design*).

R.L. Hardy (1990), "Theory and applications of the multiquadric-biharmonic method", *Comput. Math. Applic.*, Vol. 19, pp. 163–208.

C.A. Micchelli (1986), "Interpolation of scattered data: distance matrices and conditionally positive definite functions", *Constr. Approx.*, Vol. 2, pp. 11–22.

M.J.D. Powell (1992), "The theory of radial basis function approximation in 1990", in *Advances in Numerical Analysis Volume II: Wavelets, Subdivision Algorithms, and Radial Basis Functions*, ed. W. Light, Clarendon Press (Oxford), pp. 105–210.

M.J.D. Powell (1997), "A new iterative method for thin plate spline interpolation in two dimensions", *Annals Numer. Math.*, Vol. 4, pp. 519–527.

Y. Saad and M.H. Schultz (1986), "GMRES: A generalized minimum residual algorithm for solving nonsymmetric linear systems", *SIAM J. Sci. Statist. Comput.*, Vol. 7, pp. 856–869.

R. Schaback (1993), "Comparison of radial basis function interpolants", in *Multivariate Approximations: From CAGD to Wavelets*, eds. K. Jetter and F. Utreras, World Scientific (Singapore), pp. 293–305.

R. Sibson and G. Stone (1991), "Computation of thin plate splines", *SIAM J. Sci. Statist. Comput.*, Vol. 12, pp. 1304–1313.

Acknowledgement

The work of George Goodsell on radial basis function interpolation was of major importance to the development of the given algorithm. It led to the operator A that is recommended in Section 4. Further, he wrote most of the Fortran software that has been used for numerical calculations.

Anita C. Faul and Michael J.D. Powell
Department of Applied Mathematics and Theoretical Physics
University of Cambridge
Silver Street
Cambridge CB3 9EW
England
`acf22@damtp.cam.ac.uk`
`M.J.D.Powell@damtp.cam.ac.uk`

M. C. Ferris, T. S. Munson and D. Ralph

A homotopy method for mixed complementarity problems based on the PATH solver

Abstract Mixed complementarity problems can be recast as zero finding problems for the normal map, a function that is smooth on the interior of each of the cells of a piecewise linear manifold of $\mathbb{R}^n$, called the normal manifold. We develop a predictor-corrector, or path following, homotopy method based upon using piecewise linear approximations to the piecewise smooth normal map. A description of an implementation using technology found in the PATH solver is given along with computational experience on the MCPLIB test suite.

1 Introduction

The complementarity problem arises in many different applications [14]. The original sources of these problems were the optimality conditions of linear, quadratic, and nonlinear programs. Since that time, many other applications, including game theory, economic equilibria, and structure design/failure have been postulated and solved as complementarity problems. In this chapter, we will focus on the mixed (nonlinear) complementarity problem of finding $z \in [l, u]$ such that for all $i = 1, \ldots, n$,

$$\begin{array}{rcl} f_i(z) & \geq & 0 \quad \text{if } z_i = l_i \\ f_i(z) & \leq & 0 \quad \text{if } z_i = u_i \\ f_i(z) & = & 0 \quad \text{if } l_i < z_i < u_i, \end{array}$$

where $f : \mathbb{R}^n \to \mathbb{R}^n$ is a continuously differentiable function, l_i and u_i are fixed, possibly infinite numbers satisfying

$$-\infty \leq l_i < u_i \leq +\infty,$$

and $[l, u]$ denotes the set of $z \in \mathbb{R}^n$ such that $l_i \leq z_i \leq u_i$ for all i. We call this problem MCP(f, l, u) or MCP for short.

Many monotone equilibrium problems and convex optimization problems whose constraints C are *not* simple boxes $[l, u]$ can also be reformulated as

MCPs. For example, consider the nonlinear program

$$\min \phi(z) \text{ subject to } Az = b,\ g_j(z) \leq 0 \text{ for } j = 1, \ldots, q$$

where f and each g_j are real, differentiable, convex functions on $\mathbb{R}^n$, and $A \in \mathbb{R}^{p \times n}$, $b \in \mathbb{R}^p$. If the Slater constraint qualification holds, namely there exists $\hat{z}$ such that $A\hat{z} = b$ and $g_j(\hat{z}) < 0$ for each j, then finding a minimizer z, global or local, of this nonlinear program is equivalent to solving its stationarity or Karush-Kuhn-Tucker conditions; see [21]. The KKT conditions form an MCP in z and auxiliary variables $\mu \in \mathbb{R}^p$ and $\lambda \in \mathbb{R}^q$ called (KKT or Lagrange) multipliers. Define an MCP function $f : \mathbb{R}^{n+p+q} \to \mathbb{R}^{n+p+q}$ by

$$f(x, \mu, \lambda) \quad = \quad (\nabla\phi(x) - A^T\eta - \nabla g(x)^T\lambda,\ Ax - b,\ g(x))$$

where $\nabla\phi(x)$ is the gradient vector of ϕ at x, $g(x) = (g_1(x), \ldots, g_q(x))$ and $\nabla g(x)$ is the $q \times n$ Jacobian matrix of g at x. Also let the box $[l, u]$ be $\mathbb{R}^n \times \mathbb{R}^p \times \mathbb{R}^q_-$. Then the associated MCP is precisely the KKT conditions of the nonlinear program, hence is equivalent to the nonlinear program.

Our method will be based on a reformulation of the mixed complementarity problem. To explain this reformulation, let us first consider a nonempty polyhedral subset of $\mathbb{R}^n$, C. Associated with each of the faces of C is a (full-dimensional) polyhedral set, and the collection of these polyhedra comprise a piecewise linear manifold of $\mathbb{R}^n$ called the *normal manifold*, denoted $\mathcal{N}_C$. These polyhedra are called the *cells* of $\mathcal{N}_C$; a full description along with important properties are given in [24] (see also [22] for further investigation). For example, when $C = \mathbb{R}^n_+$, that is $C = [l, u]$ with each $l_i = 0$ and $u_i = \infty$, the cells of $\mathcal{N}_C$ are the orthants of $\mathbb{R}^n$.

We denote by $\pi_C(\cdot)$ the Euclidean projection mapping onto the set C; thus for the box $C = [l, u]$, $\pi_C(x)$ is the vector whose ith component is l_i if $x_i \leq l_i$, x_i if $l_i \leq x_i \leq u_i$, and u_i otherwise. The normal map [24] induced by (f, C) is the function $f_C : \mathbb{R}^n \to \mathbb{R}^n$ given by

$$f_C(x) \quad = \quad f(\pi_C(x)) + x - \pi_C(x).$$

It is well known that the projection map $\pi_C(x)$ and hence the normal map $f_C(x) = f(\pi_C(x)) + x - \pi_C(x)$ are smooth in the interior of each of the cells of $\mathcal{N}_C$. The key point is that for $C = [l, u]$, if x satisfies $f_C(x) = 0$, then $z = \pi_C(x)$ solves MCP. Furthermore, if z solves MCP then $x = z - f(z)$ is a zero of f_C.

We propose and implement a homotopy (or continuation) method for finding zeroes of the normal map $f_C(x)$ when $C = [\ell, u]$ based on classical predictor-corrector ideas. Previously proposed algorithms [27, 29, 30] use linear approximations to single (smooth) pieces of a piecewise smooth mapping, see also

[16, 17]. The method proposed in this chapter is based on piecewise linear approximations of a piecewise smooth mapping. Both our method and that of [30] are based upon the theoretical foundations developed in [1, 2, 3].

An alternative homotopy approach investigated recently [5] is to form a homotopy that is smooth except in the limit as the homotopy parameter reaches its final value. This makes application of smooth homotopy codes attractive though some care near the end of the homotopy path, when the problem is tending toward nonsmoothness, may be needed.

Section 2 gives some background material on homotopy methods. Section 3 outlines the theoretical underpinnings of our method. Section 4 describes the implementation of a large scale code based on this theory and Section 5 gives some computational results on problems from MCPLIB [7].

2 Background

We give the notation and background for homotopy methods, in particular for normal maps induced by smooth functions and polyhedral convex sets.

2.1 Homotopy mappings and paths

Continuation or homotopy methods are a technique to trace the zeroes of a homotopy mapping, starting from an easily constructed zero and moving towards the solution to the problem of interest. To describe such methods, we first need to understand the homotopy mapping.

For a fixed vector $a \in \mathbb{R}^n$, we construct a homotopy mapping $H_C^a : \mathbb{R}^{n+1} \to \mathbb{R}$ from the function $x \mapsto x - a$ to the function f_C by interpolation: for $(x, t) \in \mathbb{R}^n \times [0, 1]$,

$$H_C^a(x,t) \;=\; (1-t)(x-a) + t f_C(x).$$

Note that $x = a$ is the unique solution of $H_C^a(x, 0) = 0$, and $H_C^a(x, 1) = 0$ if and only if x is a zero of the normal map. Hence if we can calculate the endpoint $x^a(1)$ of the homotopy path, then we have solved the original problem. The idea of a continuation method is to analytically or numerically determine a path of solutions $x^a(t)$ of the equation $H_C^a(\cdot, t) = 0$ from $t = 0$, where $x^a(0) = a$ is the unique solution, to $t = 1$, where as mentioned, $x^a(1)$ is a zero of the normal map.

A standard list of properties of the homotopy path $x^a(t)$ follows. We only state these for the case of H_C^a defined above, though the class of homotopies with these properties is considerably larger [3] as we explain after Theorem 1.

Parts 1 and 2 below summarize results that are due, in essence, to [3], while Part 3 is due to [27]; see the proof for details. Parts 4 and 5 seem to be new. Part 5 refers to subanalytic functions [15, 20] which we define, for completeness, following the statement of the theorem. However we refer to [11, Section 5.2] for a succinct introduction to the properties of subanalytic sets and functions, and additional references. Also, the notation $t \to 1^-$ means $t \to 1$, $t < 1$.

Theorem 1 *Let $f : \mathbb{R}^n \to \mathbb{R}^n$ be C^1 and C be a nonempty polyhedral convex set in $\mathbb{R}^n$. For almost all $a \in \mathbb{R}^n$:*

1. *There exist $T > 0$ and a piecewise smooth path $x^a : [0, T) \to \mathbb{R}^n$ such that $x^a(0) = a$, and $\{(x^a(t), t) \,;\, t \in [0, T)\}$ is a connected component of the set $\{(x, t) : t \in [0, T),\ H_C^a(x, t) = 0\}$.*
2. *The function $x^a(\cdot)$ is nondegenerate in the sense that for each $t \in [0, T)$, first, $x^a(t)$ lies either in the interior of a unique cell of the normal manifold $\mathcal{N}_C$ or in the relative interior of a facet (an $(n-1)$-dimensional face of a cell) that is the intersection of two cells of $\mathcal{N}_C$; and, second, if $x^a(t)$ lies in a cell σ and g_σ denotes the smooth mapping that represents H_C^a on σ, then the Jacobian (derivative) matrix $\nabla g_\sigma(x^a(t), t)$ has full rank.*
3. *If f is C^2 and there exists a point $u \in C$ and a scalar $\rho > \|u\|$ such that for each $c \in C$ with $\|c\| = \rho$ we have $\langle f(c), c - u \rangle \geq 0$, then $\{x^a(t) : t \in [0, 1)\}$ is bounded and its closure contains a point $\hat{x}$ with $f_C(\hat{x}) = 0$.*
4. *If f is C^2, $T \geq 1$, and there exists a limit point $\hat{x}$ of $p(t)$ as $t \to 1^-$ such that the directional derivative $f_C'(\hat{x}; \cdot)$ is invertible, then the arc-length of the path $x^a : [0, 1) \to \mathbb{R}^n$ is finite and $x^a(t) \to \hat{x}$ as $t \to 1^-$.*
5. *If $T \geq 1$ and $\hat{x}$ is a limit point of $x^a(t)$ as $t \to 1^-$, then $f_C(\hat{x}) = 0$. If, in addition, either $T > 1$, or f is subanalytic then $x^a(t) \to \hat{x}$ as $t \to 1^-$.*

Proof Parts 1 and 2 are essentially due to [3, Theorem 1]. The fact that the graph of x^a is a connected component of $\{(x, t) : t \in [0, T),\ H_C^a(x, t) = 0\}$ is not stated but can be easily deduced from the discussion on transversality prior to [3, Theorem 1]. See also [30, Theorem 1] and subsequent discussion of these transversality properties. Part 3 is quoted from [28, Proposition 5.1].

For Part 4, suppose $\hat{x}$ is a limit point of $x^a(t)$ as $t \to 1^-$ such that $f_C'(\hat{x}; \cdot)$ is invertible. The implicit function theorem of [23] says that the path $x^a(t)$ is locally uniquely defined for t near 1^-: for some $\epsilon > 0$ and all t near 1 with $t < 1$, the homotopy equation $H_C^a(\cdot, t) = 0$ has a unique solution x within distance ϵ of $\hat{x}$. It follows that the path $\{(x^a(t), t) : t \in [0, 1)\}$ is bounded and the only limit point of $x^a(t)$ as $t \to 1^-$ is $\hat{x}$ itself; i.e., $x^a(t)$ converges to $\hat{x}$ as $t \to 1^-$. Now invertibility of $f'_C(\hat{x};\cdot)$ is equivalent to other properties such as strong regularity of the associated variational inequality at $\pi_C(\hat{x})$, [30, Definition 2.2], and coherent orientation of the normal map $\nabla f(\hat{x})_{K(\hat{x})}$, [24], where $K(\hat{x})$ is the critical cone to C at $\hat{x}$, and this normal map actually coincides with $f'_C(\hat{x};\cdot)$; see [27, Section 5] and [30, Section 2] for discussion and related results. Therefore the analysis of [27, Proposition 5.6] can be adapted to show finite arc-length, using boundedness of the path and the fact that $\hat{x}$ is the only limit point of $x^a(t)$ as $t \to 1^-$.

For Part 5, it is clear by continuity of f_C that $f_C(\hat{x}) = 0$ if $(\hat{x}, 1)$ lies in the closure of $\{(x^a(t), t) : t \in [0,1)\}$. If $T > 1$, continuity of $x^a(\cdot)$ immediately yields convergence of $x^a(t)$ to $\hat{x}$ as $t \to 1$. Suppose instead that f is subanalytic. We will show convergence of $x^a(t)$ to $\hat{x}$ as $t \to 1^-$ by refining an argument in the proof of [11, Theorem 5.10]. All polyhedral convex sets are subanalytic so that the projection π_C mapping is also subanalytic [11, Lemma 5.9]. It follows that H^a_C is also subanalytic since the composition of subanalytic mappings is also subanalytic, hence that $(H^a_C)^{-1}(0)$ is a subanalytic set. Let $P^a = \{(x^a(t), t) : t \in [0,1)\}$; from Part 1, taking $T = 1$, there is a neighborhood U of P^a such that $U \cap (H^a_C)^{-1}(0) = P^a$. Without loss of generality assume U is subanalytic; so then P^a, the intersection of subanalytic sets, is also subanalytic. This shows that $x^a : [0,T) \to \mathbb{R}^n$ is subanalytic. The lemma in the Appendix concludes the proof. □

With regard to Part 5: a set S in $\mathbb{R}^n$ is *subanalytic* if there exists a semianalytic set T in a higher dimensional space $\mathbb{R}^N$ such that $S = \{x \in \mathbb{R}^n : (x,y) \in T\}$. A set T in $\mathbb{R}^N$ is *semianalytic* if for each $\bar{z} \in T$ there is a neighborhood V of $\bar{z}$ such that $T \cap V$ can be written as the finite union of sets of the form

$$V \cap \{z \in \mathbb{R}^N \; : \; f_i(z) = 0, i = 1, \ldots, I; \; g_j(z) < 0, j = 1, \ldots, J\}$$

where I and J are nonnegative integers, and each f_i and g_j is a real analytic mapping on $\mathbb{R}^N$. A function is subanalytic if its graph is a subanalytic set. Thus the class of subanalytic functions is rather broad.

Much of Theorem 1 holds if we replace the homotopy mapping H^a_C by one of the form

$$(1-t)\phi(a,x) + t f_C(x)$$

where $\phi : \mathbb{R}^{m+n} \to \mathbb{R}^n$ is a *sufficient* mapping [3], that is ϕ is smooth such that its partial Jacobian matrix with respect to $a \in \mathbb{R}^m$, $\nabla_a \phi(a,x)$, has full rank for all $(a,x) \in \mathbb{R}^m \times \mathbb{R}^n$. (For example, the function $\phi(a,x) = x - a$ used in the definition of H_C^a is sufficient.) To be precise, in order to generalize Parts 1, 2, and 5 of Theorem 1 using a homotopy with a sufficient mapping as described, we only need to replace the qualifier $a \in \mathbb{R}^n$ by $a \in \mathbb{R}^m$ and, in the statement of Part 5, the condition "f is subanalytic" by "f and ϕ are subanalytic".

For future reference, we define the homotopy path as the set

$$P^a = \{(x^a(t), t) : t \in [0, T)\}$$

where T is the maximum value in $(0, 1]$ such that Part 1 of Theorem 1 holds.

2.2 Predictor-corrector methods

Predictor-corrector methods are a class of algorithms that attempt to numerically traverse the homotopy path by a sequence of predictor-corrector iterations. Suppose for the moment that H_C^a is smooth, e.g., $C = \mathbb{R}^n$ and $f_C = f$. At step k we are given an iterate (x^k, t^k) that is approximately on the path, that is $x^k \approx x^a(t^k)$. The predictor step identifies a nonzero vector $d = (d_x, d_t) \in \mathbb{R}^n \times \mathbb{R}^1$ tangent to the path at the current point and estimates a new point further along the path $(x', t') = (x^k, t^k) + hd$ for some small $h > 0$. The direction d is found as a vector in the kernel of the Jacobian matrix $\nabla H_C^a(x^k, t^k)$ that maintains the correct orientation, that is we want to move tangent to the curve in the direction that increases arc-length. The corrector step then tries to identify a point $(x^a(s), s)$ on the path near to (x', t'); this is used to define the next iterate, $(x^{k+1}, t^{k+1}) = (x^a(s), s)$. The corrector step is usually carried out by a version of Newton's method that uses the Moore-Penrose inverse of the $n \times (n+1)$ matrix $\nabla H_C^a(x, t)$ starting with $(x, t) = (x', t')$ and proceeding until $H_C^a(x, t)$ is approximately zero; see [4].

When H_C^a is piecewise smooth, as it is when f is C^1 and C is polyhedral convex, it is possible to mimic the predictor and corrector steps taken in the smooth case by either staying within one cell, or identifying an adjacent cell and moving into it. This means that at each predictor step, a cell σ_k containing the current iterate (x^k, t^k) is identified, and the predictor direction d is chosen using the Jacobian matrix of the smooth mapping corresponding to H_C^a on σ^k. The Moore-Penrose-Newton iteration is undertaken in a similar way. This is the line of development in [30]. We note that these algorithms, and our proposed method to follow, include some nontrivial technical details that are necessary to make it theoretically viable. In particular, it is assumed that the chosen starting point a yields a path $x^a(t)$ that is nondegenerate as described in Theorem 1, Part 2. The implementation must take special care in the near-degenerate case for which the path passes through or near a point that is contained in three or more cells.

We prefer to use a piecewise linear (PL) predictor approximation, that is to generate a piecewise linear tangent path to the piecewise smooth homotopy path. An advantage of the PL predictor is that, in principle, it allows the near-degenerate situation to be handled in the same way as the nondegenerate case. Another advantage of the PL tangent path approach is that we can use the code developed over several versions of the PATH solver [8] for MCP. For example, we rely on this code to handle numerical degeneracy in the PL path.

Consistency would dictate that each step of the corrector also be taken with respect to a PL model of the actual path. However for simplicity of implementation we have chosen a method more like that of [30]: we pick a cell containing the current point and apply a Moore-Penrose-Newton iteration as a heuristic for decreasing the distance to the path. Details of both the predictor and corrector steps are given below.

3 Homotopy method

3.1 Predictor step

To describe the predictor step, we need to define a piecewise linear path that is tangent to the piecewise smooth homotopy path P^a at a given point. To do this, we need to describe a linearization of H_C^a.

For convenience in approximating H_C^a, we write it in another way. Let $H^a : \mathbb{R}^{n+1} \to \mathbb{R}^n$ be the C^1 homotopy from the mapping $z \mapsto z - a$ to $f(z)$,

$$H^a(z,t) = (1-t)(z-a) + tf(z),$$

and observe that

$$H_C^a(x,t) = H^a(\pi_C(x),t) + x - \pi_C(x).$$

We define the linearization of the smooth mapping H^a at any point $(z,t) \in \mathbb{R}^{n+1}$ using the first two terms of the Taylor series: for $(z',t') \in \mathbb{R}^{n+1}$,

$$\mathcal{L}H^a(z,t)(z',t') = H^a(z,t) + [(1-t)I + t\nabla f(z)](z'-z) + [f(z) - (z-a)](t'-t).$$

Now we define the linearization of the piecewise smooth mapping H_C^a at (x,t) as a function of $(x',t') \in \mathbb{R}^{n+1}$,

$$\mathcal{L}H_C^a(x,t)(x',t') = \mathcal{L}H^a(\pi_C(x),t)(\pi_C(x'),t') + x' - \pi_C(x').$$

This is a piecewise linear mapping in (x', t'). It is a "point based approximation" to H_C^a at (x, t) in the terminology of [25].

By the "piecewise linear path that is tangent to the piecewise smooth homotopy path P^a at (x^k, t^k)" we mean the set

$$P_k^a = \{(x,t) : t \text{ near } t^k,\ \mathcal{L}H_C^a(x^k, t^k)(x,t) = 0\}. \tag{3.1}$$

Note $(x^k, t^k) \in P_k^a$. Given a step size $h_k > 0$ our task is to partially construct P_k^a, starting from (x^k, t^k) and moving in the direction that is associated with increasing the arc-length of the actual homotopy path P^a. This "one-sided" path is denoted Q_k^a. We keep moving along Q_k^a until we determine a point (x', t') at distance h_k from (x^k, t^k). The new point (x', t') is our prediction for the next point on the path P^a. We need a corrector scheme to move from (x', t') to a nearby point in P^a; this will be discussed in Section 3.2.

For the subsequent discussion we assume for all $t \in [0, 1)$ that either $x^a(t)$ lies in the interior of a cell of $\mathcal{N}_C$ or in the relative interior of one of its facets, and that the Jacobian matrices at $(x^a(t), t)$ of the smooth functions representing H_C^a on these cells have full rank. These properties hold for almost all a by Theorem 1.

We note that the full rank of the Jacobians of H_C^a near (x^k, t^k), together with nondegeneracy of x^k, is enough to show that the set P_k^a does indeed define a piecewise linear path, i.e., a one-dimensional piecewise affine manifold, for t near t^k. This PL path is tangent to the piecewise smooth homotopy path P^a at (x^k, t^k) if $(x^k, t^k) \in P^a$. It will be necessary to generate $Q_k^a \subset P_k^a$ starting from (x^k, t^k) and moving in one of the two possible directions as determined by the orientation calculation presented next. In practice we allow for (x^k, t^k) to be near rather than in P^a — c.f. the tolerance ϵ_c used in the corrector step to follow — so that P_k^a is approximately tangent to P^a. Such details, including the invariance of orientation on the homotopy path, are spelled out in [1, 2, 3, 4].

Orientation of the path

An orientation parameter $\eta = \pm 1$ is used to decide which way to go on the path P_k^a when starting at (x^k, t^k). The orientation parameter is defined prior to the first predictor step of the homotopy method as

$$\eta = \operatorname{sgn} \det \begin{bmatrix} \nabla H_C^a(a, 0) \\ (d^0)^T \end{bmatrix},$$

where $d^0 = (-f(a), 1)$. Note H_C^a is differentiable at any $(x, 0)$. Initially $t^0 = 0$, we have $x^0 = a = x^a(0)$, and $\nabla H_C^a(a, 0) = [I\ f(a)]$, so that the chosen vector

d^0 is in the kernel of $\nabla H_C^a(x^0, t^0)$ as in the smooth case. Therefore

$$\begin{aligned} \eta &= \operatorname{sgn}\det \begin{bmatrix} I & f(a) \\ -f(a)^T & 1 \end{bmatrix} \\ &= \operatorname{sgn}\det \left(\begin{bmatrix} I & 0 \\ -f(a)^T & 1 + \|f(a)\|^2 \end{bmatrix} \begin{bmatrix} I & f(a) \\ 0 & 1 \end{bmatrix} \right) = 1, \end{aligned}$$

where we are using the fact that the determinant of a product of two matrices is the product of the determinants of the matrices. Writing $d^0 = (d_x, d_t) \in \mathbb{R}^n \times \mathbb{R}$, we have chosen $d_t = 1 > 0$ since initially we plan to move along the path P^a by increasing t, i.e., our first prediction will be a point $(a, 0) + h_0 d^0$, for some $h_0 > 0$, which we believe will be close to a point $(x^a(t), t)$ with $t \approx hd_t$.

Let $(x^k, t^k) \in P^a$, σ be a cell containing x^k, and $\mathcal{L}H_C^a(x^k, t^k)$ be represented on σ by an affine map whose Jacobian $J \in \mathbb{R}^{n \times (n+1)}$ at (x^k, t^k) has full rank. So the kernel of J is a one-dimensional space. Let ker J be spanned by a nonzero vector $d = (d_x, d_t) \in \mathbb{R}^n \times \mathbb{R}$, where we also ensure $x^k + sd_x \in \sigma$ for small $s > 0$ by using $-d$ instead of d if necessary (this is possible since x^k is interior either to the cell σ or to the union of two cells). Now calculate the sign η_k of the determinant of the $(n+1) \times (n+1)$ matrix consisting of the matrix J augmented by the row d^T (see Section 4.1).

If $\eta_k = \eta$ then we generate the first part of the path Q_k^a starting at (x^k, t^k) and moving in the nonzero direction $d^0 = d$ specified by the (kernel of the) Jacobian of $\mathcal{L}H_C^a(x^k, t^k)$ in σ. Otherwise $\eta_k = -\eta$, and we move in the "opposite direction" which is either $d^0 = -d$ if $x^k \in \operatorname{int}\sigma$ or, if x^k is in the intersection of σ and another cell σ', a nonzero direction d^0 constructed with reference to the cell σ' instead of σ_k. (Invariance of orientation is demonstrated, in part, by the result that if σ^0 denotes the cell σ or σ' associated with d^0, depending on which situation occurs, and J^0 is the Jacobian at (x^k, t^k) of $\mathcal{L}H_C^a(x^k, t^k)$ restricted to σ^0, then the matrix consisting of J^0 augmented with the row $(d^0)^T$ has determinantal sign equal to η.)

Generating the path Q_k^a

The PL path $Q_k^a \subset P_k^a$ is generated, one line segment at a time, in the following way. The path generation procedure is initialized with $(\xi^0, \tau^0) = (x^k, t^k)$, the nonzero direction $d = d^0$ as described in Section 3.1, and an associated cell σ^0. In fact, at each step we are given a point $(\xi, \tau) \in P_k^a$, a cell σ of the normal manifold $\mathcal{N}_C$ that contains ξ, and a nonzero direction $d = (d_x, d_t) \in \mathbb{R}^n \times \mathbb{R}$ such that $\xi + sd_x \in \sigma$ for all small $s > 0$. The next line segment on the path is generated by finding the maximum value, possibly infinite, of $s \geq 0$ such that $\xi + sd_x \in \sigma$; denote this value by s'. If s' is finite and the path Q_k^a is nondegenerate (as it must be for almost all a) then the new point is $(\xi', \tau') = (\xi, \tau) + s'd \in Q_k^a$ and there is a cell σ' distinct from σ such that (ξ', τ') lies in the relative interior of a common facet of σ' and σ. It is now possible, as in Section 3.1, to choose a unit direction $d' = (d'_x, d'_t)$ in the kernel of the Jacobian J' of $\mathcal{L}H_C^a(x^k, t^k)$ when restricted to σ', such that $\xi' + sd'_x \in \sigma'$ for small $s \geq 0$. The remainder of the path is generated inductively.

A more formal justification of this path generating procedure can be given in terms of [10] where it is also shown how the degenerate case can be handled. It follows from this paper that Q_k^a is composed of finitely many line segments, the last of which is either a ray or contains the starting point (x^k, t^k).

The path generating procedure may be terminated in any step that generates (d_x, d_t) and s' from (ξ, τ) and σ as follows. Recall h_k is a positive parameter bounding the distance of points on Q_k^a to the starting point (x^k, t^k). Termination occurs at the point $(x', t') = (\xi, \tau) + s(d_x, d_t)$ if there exists $0 < s \leq s'$ such that any one of the following conditions holds:

(i) $\|(x', t') - (\xi^0, \tau^0)\| = h_k$

(ii) $t' = 1$

(iii) $t' = 0$

(iv) $(x', t') = (\xi^0, \tau^0)$.

Stopping condition (iii) is used since there is only one solution $x = x^0 = a$ to the equation $H_C^a(x, 0) = 0$, and we are not interested in revisiting $(x^0, 0)$. Stopping condition (iv) is used to prevent a kind of cycling, that is traveling in an endless loop back.

3.2 Corrector step

We are given a corrector tolerance $\epsilon_c > 0$, that is we expect every iterate (x^k, t^k) to satisfy

$$\|H_C^a(x^k, t^k)\| \leq \epsilon_c.$$

Therefore if the point (x', t') is the result of a predictor step from (x^k, t^k), then the role of the corrector step is to find the next iterate (x^{k+1}, t^{k+1}) somewhat near (x', t') such that $\|H_C^a(x^{k+1}, t^{k+1})\| \leq \epsilon_c$. If an iterative corrector scheme is unable to provide such a point within J_c iterations, where J_c is a positive integer, then the corrector step is deemed to have failed, and a new predicted point (x', t') must be provided.

A theoretically robust corrector scheme

The corrector step can be extended from the Moore-Penrose-Newton method [4] for smooth operators, which we paraphrase as follows. Suppose $H : \mathbb{R}^n \to \mathbb{R}^{n+k}$ is a smooth function, and $z^0 \in \mathbb{R}^{n+k}$ is given; for continuation methods it is enough to take $k = 1$. For each $j = 0, 1, 2, \ldots$, we assume the Jacobian matrix $\nabla H(z^j) \in \mathbb{R}^{(n+k)\times n}$ has full rank, and define z^{j+1} as the nearest point, in Euclidean distance, to z^j in $\{z : H(z^j) + \nabla H(z^j)(z - z^j) = 0\}$. This geometrically motivated algorithm is usually stated, equivalently, as $z^{j+1} = z^j - \nabla H(z^j)^\dagger H(z^j)$ where $\nabla H(z^j)^\dagger$ is the Moore-Penrose inverse of $\nabla H(z^j)$. It converges locally at a Q-quadratic rate to a zero z^* of H if $\nabla H(z^*)$ has full rank and ∇H is Lipschitz near z^*.

The paper [9] provides an extension of this algorithm, designed for solving generalized equations, that can be easily adapted to finding zeroes of the nonsmooth function H_C^a. This approach has also been studied in some detail in [19]. We give a version of the Newton method of [9] for the case of the corrector step. It uses the parameters $\epsilon_c > 0$ and $J_c \in \mathbb{N}$.

> **Corrector Step A**, a function of (x', t').
>
> Let $(y^0, s^0) = (x', t')$ and $j = 0$.
>
> While $\|H_C^a(y^j, s^j)\| > \epsilon_c$ and $j < J_c$
>
> > Find a globally nearest point (y^{j+1}, s^{j+1}) to (x', t') in the set
> >
> > $$\{(y, s) : \mathcal{L}H_C^a(y^j, s^j)(y, s) = 0\}. \tag{3.2}$$
> >
> > Let $j = j + 1$.
>
> (end While)
>
> Let $(x^\dagger, t^\dagger) = (y^j, s^j)$.

Using Theorem 1 it follows that for almost every a, if (y^j, s^j) is near enough to the path P^a, then the linearized set defined in (3.2) is a piecewise linear path at least in a neighborhood of (y^j, s^j). Thus determining a globally nearest point to this path (within a closed neighborhood of (y^j, s^j)) is made computationally possible by examining each of the finitely many line segments of the path.

Suppose f is a C^2 function. It is known [9, 19] that if $\mathcal{L}H_C^a$ satisfies certain regularity conditions, then the above Newton method converges Q-quadratically to a zero of H_C^a. Suppose further that the homotopy path P^a is bounded and nondegenerate such that, for some point $(\hat{x}, \hat{t})$ in its closure, the partial directional derivative of H_C^a with respect to x, $(H_C^a)'_x(\hat{x}, \hat{t}; \cdot)$, is invertible. Then we claim the following statement can be established using the Newton convergence result, Theorem 1 and compactness arguments: there exist $h > 0$, $\epsilon_c > 0$, and $J_c \in \mathbb{N}$ such that for any $t \in [0, \hat{t})$ and x with $\|H_C(x,t)\| \leq \epsilon_c$, if the predictor step generates (x', t') from (x, t, h), then Corrector Step A terminates in one step and the distance of (y^1, s^1) to the path P^a is a small order of the distance from $(y^0, s^0) = (x', t')$ to P^a.

We comment that the choice of a viable ϵ_c seems to depend on "how far the path is from degenerate points", hence on a, so it is not clear how to choose ϵ_c appropriately other than as a small positive number.

A heuristic corrector scheme

To simplify our implementation, that is to avoid computation of a globally nearest point to the generally nonconvex set (3.2), it is easy to extend the approach of [18] for piecewise smooth systems in the same number of variables as equations to underdetermined piecewise smooth equations. At the jth iterate, the idea is to identify any cell σ_j containing the current point $(y^j, s^j) \in \mathbb{R}^{n+1}$, and apply one step of the Moore-Penrose-Newton method to the smooth mapping representing H_C^a on σ_j.

Corrector Step B, a function of (x', t').

Let $(y^0, s^0) = (x', t')$ and $j = 0$.

While $\left\|H_C^a(y^j, s^j)\right\| > \epsilon_c$ and $j < J_c$

Let $(y^{j+1}, s^{j+1}) = (y^j, s^j) - \nabla H_j(y^j, s^j)^\dagger H_C^a(y^j, s^j)$
where σ_j is a cell of $\mathcal{N}_C$ containing (y^j, s^j), and H_j is the C^1 representation of H_C^a on σ_j.
Let $j = j + 1$.

(end While)

Let $(x^\dagger, t^\dagger) = (y^j, s^j)$.

We stress that Corrector Step B is only a heuristic for piecewise smooth systems because the convergence analysis is not as strong as it is for the classical Newton's method or its extensions in [9, 19]. See [6] for a useful but limited convergence theory of this method and some generalizations.

3.3 Formal homotopy algorithm

As described above, if a predictor step produces a point that cannot be sufficiently corrected in J_c or fewer iterations of the corrector step, then a revised, more conservative prediction is made, and the correction step is attempted again. We define this formally using the constants $\epsilon_t \geq 0$, $\epsilon_f > 0$ and $\lambda \in (0,1)$ at the outer level of the algorithm; $\eta \in \{\pm 1\}$ in the predictor step; and $\epsilon_c > 0$ and $J_c \in \mathbb{N}$ in the corrector step.

Homotopy algorithm for solving $H_C^a(x, 1) = 0$.

Let $(x^0, t^0) = (a, 0) \in P$, $k = 0$, $\hat{h} \in (0, 1]$, $\lambda \in (0, 1)$.

While $|t^k - 1| > \epsilon_t$ or $\|f_C(x^k)\| > \epsilon_f$

Let $h^k = \hat{h}$.

Repeat:

$(x', t', h^k) \longleftarrow$ **Predictor Step**(x^k, t^k, h^k).

$(x^\dagger, t^\dagger) \longleftarrow$ **Corrector Step B**(x', t').

Let $h_k = \lambda h_k$.

Until $\|H_C(x^\dagger, t^\dagger)\| \leq \epsilon_c$.

Let $(x^{k+1}, t^{k+1}) = (x^\dagger, t^\dagger)$, $k = k + 1$.

(end While)

4 Implementation

The implementation of the algorithm is presented in two parts. The predictor step uses a suitable modification of the linear complementarity problem solver contained in the PATH code. The corrector uses a heuristic based on the Moore-Penrose idea. We assume hereafter that $C = [\ell, u]$.

4.1 Prediction

PATH [8] is an implementation of a nonsmooth Newton method for solving mixed complementarity problems. At each iteration, a linear complementarity problem is solved using a homotopy method, i.e., it follows the zero curve of a homotopy from $(x^k, 0)$ to $(x', 1)$. The technique employed to generate this path is the same as that developed in Section 3.1. The code has special rules to deal with the cases where the Jacobian of the affine map on σ does not have full row rank and where there is degeneracy in the path [12].

The predictor step of our PL homotopy algorithm uses the same code to generate Q_k^a for the homotopy $\mathcal{L}H_C^a(x^k,t^k)(x,t)$ starting in a cell σ. We explain the implementation of the algorithm of Section 3.1 by generating a representation of this linearized normal map, an indexing scheme that determines σ, and the Jacobian, J, of the affine map restricted to σ. We then discuss how a set of columns spanning J and the orientation is calculated. We continue with a presentation of how the cells of the normal manifold are traversed and the termination rules used by the code.

Representation of the linearized normal map

We will describe the linearized normal map $\mathcal{L}H_C^a(x^k,t^k)(\cdot,\cdot\cdot)$ by giving a formula for the affine mapping that defines it on each cell. We first decompose $x \in \mathbb{R}^n$ into a triple $(z,w,v) \in \mathbb{R}^{3n}$, then give a simple indexing scheme for representing each cell of $\mathcal{N}_C$, and finally combine these two to describe $\mathcal{L}H_C^a(x^k,t^k)(\cdot,\cdot\cdot)$ on each cell.

Represent each x as $z - w + v$ where

$$z = \pi_C(x),\ w = [\pi_C(x) - x]_+,\ v = [x - \pi_C(x)]_+ \tag{4.1}$$

and, for any vector c, c_+ is the vector whose ith component is $\max\{c_i, 0\}$. So $z \in C$ and w, v satisfy

$$w, v \geq 0 \tag{4.2}$$

and the complementarity conditions

$$\langle z - l, w\rangle = 0,\ \langle u - z, v\rangle = 0. \tag{4.3}$$

Conversely, for any $z \in C$ and nonnegative w and v satisfying (4.3), the vector given by $x = z - w + v$ is such that (4.1) holds.

Define a *representation* $\mathcal{A} = \{\mathcal{A}_i\}$ by $\mathcal{A}_i = i$, $n+i$ or $n+2i$, corresponding to z_i, w_i or v_i, respectively, where $\mathcal{A}_i \neq n+i$ if $l_i = -\infty$ and $\mathcal{A}_i \neq n+2i$ if $u_i = \infty$. We abuse notation by writing, say, $w_i \in \mathcal{A}$ to indicate that $n+i \in \mathcal{A}$. It is clear that there is a one-to-one correspondence between the cells

$$\begin{array}{rcllll} \sigma_{\mathcal{A}} & = & \{x : & x_i \leq l_i & \text{if } w_i \in \mathcal{A} \\ & & & l_i \leq x_i \leq u_i & \text{if } z_i \in \mathcal{A} \\ & & & u_i \leq x_i & \text{if } v_i \in \mathcal{A}\} \end{array}$$

and the representations $\mathcal{A}$, and that $\mathcal{N}_C$ is precisely the family of cells indexed by representations $\mathcal{A}$.

Let $\sigma_{\mathcal{A}}$ be a cell and $\mathcal{A}$ be its representation. Then for $x \in \sigma_{\mathcal{A}}$ and (z, w, v) given by (4.1), the relationships (4.3) and (4.2) imply

$$\begin{array}{llllll} z_i = l_i & \text{and} & w_i = l_i - x_i & \text{and} & v_i = 0 & \text{if } w_i \in \mathcal{A}, \\ z_i = x_i & \text{and} & w_i = 0 & \text{and} & v_i = 0 & \text{if } z_i \in \mathcal{A}, \\ z_i = u_i & \text{and} & w_i = 0 & \text{and} & v_i = x_i - u_i & \text{if } v_i \in \mathcal{A}. \end{array}$$

Thus as x varies within $\sigma_{\mathcal{A}}$, only the components of (z, w, v) in the representation $\mathcal{A}$ can change and, furthermore, the transformation $x \mapsto (z, w, v)$ is affine.

Given x^k let $z^k = \pi_C(x^k)$, $w^k = [\pi_C(x^k) - x^k]_+$ and $v^k = [x^k - \pi_C(x^k)]_+$. Then rewrite $\mathcal{L}H_C^a(x^k, t^k)(x, t) = 0$ as the equivalent system consisting of (4.2), (4.3) together with

$$N(z, w, v, t) + q \;=\; 0, \tag{4.4}$$

where $I \in \mathbb{R}^{n \times n}$ is the identity,

$$\begin{array}{rcllcl} N & = & [A \; -I \; I \; r] & & \in & \mathbb{R}^{(3n+1)\times n} \\ A & = & t^k \nabla f(z^k) + (1 - t^k) I & & \in & \mathbb{R}^{n \times n} \\ r & = & f(z^k) - (z^k - a) & & \in & \mathbb{R}^n \\ q & = & t^k f(z^k) + (1 - t^k)(z^k - a) - M z^k - t^k r & & & \\ & = & t^k z^k - a - t^k \nabla f(z^k) z^k & & \in & \mathbb{R}^n. \end{array}$$

So a solution (z, w, v, t) of (4.2)–(4.4) yields $\mathcal{L}H_C^a(x^k, t^k)(z - w + v, t) = 0$.

Since only the components in $\mathcal{A}$ change when $x \in \sigma_{\mathcal{A}}$, it is clear that for all $x \in \sigma_{\mathcal{A}}$ and (z, w, v) representing x

$$\mathcal{L}H_C^a(x^k, t^k)(x, t) \;=\; N_{\mathcal{A}}\,(z, w, v)_{\mathcal{A}} \;+\; rt \;+\; q^{\sigma}.$$

Here $q^{\sigma} = q + N_{\bar{\mathcal{A}}}\,(z, w, v)_{\bar{\mathcal{A}}}$ where $\bar{\mathcal{A}} = \{1, \ldots, 3n\} \setminus \mathcal{A}$ so that the subvector $(z, w, v)_{\bar{\mathcal{A}}}$ is constant in σ.

The PATH code works with the triplet (z, w, v) and $N_{\mathcal{A}}$. However, in x-space the Jacobian, J, of $\mathcal{L}H_C^a(x^k, t^k)(x, t)$ restricted to $\sigma_{\mathcal{A}}$ is given by $\begin{bmatrix} J_x & J_t \end{bmatrix} = \begin{bmatrix} N_{\mathcal{A}}\Delta & r \end{bmatrix}$ where $\Delta \in \Re^{n \times n}$ is the diagonal matrix

$$\Delta_{i,i} = \begin{cases} -1 & \text{if } w_i \in \mathcal{A} \\ 1 & \text{otherwise.} \end{cases}$$

This fact will be used in the calculation of the orientation.

Bases

At the start of the predictor step at iteration k we are given $x^k \in \mathbb{R}^k$, and a representation $\mathcal{A}$ whose corresponding cell σ contains x^k. If x^k is in the intersection of two or more cells we use $\mathcal{A}$ to describe the cell we are interested in. We have seen that the linear part of $\mathcal{L}H_C^a(x^k, t^k)$ on σ can be described using the matrix $\begin{bmatrix} N_{\mathcal{A}} & r \end{bmatrix} \in \mathbb{R}^{n\times(n+1)}$. To determine our initial search direction, we need to construct a basis for the range space and kernel of this map. More formally, given a cell σ and its representation $\mathcal{A}$, define a *basis* $\mathcal{B}$ as a list of n indices taken from $\mathcal{A} \cup \{3n+1\}$, where $3n+1$ is the index of t, such that $N_{\mathcal{B}}$ is invertible.

The linear complementarity code is initially given a candidate basis, $\tilde{\mathcal{B}}$. It is possible that $N_{\tilde{\mathcal{B}}}$ is not invertible. In this case we attempt to uncover a basis. We apply a LU factorization to the $n \times (n+1)$ matrix $N_{\mathcal{A}\cup\{3n+1\}}$ to determine the linearly dependent column. The remaining columns form a basis, $\mathcal{B}$.

If more than one linearly dependent column is identified, the theory breaks down and this procedure fails. If x^k is in the intersection of two or more cells, we try to choose an adjacent cell with full row rank. Specifically, the adjacent cell attempted is $\bar{\sigma}$ with representation:

$$\bar{\mathcal{A}}_i = \begin{cases} w_i & \text{if } z_i^k = l_i \\ v_i & \text{if } z_i^k = u_i \\ z_i & \text{otherwise.} \end{cases}$$

We choose a candidate basis as $\bar{\mathcal{B}} = \bar{\mathcal{A}}$. We then apply the above procedure with $\bar{\mathcal{B}}$ replacing $\tilde{\mathcal{B}}$. In the event this fails, an error is reported and the code halts.

The reason we use a candidate basis as our initial guess as to the invertible basis as opposed to always factoring the larger $n \times (n+1)$ matrix is that the code written for PATH only uses the $n \times n$ basis matrix for computations. We did not want to rewrite the code to use the $n \times (n+1)$ matrix. Typically the candidate basis is invertible and we save a factorization.

Orientation

We now have an initial basis $\mathcal{B}^k$ and a linearly dependent column $\beta^k = (\mathcal{A} \cup \{3n+1\}) \setminus \mathcal{B}^k$. We need to determine the initial direction of the path Q_k^a, that is whether the variable $(z, w, v, t)_{\beta^k}$ is going to initially increase or decrease. To do this, we need to find the orientation.

For the orientation, we determine the sign of the determinant of the following matrix:

$$\begin{bmatrix} J_x & J_t \\ d_x^T & d_t \end{bmatrix} = \begin{bmatrix} N_{\mathcal{A}}\Delta & r \\ d_x^T\Delta & d_t \end{bmatrix} = \begin{bmatrix} B & b \\ d_B^T & 1 \end{bmatrix} P$$

where $\begin{bmatrix} d_x^T & d_t \end{bmatrix}$ spans the kernel of $\begin{bmatrix} J_x & J_t \end{bmatrix}$, $B = \begin{bmatrix} N_{\mathcal{A}}\Delta & r \end{bmatrix}_{\mathcal{B}^k}$, $b = \begin{bmatrix} N_{\mathcal{A}}\Delta & r \end{bmatrix}_{\beta^k}$, and P is a permutation matrix. Let LU be the decomposition of B into an invertible lower triangular matrix, L, and an invertible upper triangular matrix, U. Letting $d_B^T = -U^{-1}L^{-1}b = -B^{-1}b$, we can see that $\begin{bmatrix} d_B^T & 1 \end{bmatrix}$ spans the kernel of $\begin{bmatrix} B & b \end{bmatrix}$. Combining these facts, we have

$$\begin{aligned} \begin{bmatrix} B & b \\ d_B^T & 1 \end{bmatrix} P &= \begin{bmatrix} LU & b \\ d_B^T & 1 \end{bmatrix} P \\ &= \begin{bmatrix} L & 0 \\ d_B^T U^{-1} & 1 - d_B^T U^{-1} L^{-1} b \end{bmatrix} \begin{bmatrix} U & L^{-1}b \\ 0 & 1 \end{bmatrix} P \\ &= \begin{bmatrix} L & 0 \\ d_B^T U^{-1} & 1 + \|d_B\|^2 \end{bmatrix} \begin{bmatrix} U & L^{-1}b \\ 0 & 1 \end{bmatrix} P. \end{aligned}$$

Therefore, the orientation is just the sign of the determinant of L times the sign of the determinant of U multiplied by the sign of determinant of P. We will initially increase the entering variable if the orientation is η and decrease it otherwise. We then traverse the cells using a complementary pivoting strategy.

Complementary pivoting

Given the basis, $\mathcal{B}^k$, and the entering variable, β^k, we calculate a direction, $d = -N_{\mathcal{B}^k}^{-1} N_{\beta^k}$. We will either increase or decrease variable $(z, w, v, t)_{\beta^k}$ along this direction depending upon the orientation calculation above. A standard ratio test (on (z, w, v, t)) determines when we run into the boundary of a cell and hence gives a leaving variable and steplength. We then perform a pivot by updating $\mathcal{B}^k$ to include the entering variable and exclude the leaving variable. The new entering variable is chosen as follows:

- If z_i leaves at its lower bound, w_i is the new entering variable.
- If z_i leaves at its upper bound, v_i is the new entering variable.

- If w_i leaves at its lower bound, z_i is the new entering variable at its lower bound.
- If v_i leaves at its lower bound, z_i is the new entering variable at its upper bound.
- If t leaves, we are done.

This defines the new cell along with a corresponding basis, entering variable, and direction. The code uses a rank-1 update of the LU factorization of the old basis to find the needed decomposition of the new basis.

Termination

The four conditions for termination presented in Section 3.1 are used in the code. If t leaves the basis we terminate according to the complementary pivoting rules. This corresponds to $t = 0$ or $t = 1$. In the code, we relax the distance based termination rule to be

$$\left\|(\pi_C(x), t) - (\pi_C(x^k), t^k)\right\| = h_k$$

because the approximation error, i.e., the difference between $\mathcal{L}H_C^a(x^k, t^k)(x,t)$ and $H_C^a(x,t)$ depends more on $\pi_C(x) - \pi_C(x^k)$ than on $[x - \pi_C(x)] - [x^k - \pi_C(x^k)]$.

Special rules

The corrector can lead us to a place where $t^k > 1$. In this case we reverse the direction we are traveling. We put a lower bound on the t variable of 1 and an upper bound on t of ∞. We then use the above direction calculation, but take the opposite direction in which to initially travel. Everything then continues as normal, with the termination being that t leaves at its lower bound or we take a step of the maximum allowed distance.

4.2 Correction

The corrector code is given an initial (x', t') from the predictor. The implementation of the corrector uses spacer steps which move the current iterate closer to P^a. They are defined as follows:

$$\hat{y}_i^0 = \begin{cases} (\pi_C(x') - [H^a(x', t')]_+)_i & \text{if variable } \pi_C(x') = l_i \\ (\pi_C(x') + [-H^a(x', t')]_+)_i & \text{if variable } \pi_C(x') = u_i \\ x_i' & \text{otherwise.} \end{cases}$$

Let $s^0 = t'$. We then calculate $\left\| H_C^a(y^0, s^0) \right\|$ and check to see if it is zero as detailed in Section 3.2. If it is not, we use the Moore-Penrose idea to move back onto the zero curve of the homotopy.

The Moore-Penrose iterate solves a minimization problem:

$$\begin{array}{ll} \min & \left\| (x,t) - (y^j, s^j) \right\|^2 \\ \text{subject to} & \begin{bmatrix} J_x & J_t \end{bmatrix} \begin{bmatrix} x \\ t \end{bmatrix} = -H_C^a(\hat{x}, \hat{t}) \end{array}$$

where $\begin{bmatrix} J_x & J_t \end{bmatrix}$ is the Jacobian of the affine map $\mathcal{L}H_C^a(y^j, s^j)(\cdot,\cdot)$ on σ, a cell containing y^j, (see Section 4.1 for details on how this is calculated).

To solve this problem, we solve the following system of equations:

$$\begin{bmatrix} I & 0 & -J_x^T \\ 0 & 1 & -J_t^T \\ J_x & J_t & 0 \end{bmatrix} \begin{bmatrix} d_x \\ d_t \\ d_u \end{bmatrix} = \begin{bmatrix} 0 \\ -H_C^a(y^j, s^j) \end{bmatrix}.$$

The direction (d_x, d_t) is nonzero since $H_C^a(y^j, s^j) \neq 0$. We move in the direction (d_x, d_t) until either the full step is taken or we encounter the boundary of σ to obtain y^{j+1} and s^{j+1} and an adjacent cell σ'. We then check for termination, and if necessary perform another iteration.

The code does not allow the residual to increase from one corrector step to the next and returns an error code if at the end the homotopy parameter is negative. If the corrector fails, we take a half predictor step and attempt the corrector again. A half predictor step is quickly found using the reconstruction technique documented in [12].

4.3 Updates

The parameters in the code are updated after each major iteration (predictor, corrector sequence). We modify the maximum distance allowed to travel to become larger if we are doing well and make it smaller if we are doing poorly.

5 Results

The algorithm was implemented using the current version of the linear complementarity problem solver found in PATH. The framework reported in [12] was used so that we could easily access problems generated by the GAMS modeling language.

We ran the code on the MCPLIB [7] test suite of problems. The MPSGE [26] models in the collection were omitted due to the fact that function and Jacobian evaluations for these models have the undesirable side-effect of changing the evaluation point, thereby making corrector steps impossible to perform.

The model name, dimension of the problem, number of function evaluations (Func), and solution time (Time) for the first starting point of the remaining models in this test set are reported in Table 5.1. The test was carried out on a Sun Ultrasparc machine that has 768 megabytes of available memory.

We note that most of the academic test problems which cause problems for other algorithms (e.g., billups, dirksel, ralph, simple-ex) are easily solved using the predictor-corrector code. Over the entire set of 77 models with 436 total starting points, we achieved a 67.7% success rate. Of the 141 failures, 83 of these occurred for starting points associated with just two problems, namely `games` and `tinsmall`.

The current implementation demonstrates that homotopy methods can be generated for large scale complementarity problems. The choice of homotopy function (in our case $x - a$) is critical for the success of these methods. We were disappointed with the code's robustness; in particular the implementation is very sensitive to how the original problem is formulated (e.g., where $f_1(x) = 0$ or $-f_1(x) = 0$.

Three approaches may be useful in improving these results. Firstly, a different choice of homotopy could be used instead of $x - a$ to attempt to match it more closely to the underlying problem. Secondly, the preprocessor for MCP described in [13] could be used to exploit more fully the underlying problem structure. Finally, this preprocessor could be adapted to automatically identify poorly posed problems and reformulate appropriately. This is the subject of future research.

References

[1] J. C. Alexander. The topological theory of an embedding method. In H. Wacker, editor, *Continuation Methods*, pages 37–68. Academic Press, New York, 1978.

[2] J. C. Alexander, R. B. Kellog, T.-Y. Li, and J .A. Yorke. Piecewise smooth continuation. Manuscript, 1979.

[3] J. C. Alexander, T.-Y. Li, and J .A. Yorke. Piecewise smooth homotopies. In B. C. Eaves, F. J. Gould, H.-O. Peitgen, and M. J. Todd, editors,

Table 5.1: Results on MCPLIB Problems

Problem	Size	Func	Time	Problem	Size	Func	Time
badfree	5	44	0.03	mr5mcf	350	-	fail
bert_oc	5000	-	fail	munson3	2	-	fail
bertsekas	15	441	0.66	munson4	1	-	fail
billups	1	32	0.02	nash	10	203	0.71
bratu	5625	181	13208.96	ne-hard	3	111	0.08
cammcf	242	-	fail	obstacle	2500	-	fail
choi	13	21	2.66	opt_cont	288	173	12.10
colvdual	20	269	0.45	opt_cont127	4096	641	1726.96
colvnlp	15	275	0.33	opt_cont255	8192	1888	12470.15
cycle	1	23	0.01	opt_cont31	1024	323	114.73
degen	2	27	0.02	opt_cont511	16384	-	fail
dirkse1	2	135	0.08	pgvon105	105	619	12.37
duopoly	63	-	fail	pgvon106	106	-	fail
eckstein	1	21	0.01	pies	42	1595	4.91
ehl_k40	41	195	6.68	powell	16	61	0.20
ehl_k60	61	233	18.26	powell_mcp	8	75	0.11
ehl_k80	81	265	38.50	qp	4	39	0.03
ehl_kost	101	291	74.67	ralph	2	33	0.02
electric	158	-	fail	runge	1	13	0.01
explcp	16	98	0.19	scarfanum	13	88	0.23
fixedpt	406	-	fail	scarfasum	14	46	0.13
forcebsm	184	-	fail	scarfbnum	39	653	2.46
forcedsa	186	-	fail	scarfbsum	40	621	3.28
freebert	15	439	0.65	shubik	30	1534	2.78
gafni	5	48	0.07	simple-ex	17	173	0.67
games	16	-	fail	simple-red	13	121	0.65
hanskoop	14	58	0.10	sppe	27	447	0.80
hansmcf	43	139	0.74	tinloi	146	54	1.16
hydroc06	29	-	fail	tinsmall	42	647	3.58
hydroc20	99	-	fail	tobin	42	305	0.79
jel	6	317	0.37	trafelas	2376	-	fail
josephy	4	43	0.03	trig	10	-	fail
kojshin	4	61	0.04	vonthmcf	125	-	fail
kyh-scale	28	-	fail	xu1	8	73	0.07
kyh	28	-	fail	xu2	8	73	0.07
lincont	419	-	fail	xu3	11	63	0.08
mathinum	3	47	0.04	xu4	11	39	0.04
mathisum	4	35	0.03	xu5	20	70	0.12
methan08	31	-	fail				

Homotopy Methods and Global Convergence, pages 1–14. Plenum Press, New York, 1983.

[4] E. L. Allgower and K. Georg. *Numerical Continuation Methods, An Introduction.* Springer-Verlag, Berlin, 1990.

[5] S. C. Billups, A. L. Speight, and L. T. Watson. Nonmonotone path following methods for nonsmooth equations and complementarity problems. Technical report, Department of Mathematics, University of Colorado, Denver, Colorado, 1999.

[6] X. Chen, Z. Nashed, and L. Qi. Convergence of Newton's method for singular smooth and nonsmooth equations using adaptive outer inverses. *SIAM Journal on Optimization*, 7:445–462, 1997.

[7] S. P. Dirkse and M. C. Ferris. MCPLIB: A collection of nonlinear mixed complementarity problems. *Optimization Methods and Software*, 5:319–345, 1995.

[8] S. P. Dirkse and M. C. Ferris. The PATH solver: A non-monotone stabilization scheme for mixed complementarity problems. *Optimization Methods and Software*, 5:123–156, 1995.

[9] A. L. Dontchev. Local convergence of the Newton method for generalized equations. *Comptes rendus de l'Acadmie des sciences. Srie 1, Mathmatique*, 322:327–331, 1996.

[10] B. C. Eaves. A short course in solving equations with PL homotopies. In R. W. Cottle and C. E. Lemke, editors, *Nonlinear Programming*, pages 73–143, Providence, RI, 1976. American Mathematical Society, SIAM–AMS Proceedings.

[11] F. Facchinei and J. S. Pang. Total stability of variational inequalities. Manuscript, Universitá di Roma "La Sapienza", Dipartimento di Informatica e Sistemistica, Roma, Italia, 1988.

[12] M. C. Ferris and T. S. Munson. Interfaces to PATH 3.0: Design, implementation and usage. *Computational Optimization and Applications*, 12:207–227, 1999.

[13] M. C. Ferris and T. S. Munson. Preprocessing complementarity problems. Mathematical Programming Technical Report 99-07, Computer Sciences Department, University of Wisconsin, Madison, Wisconsin, 1999.

[14] M. C. Ferris and J. S. Pang. Engineering and economic applications of complementarity problems. *SIAM Review*, 39:669–713, 1997.

[15] H. Hironaka. Introduction to real-analytic sets and real-analytic maps. Technical report, Instituto Matematico "L. Tonelli", Università di Pisa, Pisa, Italy, 1973.

[16] M. Kojima. Recent advances in mathematical programming. X. Computation of fixed points by continuation methods. *Systems and Control*, 25:421–430, 1981.

[17] M. Kojima and R. Hirabayashi. Continuous deformation of nonlinear programs. *Mathematical Programming Study*, 21:150–198, 1984.

[18] M. Kojima and S. Shindo. Extensions of Newton and quasi-Newton methods to systems of PC^1 equations. *Journal of Operations Research Society of Japan*, 29:352–374, 1986.

[19] B. Kummer. Metric regularity: characterizations, nonsmooth variations, and successive approximation. *Journal of Mathematical Analysis and its Applications*, forthcoming.

[20] M. S. Lojasiewicz. *Ensembles semi-analytiques*. Institut des Hautes Etudes Scientifiques, Bures-sur-Yvette, 1964.

[21] O. L. Mangasarian. *Nonlinear Programming*. McGraw–Hill, New York, 1969. SIAM Classics in Applied Mathematics 10, SIAM, Philadelphia, 1994.

[22] D. Ralph. On branching numbers of normal manifolds. *Nonlinear Analysis, Theory, Methods and Applications*, 22:1041–1050, 1994.

[23] S. M. Robinson. An implicit–function theorem for a class of nonsmooth functions. *Mathematics of Operations Research*, 16:292–309, 1991.

[24] S. M. Robinson. Normal maps induced by linear transformations. *Mathematics of Operations Research*, 17:691–714, 1992.

[25] S. M. Robinson. Newton's method for a class of nonsmooth functions. *Set Valued Analysis*, 2:291–305, 1994.

[26] T. F. Rutherford. Applied general equilibrium modeling with MPSGE as a GAMS subsystem: An overview of the modeling framework and syntax. *Computational Economics*, forthcoming.

[27] H. Sellami. *A Continuation Method for Normal Maps.* PhD thesis, University of Wisconsin, Madison, Wisconsin, 1994.

[28] H. Sellami. A homotopy continutation method for normal maps. *Mathematical Programming*, 82:317–337, 1998.

[29] H. Sellami and S. M. Robinson. Homotopies based on nonsmooth equations for solving nonlinear variational inequalities. In G. Di Pillo and F. Giannessi, editors, *Nonlinear Optimization and Applications*, pages 327–343. Plenum Press, New York, 1996.

[30] H. Sellami and S. M. Robinson. Implementation of a continutation method for normal maps. *Mathematical Programming*, 76:563–578, 1997.

Appendix

We use a property of subanalytic sets [15, 20] in the proof of the next result: If S is a subanalytic set in $\mathbb{R}^n$ then for any point $\hat{x}$ in its closure there is a continuous function $g : [0,1] \to \mathbb{R}^n$ with $g(1) = \hat{x}$ and $g(s) \in S$ for $s \in [0,1)$.

Lemma 2 *If $p : [0,1) \to \mathbb{R}^n$ is a subanalytic function then either there exists the limit $\lim_{t\to 1^-} p(t)$, or $\|p(t)\| \to \infty$ as $t \to 1^-$.*

Proof Let $p : [0,1) \to \mathbb{R}^n$ be subanalytic, and $\hat{p}$ be a limit point of $p(t)$ as $t \to 1^-$. It is sufficient to show that the limit $\lim_{t\to 1^-} p(t)$ exists and equals $\hat{p}$.

The graph of p, $P = \{(t, p(t)) : t \in [0,1)\}$, is subanalytic by definition and $(1,\hat{p})$ lies in its closure. Let $g : [0,1] \to \mathbb{R}^n$ be a continuous function with $g(1) = (1,\hat{p})$ and $g(s) \in P$ for $s \in [0,1)$. Denote the first component function of $g(s)$ by $\gamma(s) \in [0,1]$. Hence for $s \in [0,1)$, $\gamma(s) \in [0,1)$ and $g(s) = (\gamma(s), p\circ\gamma(s))$.

Let $\epsilon > 0$. We will complete the proof by providing $\delta \in (0,1]$ such that

$$\|p(t) - \hat{p}\| < \epsilon \quad \text{for all } t \in [1-\delta, 1). \tag{5.1}$$

Continuity of g at $s = 1$ yields $\delta_g \in (0,1]$ such

$$\|p \circ \gamma(s) - \hat{p}\| < \epsilon \quad \text{for all } s \in [1-\delta_s, 1).$$

Now $\gamma(1-\delta_g) < 1 = \gamma(1)$ so continuity of γ ensures (by the intermediate value theorem) that $\gamma([1-\delta_g, 1)) \supset [\gamma(1-\delta_g), 1)$. Let $\delta = 1 - \gamma(1-\delta_g) \in (0,1]$ and deduce (5.1) from the previous bound. □

An easy corollary extends a result of [11, Theorem 5.10], in the context of a homotopy method applied to variational inequalities, in which Parts 2 and 3 were shown to be equivalent.

Corollary 3 *For a subanalytic function* $p : [0, 1) \to \mathbb{R}^n$ *the following statements are equivalent:*

1. $\liminf_{t \to 1^-} \|p(t)\| < \infty$.
2. $\limsup_{t \to 1^-} \|p(t)\| < \infty$.
3. $\lim_{t \to 1^-} p(t)$ *exists.*

Acknowledgements

We are grateful to Jeremy Glick for preliminary work on smooth homotopy methods and to Stephen M. Robinson for discussions on homotopy methods applied to variational inequalities.

The work of Ferris and Munson was based on research supported by the National Science Foundation grant CCR-9972372 and the Air Force Office of Scientific Research grant F49620-98-1-0417. The work of Ralph was based on research supported by the Australian Research Council.

Michael C. Ferris, Todd S. Munson
Computer Sciences Department
University of Wisconsin
Madison, WI 53706, USA

Daniel Ralph
Department of Mathematics
University of Melbourne
Melbourne, Australia

E. Hairer and Ch. Lubich

Energy conservation by Störmer-type numerical integrators

Abstract For the numerical solution of second-order, highly oscillatory differential equations we study a class of symmetric methods that includes the Störmer/Verlet method, the trapezoidal rule and the Numerov method. We consider Hamiltonian systems where high oscillations are generated by a single frequency that is well separated from the lower frequencies. We apply the numerical methods with step sizes whose product with the high frequency is in the range of linear stability, but is not assumed to be small. As a main result of this chapter, we show long-time conservation of the time averages of the total energy and the oscillatory energy, and pointwise near-preservation of modified energies.

1 Introduction

The Störmer/Verlet method is the most widely used numerical integrator in molecular dynamics, for a good part because of its long-time conservation of the total energy and also of adiabatic invariants such as the oscillatory energy. The long-time behaviour has previously been explained by a backward analysis which interprets the numerical solution as the "almost" exact solution of a perturbed Hamiltonian system [BG94, HaL97, Rei98, Rei99]. This explanation applies only when the product $h\omega$ of the time step h with the highest frequency ω is very small, a situation that is often not met in practical computations.

Here we consider the numerical energy behaviour for fixed positive $h\omega$ in the interval of linear stability, for a class of nonlinear model problems with a single, constant high frequency ω. This problem class includes the Fermi-Pasta-Ulam model of alternating soft nonlinear and stiff linear springs. Although oscillations in the energy are of a size roughly proportional to $(h\omega)^2$, the averages of the energy over finite time windows are preserved over very long time intervals. This is explained, and the deviation of the finite-time averages from the initial energy is explicitly determined in terms of certain modified energies which are well preserved over long times. The analysis uses only the symmetry of the method, not its symplecticness.

We consider a class of symmetric methods that includes the Störmer/Verlet method, the trapezoidal rule and the Numerov method. In Section 2 we begin by studying the long-time conservation of the total energy when the numerical methods are applied with small step sizes ($h\omega \to 0$).

Section 3 presents numerical experiments with a Fermi-Pasta-Ulam chain. The numerical phenomena observed for $h\omega$ bounded away from 0 are completely explained by the theoretical results of Section 4, the main results of this chapter. For the methods of Section 2, it is shown that certain modified total and oscillatory energies, and also the time averages of the original total and oscillatory energies, are preserved up to $O(h^2)$ over time intervals of length h^{-N}, where N may become arbitrarily large.

Section 5 reviews and refines results from [HaL99] concerning the long-time energy conservation by a class of methods which integrate the linear part of oscillatory differential equations exactly. That class includes Gautschi-type methods [Gau61, HoL99] and the mollified impulse method [GSS99]. Section 6 contains the proofs of the theorems of Section 4, which are based on the results of Section 5.

2 Energy conservation by symmetric methods used with small step size

For the numerical solution of second-order differential equations

$$\ddot{x} = f(x), \qquad x(0), \dot{x}(0) \text{ given}, \tag{2.1}$$

with smooth nonlinearity in gradient form $f(x) = -\nabla U(x)$, we consider the following class of symmetric methods (cf. [New59, SZS97]):

$$\begin{aligned} x_{n+1} - 2x_n + x_{n-1} &= h^2 f_n + \alpha h^2 (f_{n+1} - 2f_n + f_{n-1}) \\ 2h\dot{x}_n &= (x_{n+1} - x_{n-1}) - \beta h^2 (f_{n+1} - f_{n-1}). \end{aligned} \tag{2.2}$$

Here, $f_n = f(x_n)$, and α, β are real parameters determining the method. This class includes the Störmer/Verlet method ($\alpha = \beta = 0$), the trapezoidal rule ($\alpha = \beta = 1/4$), and the fourth-order Numerov method ($\alpha = 1/12$, $\beta = 1/6$). If $\alpha = \beta$, the method results from the symmetric one-step scheme

$$\begin{aligned} x_{n+1} &= x_n + h\dot{x}_n + \tfrac{1}{2}h^2 f_n + \alpha h^2 (f_{n+1} - f_n) \\ \dot{x}_{n+1} &= \dot{x}_n + \tfrac{1}{2}h f_{n+1} + \tfrac{1}{2}h f_n. \end{aligned} \tag{2.3}$$

In the general case, for given $(x_n, \dot{x}_n)$, the equations (2.2) uniquely define x_{n-1} and x_{n+1}. As soon as we have x_n and x_{n+1}, we can compute x_{n+2} and $\dot{x}_{n+1}$

again from (2.2). This shows that (2.2) can be interpreted as a one-step method $(x_n, \dot{x}_n) \mapsto (x_{n+1}, \dot{x}_{n+1})$.

We show that for all values of α, β the symmetric method (2.2) approximately conserves the total energy

$$H(x, \dot{x}) = \tfrac{1}{2}|\dot{x}|^2 + U(x) \tag{2.4}$$

over very long time intervals.

Theorem 2.1 *The total energy along the numerical solution of method (2.2) satisfies*

$$|H(x_n, \dot{x}_n) - H(x_0, \dot{x}_0)| \leq Ch^2 + C_N h^N t \qquad \text{for } 0 \leq t = nh \leq h^{-N}$$

for arbitrary positive integer N. The constants C and C_N are independent of t. C_N depends on bounds of derivatives of f up to Nth order in a region that contains the numerical solution values (x_n). For the Numerov method the estimate holds with Ch^4 instead of Ch^2.

Theorem 2.1 gives long-time estimates for *small* step sizes, i.e., for $h^2 L \to 0$, where L is a Lipschitz constant of f. We give two different proofs of this result.

The first proof uses only the symmetry of the methods and makes no appeal to symplecticness. It is this type of argument which extends to energy conservation in oscillatory systems when the product of the step size with the highest frequencies is bounded away from 0.

The second proof is based on the observation that the methods (2.2), though not themselves symplectic with the exception of the Störmer/Verlet method ($\alpha = 0$, $\beta = 0$), are closely related to symplectic methods. Backward analysis, which interprets the numerical solution of a symplectic one-step method as the almost-exact solution of a modified Hamiltonian system, then yields the result.

When the nonlinearity f is analytic, both proofs can be refined to yield an estimate $Ch^2 + C_0 e^{-c/h} t$ over exponentially long times $t \lesssim e^{c/h}$, with c proportional to $1/\sqrt{L}$. Compare [BG94, HaL97, Rei98] where such exponential estimates are proved for symplectic methods.

First proof. (a) The first step consists in constructing a smooth function $y(t)$, depending on the step size h, such that it gives a small defect – of size $O(h^{N+2})$ – when inserted into the numerical scheme:

$$\begin{aligned} y(t+h) - 2y(t) + y(t-h) &= h^2 f(y(t)) \\ &\quad + \alpha h^2 \big(f(y(t+h)) - 2f(y(t)) + f(y(t-h))\big) + O(h^{N+2}) \end{aligned} \tag{2.5}$$

and which satisfies $y(0) = x_0$, $y(h) = x_1$. The function $y(t)$ is needed on a bounded time interval, $0 \le t \le T$, say. It can be constructed in a standard way, either by an asymptotic expansion of the numerical solution in powers of the step size, or by backward analysis where the numerical solution is interpreted as the formally exact (up to $O(h^N)$) solution of a modified differential equation. See [HaL98] for a review and comparison of these two techniques and for detailed references to the pertinent literature.

Since $y(t+h) = \sum_{k=0}^{N-1} y^{(k)}(t)h^k/k! + O(h^N)$, the relation (2.5) is equivalent to

$$\sum_{l=1}^{N/2} c_l\, y^{(2l)}(t)\, h^{2l-2} = f(y(t)) + O(h^N), \tag{2.6}$$

where c_l are the coefficients of the expansion

$$\frac{e^z - 2 + e^{-z}}{1 + \alpha(e^z - 2 + e^{-z})} = \sum_{l=1}^{\infty} c_l\, z^{2l}.$$

We note that $c_1 = 1$ and $c_2 = 1/12 - \alpha$, so that $c_2 = 0$ for the Numerov method. The symmetry of the method manifests itself through the fact that only even-order derivatives of $y(t)$ (and even powers of the step size) are present in (2.6).

(b) We multiply (2.6) with $\dot{y}(t)^T$ and integrate over t. The crucial observation is now that the product of $\dot{y}(t)$ with an *even*-order derivative of $y(t)$ is a total differential:

$$\dot{y}^T y^{(2l)} = \frac{d}{dt}\Big(\dot{y}^T y^{(2l-1)} - \ddot{y}^T y^{(2l-2)} + \ldots \mp (y^{(l-1)})^T y^{(l+1)} \pm \frac{1}{2}\,(y^{(l)})^T y^{(l)}\Big) =: \frac{d}{dt} a_l\,.$$

In particular, $a_1 = |\dot{y}|^2$. Moreover, for $f(y) = -\nabla U(y)$ we clearly have $\dot{y}^T f(y) = -(d/dt)U(y)$. Setting

$$\widetilde{H}[y](t) = \sum_{l=1}^{N/2} c_l\, a_l(t)\, h^{2l-2} + U(y(t)),$$

we thus obtain for $0 \le t \le 1$

$$\widetilde{H}[y](t) - \widetilde{H}[y](0) = O(h^N) \tag{2.7}$$

and

$$\widetilde{H}[y](t) = H(y(t), \dot{y}(t)) + O(h^p), \tag{2.8}$$

where $p = 4$ for the Numerov method (because $c_2 = 0$), and $p = 2$ otherwise.

(c) We define the function $y'(t)$ by the formula corresponding to $\dot{x}_n$ in (2.2):

$$2h\, y'(t) = y(t+h) - y(t-h) - \beta h^2 \big(f(y(t+h)) - f(y(t-h))\big). \tag{2.9}$$

We then have $y'(t) = \dot{y}(t) + O(h^p)$ and hence, by (2.8),

$$H(y(t), y'(t)) = H(y(t), \dot{y}(t)) + O(h^p). \tag{2.10}$$

(d) The proof is now completed by a standard argument: We fix a positive T and choose the integer M such that $Mh \leq T < (M+1)h$. We denote by $y_k(t)$ the smooth function, unique up to $O(h^N)$, which satisfies (2.5) and starts from x_{kM} and x_{kM+1}. Similarly, we write $y_k'(t)$ for the corresponding function constructed by (2.9). The finite-time stability of the numerical method then yields

$$\begin{aligned} y_k(t) &= x_{kM+m} + O(h^N) \\ y_k'(t) &= \dot{x}_{kM+m} + O(h^N) \end{aligned} \qquad \text{for } \; 0 \leq t = mh \leq T \tag{2.11}$$

and hence

$$H(x_{kM+m}, \dot{x}_{kM+m}) = H(y_k(t), y_k'(t)) + O(h^N), \qquad 0 \leq t = mh \leq T. \tag{2.12}$$

We obtain equally

$$\widetilde{H}[y_{k+1}](0) = \widetilde{H}[y_k](Mh) + O(h^N)$$

so that, using (2.7),

$$\widetilde{H}[y_k](t) = \widetilde{H}[y_0](0) + O(kh^N), \qquad 0 \leq t \leq T.$$

Together with (2.12), (2.10), (2.8), this yields the result stated in Theorem 2.1.

Second proof. (a) We begin with the following observation (cf. [SZS97]): *For every α, the symmetric one-step scheme (2.3) is conjugate to a symplectic one-step method.*

To see this, we write the method (2.3) as

$$\begin{aligned} x_{n+1} - \alpha h^2 f_{n+1} &= x_n - \alpha h^2 f_n + h\dot{x}_n + \tfrac{1}{2} h^2 f_n \\ \dot{x}_{n+1} - \tfrac{1}{2} h f_{n+1} &= \dot{x}_n + \tfrac{1}{2} h f_n. \end{aligned}$$

Taking the wedge product of the differentials of both equations, and using $dx_n \wedge df_n = 0$, we obtain

$$(dx_{n+1} - \alpha h^2 df_{n+1}) \wedge d\dot{x}_{n+1} = (dx_n - \alpha h^2 df_n) \wedge d\dot{x}_n.$$

Hence, with the transformation

$$q_n = x_n - \alpha h^2 f_n, \qquad p_n = \dot{x}_n,$$

the scheme (2.3) is conjugate to the symplectic method $(q_n, p_n) \mapsto (q_{n+1}, p_{n+1})$. Since the transformation only depends on h^2, this new method is also symmetric. The modified Hamiltonian of this one-step method is

$$\widetilde{H}(q,p) = H(q,p) + \frac{h^2}{24}\Big(2U''(q)p^2 - (12\alpha+1)(U'(q))^2\Big) + O(h^4), \tag{2.13}$$

where $H(q,p) = \frac{1}{2}p^2 + U(q)$.

(b) Denote by $\widetilde{q}(t)$ and $\widetilde{p}(t)$ the solution of the modified equation of the symplectic one-step method $(q_n, p_n) \mapsto (q_{n+1}, p_{n+1})$. We then define $\widetilde{x}(t)$ from $\widetilde{q}(t) = \widetilde{x}(t) - \alpha h^2 f(\widetilde{x}(t))$ and $\widetilde{x}'(t) = \widetilde{p}(t) - (\beta - \alpha)\frac{1}{2}h\big(f(\widetilde{x}(t+h)) - f(\widetilde{x}(t-h))\big)$, so that the numerical solution of (2.2) satisfies (formally)

$$x_n = \widetilde{x}(t_n), \qquad \dot{x}_n = \widetilde{x}'(t_n).$$

Using the modified differential equation for $\widetilde{q}(t)$ and $\widetilde{p}(t)$ and the definition of $\widetilde{x}(t)$, the expression $h\big(f(\widetilde{x}(t+h)) - f(\widetilde{x}(t-h))\big)$ can be written as a series in h^2 with coefficients depending on $\widetilde{x}(t)$ and $\widetilde{x}'(t)$ only. The modified Hamiltonian $\widehat{H}(\widetilde{x}, \widetilde{x}') := \widetilde{H}(\widetilde{q}, \widetilde{p})$ is then constant up to $O(h^N)$ along the numerical solution over bounded time intervals, and by the same argument as in part (d) of the previous proof, it is constant up to $O(th^N)$ for all $t = nh$. The first terms of this modified Hamiltonian are:

$$\begin{aligned}\widehat{H}(\widetilde{x}, \widetilde{x}') = H(\widetilde{x}, \widetilde{x}') \;&+\; \Big(\alpha - \beta + \frac{1}{12}\Big)h^2 U''(\widetilde{x})(\widetilde{x}')^2 \\ &+\; \Big(\alpha - \frac{1}{12}\Big)\frac{h^2}{2}(U'(\widetilde{x}))^2 + O(h^4).\end{aligned}$$

This proves the statement of Theorem 2.1.

3 Numerical experiments

We consider a chain of springs, where soft nonlinear springs alternate with stiff harmonic springs (see [GGMV92] and Fig. 3.1). The variables $x_1, \ldots, x_{2n}$ (and $x_0 = 0$, $x_{2n+1} = 0$) stand for the displacements of end-points of the springs. The movement is described by a Hamiltonian system with

$$H(x, \dot{x}) = \frac{1}{2}\sum_{i=1}^{n}(\dot{x}_{2i-1}^2 + \dot{x}_{2i}^2) + \frac{K}{2}\sum_{i=1}^{n}(x_{2i} - x_{2i-1})^2 + \sum_{i=0}^{n}(x_{2i+1} - x_{2i})^4.$$

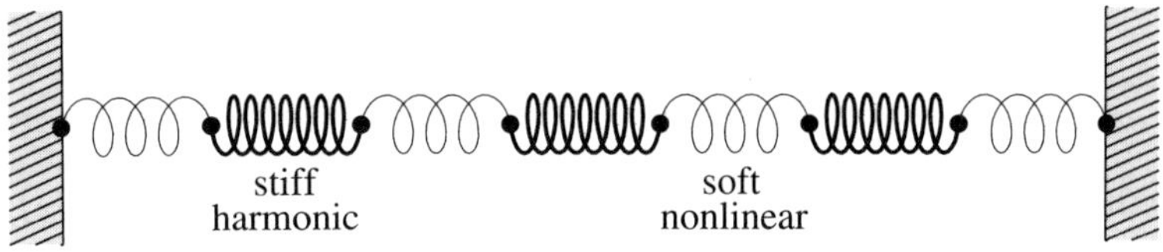

Figure 3.1: Alternating soft and stiff springs.

Using the symplectic change of variables $u_i = (x_{2i} + x_{2i-1})/\sqrt{2}$, $v_i = (x_{2i} - x_{2i-1})/\sqrt{2}$, we get a new Hamiltonian system with

$$H(u, v, \dot{u}, \dot{v}) = \frac{1}{2}\sum_{i=1}^{n}(\dot{u}_i^2 + \dot{v}_i^2) + \frac{\omega^2}{2}\sum_{i=1}^{n} v_i^2 + \frac{1}{4}\sum_{i=0}^{n}(u_{i+1} - v_{i+1} - u_i - v_i)^4, \tag{3.1}$$

where $u_0 = v_0 = u_{n+1} = v_{n+1} = 0$ and $\omega^2 = 2K$.

For our numerical experiments we consider the case $n = 3$ (as shown in Fig. 3.1) with different values of ω. As initial values we choose

$$u_1(0) = 1, \quad \dot{u}_1(0) = 1, \quad v_1(0) = \sqrt{2}/\omega, \tag{3.2}$$

and zero for the remaining initial values.

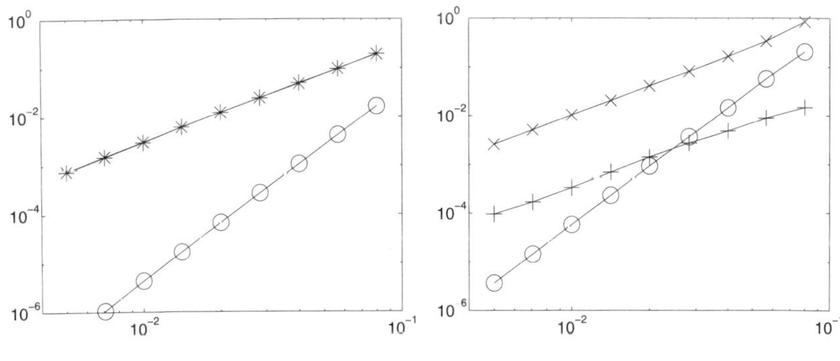

Figure 3.2: Maximum difference in total energy vs. step size. $\omega = 1$ and $\omega = 20$.

Fig. 3.2 and the first picture of Fig. 3.3 show the difference between the maximum and the minimum value of the total energy H along the numerical solution over the time interval $0 \leq t \leq 1000$, plotted as functions of the step size h. We show the results of the following numerical methods:

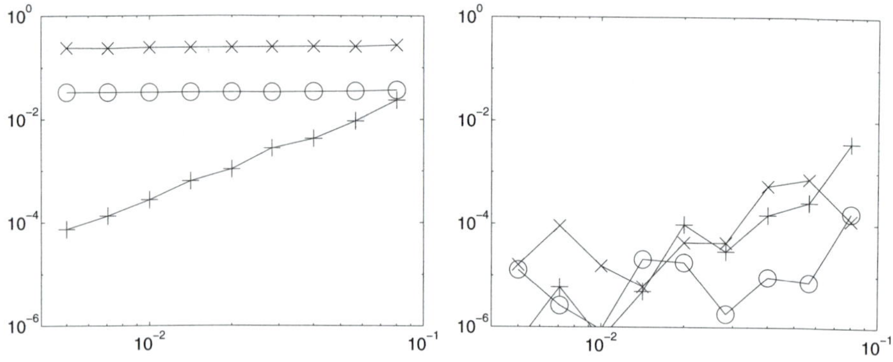

Figure 3.3: Maximum difference in total energy and difference of energy averages. $\omega = 1/h$.

- Störmer/Verlet method: marker $\times$
- trapezoidal rule: marker $+$
- Numerov method: marker $\circ$

The left-hand picture of Fig. 3.2 shows the nonstiff case $\omega = 1$, the right-hand picture gives the results for $\omega = 20$. In comparison, there is a clear deterioration for the larger ω in the Störmer/Verlet and Numerov methods, from which the trapezoidal rule does not suffer. The orders are as predicted by Theorem 2.1.

Fig. 3.3 shows the corresponding numerical results for the choice $\omega = 1/h$. Here only the energy deviation in the trapezoidal rule decreases with h, whereas it remains essentially constant for the other two methods. The right-hand picture in Fig. 3.3 makes clear that this is not due to a drift in the energy. It shows the difference between the time averages of H along the numerical solution on the intervals $[0, 10]$ and $[990, 1000]$. Here all three methods behave in a similar way.

The increasingly oscillatory behaviour of the total energy of the Störmer/Verlet method with growing values of $h\omega$ is illustrated in Fig. 3.4, which shows the energy H (or rather $H - 0.8$ for graphical reasons) as a function of time on the interval $[0, 100]$ with $\omega = 20$ and $h = 0.01$ (left) and $h = 0.04$ (right). One also observes that the average value of H is distinctly smaller for the larger step size. In addition, Fig. 3.4 shows the evolution of the oscillatory energy

$$I = I_1 + I_2 + I_3 \qquad \text{with} \qquad I_j(v_j, \dot v_j) = \tfrac{1}{2}\big(\dot v_j^2 + \omega^2 v_j^2\big).$$

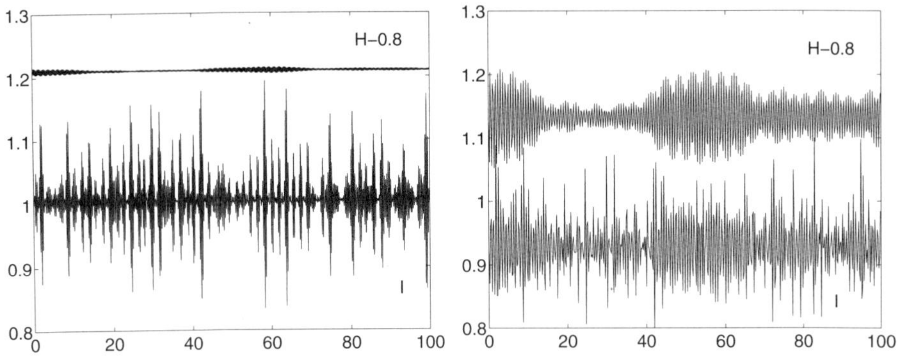

Figure 3.4: Total energy and oscillatory energy versus time. Störmer/Verlet $h = 0.01$ and $h = 0.04$, $\omega = 20$.

This quantity is an adiabatic invariant, known to remain constant up to oscillations of size $O(\omega^{-1})$ on time intervals that are exponentially long in ω, i.e., for $t \le e^{c\omega}$; see [BGG87, GGMV92, HaL99]. The numerical oscillatory energy also shows no drift. Comparing the two pictures, it is seen that the average value of I becomes smaller by about the same amount as that of the total energy H.

These numerical phenomena are explained in the following sections.

4 Energy behaviour for large step sizes

The equations of motion of the system (3.1) are of the form

$$\ddot{x} = -\Omega^2 x + g(x) \tag{4.1}$$

with smooth nonlinearity $g(x) = -\nabla V(x)$ and

$$\Omega = \begin{pmatrix} 0 & 0 \\ 0 & \omega I \end{pmatrix}, \qquad \omega \gg 1. \tag{4.2}$$

We are interested in the numerical behaviour for step sizes $h \ll 1$ with

$$h\omega \ge c > 0. \tag{4.3}$$

We always assume that $h\omega$ is such that the method (2.2) applied to the linear problem (4.1) with $g(x) \equiv 0$ is stable. That is, the linear recurrence relation

$$y_{n+1} - 2y_n + y_{n-1} = -h^2\omega^2 y_n - \alpha h^2\omega^2(y_{n+1} - 2y_n + y_{n-1})$$

has its characteristic roots on the unit circle. For $\alpha \geq 1/4$ this is the case without any restriction on h (unconditional stability), and for $\alpha < 1/4$ under the assumption

$$h\omega < \frac{2}{\sqrt{1-4\alpha}}.$$

The two characteristic roots are then $\exp(\pm ih\widetilde{\omega})$ with $h\widetilde{\omega}$ determined by the condition

$$\sin\tfrac{1}{2}h\widetilde{\omega} = \frac{\frac{1}{2}h\omega}{\sqrt{1+\alpha h^2\omega^2}}. \tag{4.4}$$

If we slightly strengthen the previous stability assumption to

$$h\omega \leq \frac{d}{\sqrt{1-4\alpha}} \qquad \text{with some} \quad d < 2$$

(and to $h\omega \leq const$ for $\alpha = 1/4$), this together with (4.3) implies the nonresonance condition

$$|\sin(\tfrac{1}{2}kh\widetilde{\omega})| \geq c > 0 \qquad \text{for} \quad k = 1, 2. \tag{4.5}$$

For the subsequent analysis we further need that

$$|\sin(\tfrac{1}{2}kh\widetilde{\omega})| \geq c\sqrt{h} > 0 \qquad \text{for} \quad k = 3, 4, \ldots, 2N. \tag{4.6}$$

Our aim is to explain the behaviour of the total and the oscillatory energies

$$\begin{aligned} H(x,\dot{x}) &= \tfrac{1}{2}|\dot{x}|^2 + \tfrac{1}{2}\omega^2|x_2|^2 + V(x) \\ I(x,\dot{x}) &= \tfrac{1}{2}|\dot{x}_2|^2 + \tfrac{1}{2}\omega^2|x_2|^2, \end{aligned} \tag{4.7}$$

where $x = (x_1, x_2)^T$ according to the partitioning of Ω. The oscillatory energy $I(x,\dot{x})$ still has $O(\omega^{-1})$ oscillations along the exact solution $x(t)$. The analysis of [BGG87] or of [HaL99] shows that these are of the form $x_2^T g_2(x) + O(\omega^{-2})$. We therefore introduce

$$J(x,\dot{x}) = I(x,\dot{x}) - x_2^T g_2(x), \tag{4.8}$$

which has only $O(\omega^{-2})$ oscillations along the exact solution on exponentially long time intervals.

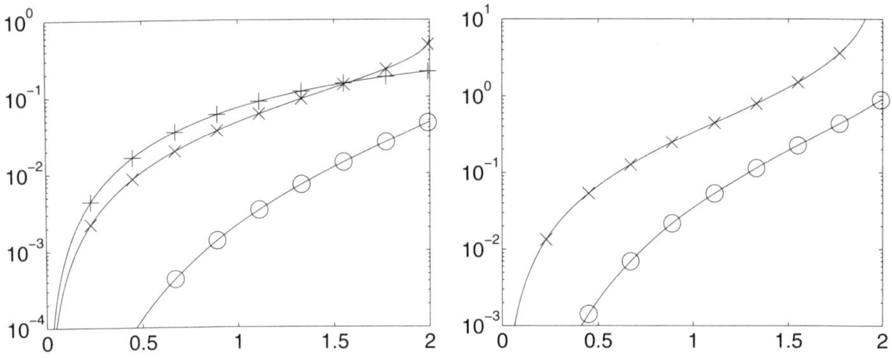

Figure 4.1: $|\widetilde{\omega}/\omega - 1|$ and γ vs. $h\omega$.

We show that the following modified energies are well conserved by the numerical scheme:

$$\begin{aligned} H^*(x,\dot{x}) &= H(x,\dot{x}) + \tfrac{1}{2}\gamma|\dot{x}_2|^2 \\ J^*(x,\dot{x}) &= J(x,\dot{x}) + \tfrac{1}{2}\gamma|\dot{x}_2|^2 \end{aligned} \tag{4.9}$$

where

$$\gamma = \frac{1}{1+\delta h^2\omega^2}\left(\frac{1+\alpha h^2\omega^2}{1+\beta h^2\omega^2}\right)^2 - 1 \qquad \text{with } \delta = \alpha^2 - (\alpha - \tfrac{1}{2})^2. \tag{4.10}$$

Note, $\gamma = 0$ for all values of $h\omega$ if and only if $\alpha = \beta = 1/4$, that is, for the trapezoidal rule. Fig. 4.1 plots the dispersion $|\widetilde{\omega}/\omega - 1|$ (left) and γ (right) as functions of $h\omega$ for the Störmer/Verlet method ($\times$), for the trapezoidal rule ($+$), and for the Numerov method ($\circ$).

Theorem 4.1 *Under the above conditions, the modified energies along the numerical solution satisfy*

$$\begin{aligned} H^*(x_n,\dot{x}_n) &= H^*(x_0,\dot{x}_0) + O(h^2) \\ J^*(x_n,\dot{x}_n) &= J^*(x_0,\dot{x}_0) + O(h^2) \end{aligned} \qquad \text{for } 0 \le nh \le h^{-N}. \tag{4.11}$$

The constants symbolized by the O-notation depend on N, on a bound of the initial energy, and on bounds of derivatives of g in a region that contains the numerical solution values (x_n), but they are independent of n and h, ω with (4.5) and (4.6), and independent of the smoothness of the solution.

Fig. 4.2 plots the difference between the maximum and minimum values of H^* and J^* along the numerical solution on the interval $0 \le t \le 1000$ versus the step size h, again for the example (3.1) with frequency $\omega = 1/h$. Compare with Fig. 3.3.

Remark. In [Rei99] it has been first suggested, based on the analysis of the linear problem $\ddot{x} = -\Omega^2 x$ and on numerical experiments, that the modified energies H^* and I^*, with $I^*(x, \dot{x}) = I(x, \dot{x}) + \frac{1}{2}\gamma|\dot{x}_2|^2$, might be well preserved over long times, even for problems with solution-dependent high frequency.

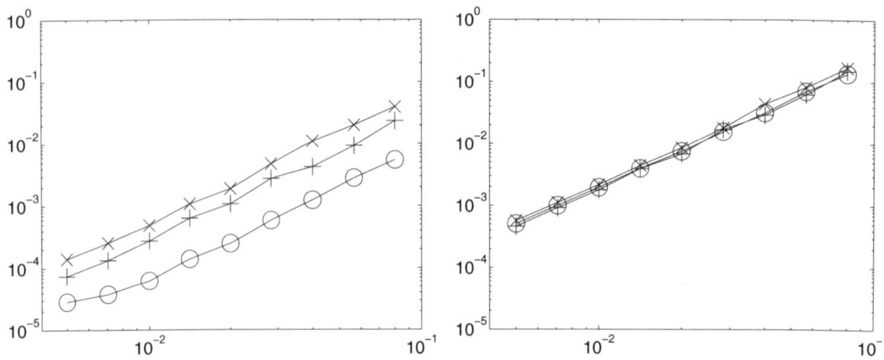

Figure 4.2: Maximum difference in H^* and J^* vs. step size, $\omega = 1/h$.

We now consider time averages of the energy over finite time intervals. Let $\varphi(t)$ be a nonnegative function with bounded support, say $[-T, T]$, whose derivative is of bounded variation. For example, take the hat function

$$\varphi(t) = 1 - |t|/T \quad \text{for } |t| \le T, \quad \varphi(t) = 0 \quad \text{else.}$$

Let its mass on the grid of width h be denoted by

$$\mu = \sum_{j=-\infty}^{\infty} \varphi(jh),$$

and consider the time averages of the total and the oscillatory energies with respect to the filter function φ:

$$\begin{aligned} \overline{H}_n &= \frac{1}{\mu} \sum_{j=-\infty}^{\infty} \varphi(jh)\, H(x_{n+j}, \dot{x}_{n+j}) \\ \overline{I}_n &= \frac{1}{\mu} \sum_{j=-\infty}^{\infty} \varphi(jh)\, I(x_{n+j}, \dot{x}_{n+j}). \end{aligned} \tag{4.12}$$

Theorem 4.2 *Under the above conditions, the time averages of the total and the oscillatory energies along the numerical solution satisfy*

$$
\begin{aligned}
\overline{H}_n &= H^*(x_n, \dot{x}_n) - \tfrac{1}{2} \frac{\gamma}{1+\gamma} J^*(x_n, \dot{x}_n) + O(h^2) \\
\overline{I}_n &= J^*(x_n, \dot{x}_n) - \tfrac{1}{2} \frac{\gamma}{1+\gamma} J^*(x_n, \dot{x}_n) + O(h^2),
\end{aligned}
\tag{4.13}
$$

and therefore

$$
\begin{aligned}
\overline{H}_n &= \overline{H}_0 + O(h^2) \\
\overline{I}_n &= \overline{I}_0 + O(h^2)
\end{aligned}
\qquad \textit{for } \; 0 \le nh \le h^{-N}.
\tag{4.14}
$$

The constants symbolized by the O-notation depend on the choice of φ in addition to the quantities of the previous theorem, but are again independent of n, h and ω.

Remark. If we choose $\varphi(t)$ as the characteristic function of the time window $[-T, T]$, we obtain $O(h)$ instead of the above $O(h^2)$ bounds. This is why we have chosen the formulation with a filter function $\varphi(t)$ as above.

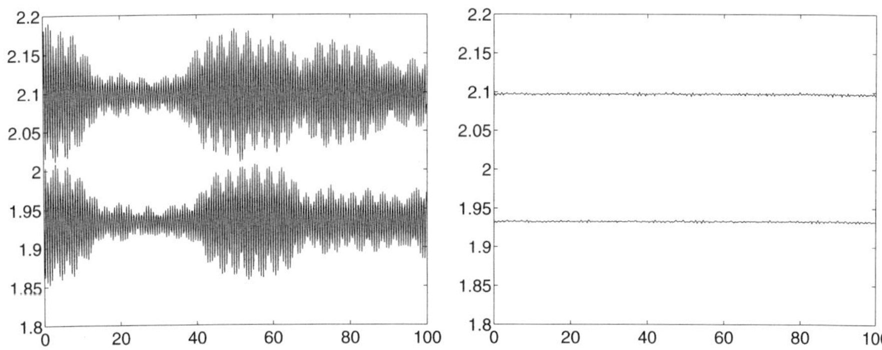

Figure 4.3: Total energies H and their predicted averages.

In the left-hand picture of Fig. 4.3 we plot the total energy versus time for the Störmer/Verlet method with $h = 0.04$, $\omega = 20$ and two different choices of initial values: one for the values (3.2), which have $v_1(0) = \sqrt{2}/\omega$, $\dot{v}_1(0) = 0$, and one for $v_1(0) = 0$, $\dot{v}_1(0) = \sqrt{2}$ instead. The total energies at the initial values are $H(x_0, \dot{x}_0) = 2.015$ and 2, respectively. In the right-hand picture we plot the quantities

$$
H^*(x, \dot{x}) - \tfrac{1}{2} \frac{\gamma}{1+\gamma} J^*(x, \dot{x})
$$

along the numerical solutions. They are almost constant and agree perfectly with the average values of the total energies, as predicted by Theorems 4.1 and 4.2.

The above two theorems therefore explain the numerical behaviour seen in Figs. 3.3 and 3.4. Their proofs rest on a reinterpretation of the methods (2.2) as almost belonging to a class of methods whose energy behaviour we recently studied in [HaL99]. Results of that article are reviewed in the following section.

5 Exponential integrators for oscillatory problems

In [HaL99] we studied energy conservation along the numerical solutions of (4.1) for a class of methods that become exact solvers for $g(x) \equiv 0$ and reduce to the Störmer/Verlet method for $\Omega = 0$:

$$\begin{aligned} x_{n+1} &= \cos h\Omega\, x_n + \Omega^{-1} \sin h\Omega\, \dot{x}_n + \tfrac{1}{2} h^2 \Psi\, g_n \\ \dot{x}_{n+1} &= -\Omega \sin h\Omega\, x_n + \cos h\Omega\, \dot{x}_n + \tfrac{1}{2} h \big(\Psi_0\, g_n + \Psi_1\, g_{n+1} \big) \end{aligned} \tag{5.1}$$

where $g_n = g(\Phi x_n)$ and $\Phi = \phi(h\Omega)$, $\Psi = \psi(h\Omega)$, $\Psi_0 = \psi_0(h\Omega)$, $\Psi_1 = \psi_1(h\Omega)$ with real functions $\phi(\xi)$, $\psi(\xi)$, $\psi_0(\xi)$, $\psi_1(\xi)$ depending smoothly on ξ^2 and satisfying $\phi(0) = \psi(0) = \psi_0(0) = \psi_1(0) = 1$.

The method is assumed to be symmetric, which holds if and only if

$$\psi(\xi) = \operatorname{sinc} \xi \cdot \psi_1(\xi), \qquad \psi_0(\xi) = \cos \xi \cdot \psi_1(\xi) \tag{5.2}$$

(where $\operatorname{sinc} \xi = \sin \xi / \xi$). Eliminating $\dot{x}_n$ yields the two-step recurrence

$$x_{n+1} - 2 \cos h\Omega\, x_n + x_{n-1} = h^2 \Psi\, g_n . \tag{5.3}$$

An approximation to the derivative is obtained by the second relation of (5.1) or, if $h\omega$ is not an integer multiple of π, equivalently by the formula

$$2h \operatorname{sinc} h\Omega\, \dot{x}_n = x_{n+1} - x_{n-1} . \tag{5.4}$$

Methods of the type (5.1) or (5.3) have been proposed by several authors for different choices of ψ and ϕ [Gau61, Deu79, GSS99, HoL99]. The papers [GSS99] and [HoL99] derive error bounds on bounded time intervals which are independent of ω and of the smoothness of the solution.

The choice of the functions ψ and ϕ becomes important when $h\omega$ is close to an integer multiple of π. Here we restate only a weaker version of the main result of [HaL99], which pertains to the nonresonant situation

$$\begin{aligned} |\sin(\tfrac{1}{2} k h\omega)| &\geq c > 0 && \text{for } k = 1, 2 \\ |\sin(\tfrac{1}{2} k h\omega)| &\geq c\sqrt{h} && \text{for } k = 3, 4, \ldots, 2N. \end{aligned} \tag{5.5}$$

Theorem 5.1 [HaL99] *Under the condition (5.5), the total and the oscillatory energies along the numerical solution of the method (5.1) satisfy*

$$\begin{aligned} H(x_n,\dot x_n) &= H(x_0,\dot x_0)+O(h) \\ I(x_n,\dot x_n) &= I(x_0,\dot x_0)+O(h) \end{aligned} \qquad \text{for } 0\le nh\le h^{-N}. \tag{5.6}$$

The constants symbolized by the O-notation depend on N, on a bound of the initial energy, and on bounds of derivatives of g in a region that contains the numerical solution values (x_n), but they are independent of n and h, ω with (5.5) and independent of the smoothness of the solution.

A key ingredient of the proof in [HaL99] is the *frequency expansion* of the numerical solution over bounded time intervals,

$$x_n = y(t) + \sum_{0<|k|\le N} e^{ik\omega t} z^k(t) + O(h^{N+2}), \qquad t=nh, \tag{5.7}$$

where the functions $y=(y_1,y_2)$ and $z^k=(z_1^k,z_2^k)$ together with all their derivatives are bounded by

$$\begin{aligned} y_1 &= O(1), & z_1^1 &= O(h\omega^{-1}), & z_1^k &= O(h\omega^{-k}), \\ y_2 &= O(\omega^{-2}), & z_2^1 &= O(\omega^{-1}), & z_2^k &= O(h\omega^{-k}) \end{aligned}$$

for $k=2,\ldots,N$, and

$$\omega^2 y_2 = \frac{\psi(h\omega)}{\mathrm{sinc}^2(\frac12 h\omega)}\, g_2(y) + O(h), \qquad \dot z_2^1 = O(h\omega^{-1}).$$

A similar expansion is valid for the derivative approximation $\dot x_n$. The energy conservation is then obtained by arguments of a type similar to those of the first proof of Theorem 2.1 applied to the system of equations that determines the coefficient functions of the frequency expansion.

Theorem 5.1 can be refined by carefully tracing the $O(\omega^{-1})$ terms in Section 5 of [HaL99]. In this way, using the above estimates of the frequency expansion coefficients, we find that the modified energies

$$\begin{aligned} \widehat H(x,\dot x) &= H(x,\dot x) - \rho\, x_2^T g_2(x) \\ \widehat J(x,\dot x) &= J(x,\dot x) - \rho\, x_2^T g_2(x), \end{aligned} \tag{5.8}$$

with

$$\rho = \frac{\psi(h\omega)}{\mathrm{sinc}^2(\frac12 h\omega)} - 1,$$

are better conserved than H and I. Note, $\rho = 0$ for the choice $\psi(\xi) = \mathrm{sinc}^2(\frac{1}{2}\xi)$ [Gau61, HoL99] which ensures also that the method (5.3) is exact for $g(x) = const.$

Theorem 5.2 *Under the condition (5.5), the modified energies (5.8) along the numerical solution satisfy*

$$\begin{aligned} \widehat{H}(x_n, \dot{x}_n) &= \widehat{H}(x_0, \dot{x}_0) + O(h^2) \\ \widehat{J}(x_n, \dot{x}_n) &= \widehat{J}(x_0, \dot{x}_0) + O(h^2) \end{aligned} \qquad \textit{for } 0 \le nh \le h^{-N}. \tag{5.9}$$

The constants symbolized by the O-notation are again independent of n and h, ω with (5.5) and independent of the smoothness of the solution.

Using Theorem 5.2 and the arguments of the proof of Theorem 4.2 in Part (c) of Section 6 below, we obtain that the $O(h^2)$ bounds (4.14) for the averages of the total, and the oscillatory energies are valid also for the method (5.1) under the nonresonance condition (5.5).

6 Proofs of Theorems 4.1 and 4.2

We give the proofs for the Störmer/Verlet method ($\alpha = \beta = 0$) and briefly discuss the other methods of the class (2.2) at the end of this section. The idea of the proof is to interpret the method as a method of the type of the previous section for a modified frequency.

(a) The Störmer/Verlet method applied to (4.1) can be rewritten as

$$x_{n+1} - 2\cos h\widetilde{\Omega}\, x_n + x_{n-1} = h^2 g(x_n),$$

where $\widetilde{\Omega} = \begin{pmatrix} 0 & 0 \\ 0 & \widetilde{\omega} I \end{pmatrix}$ with $\widetilde{\omega}$ defined by (4.4). Together with the modified velocities x'_n defined by

$$2h\,\mathrm{sinc}\, h\widetilde{\Omega}\; x'_n = x_{n+1} - x_{n-1}$$

this becomes a scheme of the type (5.3), (5.4) with $\widetilde{\omega}$ instead of ω and with $\psi(\xi) = \phi(\xi) = 1$. By Theorem 5.1, we then know that

$$\begin{aligned} \widetilde{H}(x_n, x'_n) &= \widetilde{H}(x_0, x'_0) + O(h) \\ \widetilde{I}(x_n, x'_n) &= \widetilde{I}(x_0, x'_0) + O(h) \end{aligned} \qquad \text{for } 0 \le nh \le h^{-N}, \tag{6.1}$$

where $\widetilde{H}$ and $\widetilde{I}$ are defined as in (4.7) with $\widetilde{\omega}$ in place of ω. The components of $\dot{x}_n$ and x'_n are related by (omitting the subscript n)

$$\dot{x}_1 = x'_1, \qquad \dot{x}_2 = \operatorname{sinc}(h\widetilde{\omega})\, x'_2 = \frac{\omega}{\widetilde{\omega}}\sqrt{1 - \tfrac{1}{4}h^2\omega^2}\, x'_2, \tag{6.2}$$

so that

$$\begin{aligned} \widetilde{I}(x,x') &= \tfrac{1}{2}|x'_2|^2 + \tfrac{1}{2}\widetilde{\omega}^2|x_2|^2 = \tfrac{1}{2}\frac{\widetilde{\omega}^2}{\omega^2}\frac{1}{1-\frac{1}{4}h^2\omega^2}\,|\dot{x}_2|^2 + \tfrac{1}{2}\frac{\widetilde{\omega}^2}{\omega^2}\,\omega^2|x_2|^2 \\ &= \frac{\widetilde{\omega}^2}{\omega^2}\, I^*(x,\dot{x}) \end{aligned} \tag{6.3}$$

with $I^*(x,\dot{x}) = I(x,\dot{x}) + \frac{1}{2}\gamma|\dot{x}_2|^2$. Similarly, we have

$$\begin{aligned} H^*(x,\dot{x}) &= \tfrac{1}{2}|\dot{x}_1|^2 + V(x) + I^*(x,\dot{x}) \\ &= \widetilde{H}(x,x') + \left(\frac{\omega^2}{\widetilde{\omega}^2} - 1\right)\widetilde{I}(x,x'), \end{aligned} \tag{6.4}$$

and hence (6.1) yields

$$\begin{aligned} H^*(x_n,\dot{x}_n) &= H^*(x_0,\dot{x}_0) + O(h) \\ I^*(x_n,\dot{x}_n) &= I^*(x_0,\dot{x}_0) + O(h) \end{aligned} \qquad \text{for } 0 \le nh \le h^{-N}. \tag{6.5}$$

(b) For the proof of the actually observed $O(h^2)$ estimates we need to take recourse to Theorem 5.2. By that theorem, the modified energies

$$\begin{aligned} \widehat{H}(x,x') &= \widetilde{H}(x,x') - \widetilde{\rho}\, x_2^T g_2(x) \\ \widehat{J}(x,x') &= \widetilde{J}(x,x') - \widetilde{\rho}\, x_2^T g_2(x), \end{aligned} \tag{6.6}$$

with $\widetilde{J}(x,x') = \widetilde{I}(x,x') - x_2^T g_2(x)$ and

$$\widetilde{\rho} = \frac{1}{\operatorname{sinc}\left(\frac{1}{2}h\widetilde{\omega}\right)^2} - 1 = \frac{\widetilde{\omega}^2}{\omega^2} - 1$$

by (4.4), satisfy tighter bounds than (6.1):

$$\begin{aligned} \widehat{H}(x_n,x'_n) &= \widehat{H}(x_0,x'_0) + O(h^2) \\ \widehat{J}(x_n,x'_n) &= \widehat{J}(x_0,x'_0) + O(h^2) \end{aligned} \qquad \text{for } 0 \le nh \le h^{-N}. \tag{6.7}$$

When we insert $\widetilde{H}(x,x')$ and $\widetilde{I}(x,x')$ from (6.6) into (6.4), the terms $x_2^T g_2(x)$ cancel, so that

$$H^*(x_n,\dot{x}_n) = \widehat{H}(x_n,x_n') + \left(\frac{\omega^2}{\widetilde{\omega}^2} - 1\right)\widehat{J}(x_n,x_n'). \tag{6.8}$$

In the same way we obtain from (6.3)

$$J^*(x_n,\dot{x}_n) = \frac{\omega^2}{\widetilde{\omega}^2}\,\widehat{J}(x_n,x_n'). \tag{6.9}$$

Hence, (6.7) gives the desired $O(h^2)$ bounds (4.11). This completes the proof of Theorem 4.1 for the Störmer/Verlet method.

(c) We now turn to the proof of Theorem 4.2. In view of Theorem 4.1, it is sufficient to prove (4.13), and to prove this for $n=0$. Consider the frequency expansions of x_n and x_n' for $t=nh$ in a bounded interval around 0. Theorem 5.2 of [HaL99] and its proof show that, under the nonresonance conditions (4.5) and (4.6),

$$x_{n2}' = i\widetilde{\omega}\big(v_2(t) - \overline{v}_2(t)\big) + O(h^2), \qquad t = nh,$$

where $v_2(t) = e^{i\widetilde{\omega}t} z_2^1(t)$ with $z_2^1(t)$ of the frequency expansion (5.7) (with $\widetilde{\omega}$ instead of ω). With (6.2) it follows that

$$\dot{x}_{n2} = i\omega\sqrt{1 - \tfrac{1}{4}h^2\omega^2}\,\big(v_2(t) - \overline{v}_2(t)\big) + O(h^2),$$

and therefore, recalling the definition (4.10) of γ,

$$|\dot{x}_{n2}|^2 = \omega^2 \frac{1}{1+\gamma}\left(2|z_2^1(t)|^2 - 2\,\mathrm{Re}\,e^{2i\widetilde{\omega}t} z_2^1(t)^2\right) + O(h^2). \tag{6.10}$$

Now, Theorem 6.2 of [HaL99] yields that

$$2\widetilde{\omega}^2 |z_2^1(t)|^2 = \widehat{J}(x_n,x_n') + O(h^2)$$

and hence, by (6.9),

$$2\omega^2 |z_2^1(t)|^2 = J^*(x_n,\dot{x}_n) + O(h^2).$$

With two partial summations, it follows that the average with respect to the filter function $\varphi(t)$ over the oscillatory terms $\omega^2 e^{2i\widetilde{\omega}t} z_2^1(t)^2$ of (6.10) is $O(h^2)$. This yields

$$\frac{1}{\mu}\sum_{j=-\infty}^{\infty} \varphi(jh)|\dot{x}_{j2}|^2 = \frac{1}{1+\gamma} J^*(x_0,\dot{x}_0) + O(h^2).$$

Together with (4.9) and (4.12), this gives the formula for $\overline{H}_0$ in (4.13) and

$$\overline{J}_0 = J^*(x_0, \dot{x}_0) - \tfrac{1}{2}\frac{\gamma}{1+\gamma} J^*(x_0, \dot{x}_0) + O(h^2), \tag{6.11}$$

where $\overline{J}_n$ is defined in the same way as $\overline{I}_n$ in (4.12), with J instead of I. Finally, the frequency expansion shows that the average over the oscillatory terms $x_{n2}^T g_2(x_n)$ appearing in the definition (4.8) of $J(x_n, \dot{x}_n)$ is $O(h^2)$. So we have

$$\overline{J}_0 = \overline{I}_0 + O(h^2).$$

This, together with equation (6.11), completes the proof of Theorem 4.2 for the Störmer/Verlet method.

(d) The implicit methods (2.2) can be similarly interpreted as belonging to a generalization of the class of methods of Section 5 to symmetric implicit methods. The analysis of [HaL99] can be extended to that class in a straightforward, though tedious way. The above arguments for the Störmer/Verlet method then carry over to the whole class (2.2). We omit the details of this extension.

References

[BGG87] G. Benettin, L. Galgani, A. Giorgilli, "Realization of holonomic constraints and freezing of high frequency degrees of freedom in the light of classical perturbation theory. Part I". Commun. Math. Phys. **113**, 87-103 (1987). "Part II". Commun. Math. Phys. **121**, 557-601 (1987).

[BG94] G. Benettin, A. Giorgilli, "On the Hamiltonian interpolation of near to the identity symplectic mappings with application to symplectic integration algorithms". J. Statist. Phys. **74**, 1117-1143 (1994).

[Deu79] P. Deuflhard, "A study of extrapolation methods based on multistep schemes without parasitic solutions". ZAMP **30**, 177-189 (1979).

[GGMV92] L. Galgani, A. Giorgilli, A. Martinoli, S. Vanzini, "On the problem of energy equipartition for large systems of the Fermi-Pasta-Ulam type: analytical and numerical estimates". Physica D **59**, 334-348 (1992).

[GSS99] B. García-Archilla, J.M. Sanz-Serna, R. Skeel, "Long-time-step methods for oscillatory differential equations". SIAM J. Sci. Comput. **20**, 930-963 (1999).

[Gau61] W. Gautschi, "Numerical integration of ordinary differential equations based on trigonometric polynomials". Numer. Math. **3**, 381-397 (1961).

[HaL97] E. Hairer, Ch. Lubich, "The life-span of backward error analysis for numerical integrators". Numer. Math. **76**, 441-462 (1997).

[HaL98] E. Hairer, Ch. Lubich, "Asymptotic expansions and backward analysis of numerical integrators". Report, 1998, to appear in *IMA Proceedings Dynamics of Algorithms,* Springer.

[HaL99] E. Hairer, Ch. Lubich, "Long-time energy conservation of numerical methods for oscillatory differential equations". Report, 1999, submitted to SIAM J. Numer. Anal.

[HoL99] M. Hochbruck, Ch. Lubich, "A Gautschi-type method for oscillatory second-order differential equations". Numer. Math. **83**, 403-426 (1999).

[New59] N.M. Newmark, "A method of computation for structural dynamics". ASCE J. Engrg. Mech. Division **85**, 67-94 (1959).

[Rei98] S. Reich, *Dynamical Systems, Numerical Integration, and Exponentially Small Estimates.* Habilitationsschrift, FU Berlin, 1998.

[Rei99] S. Reich, "Preservation of adiabatic invariants under symplectic discretization". Appl. Numer. Math. **29**, 45-55 (1999).

[SZS97] R.D. Skeel, G. Zhang, T. Schlick, "A family of symplectic integrators: stability, accuracy, and molecular dynamics applications". SIAM J. Sci. Comput. **18**, 203-222 (1997).

Ernst Hairer
Department de Mathématiques
Université de Genève
CH–1211 Genève 24
Switzerland
Ernst.Hairer@math.unige.ch
http://www.unige.ch/math/folks/hairer

Christian Lubich
Mathematisches Institut
Universität Tübingen
Auf der Morgenstelle 10
D–72076 Tübingen
Germany
lubich@na.uni-tuebingen.de
http://na.uni-tuebingen.de

A. Klawonn and O. B. Widlund

New results on FETI methods for elliptic problems with discontinuous coefficients

Abstract The FETI algorithms are among the most severely tested domain decomposition methods for elliptic partial differential equations. They are iterative substructuring methods and have many algorithmic components in common with the Neumann–Neumann domain decomposition methods but there are also considerable differences. The purpose of this chapter is to introduce some recent results and a new family of FETI algorithms. In particular, bounds on the rate of convergence, which are uniform with respect to the coefficients of a family of elliptic problems with heterogeneous coefficients, are presented for these new algorithms.

1 Introduction

The Finite Element Tearing and Interconnecting (FETI) methods were first introduced by Farhat and Roux [9]. An important advance, making the rate of convergence of the iteration less sensitive to the number of unknowns of the local problems, was made by Farhat, Mandel, and Roux a few years later [7]. Our own work is based on the pioneering work by Mandel and Tezaur [18], who fully analyzed a variant of that algorithm. For a detailed introduction, see [8] or [26].

The principal purpose of this chapter is to survey some recent results developed by the authors; cf. [15, 14, 16]. We introduce a new one parameter family of FETI preconditioners and show that the rate of convergence is bounded independently of possible jumps of the coefficients of an elliptic model problem previously considered in the theory of Neumann–Neumann and other iterative substructuring algorithms; see [4, 3, 17, 24, 25]. In fact, we have found it possible, see [16], to reduce the analytic core of the theory for the new class of FETI methods to a variant of an estimate which is central in the Neumann–Neumann theory. Thus, we write an arbitrary element in a product space of traces of finite element functions as the sum of two terms. One of them is central in the FETI theory, the other in the Neumann–Neumann theory. The norm of each of the two terms is bounded by a factor $C(1 + \log(H/h))$ times that of the given function. Here, and from now on, C is a generic constant, which may depend on the aspect ratios of the elements and subregions, but which is independent of the mesh parameters h and H and the coefficients of the elliptic problem; h is the diameter of a typical element into which the subregions have been divided and H is the typical diameter of a subregion. We note that $(H/h)^d, d = 2, 3,$ measures the number of degrees of freedom associated with a subregion.

The result for the new family of FETI algorithms becomes possible because of two special scalings. One of them, for the preconditioner, is closely related to an important algorithmic idea used in the best of the Neumann–Neumann methods. A proof of one of the two spectral bounds that are required is then just as elementary as for the Neumann–Neumann case. The other scaling affects the choice of the projection which is used in each step of the FETI iteration, whether preconditioned or not.

For a certain choice of the two scalings, our preconditioner results in a method that is identical to one recently tested successfully by Rixen and Farhat [22]; see also [1] in which the class of algorithms is extended and the methods further tested for very difficult and large problems. Our algorithms are also defined for this class of problems, but in our analysis we have to impose certain restrictions on the coefficients and on the geometry of the subregions. We note that, by now, many variants of the FETI algorithms have been designed and that a number of them have been tested extensively; see in particular the discussion in [23]. Our results have also already been extended to Maxwell's equation in two dimensions by Toselli and Klawonn [27].

In the final section of this chapter, we discuss a FETI algorithm with inexact subdomain solvers; see Klawonn and Widlund [15]. Here, we are able to improve the condition number estimate given in [15], using the results from [16]. We note that the FETI algorithms as previously developed require the use of exact solvers for certain subproblems directly related to the subdomains into which the domain of the given problem has been split. Our extension has required a redesign of the algorithm; the purpose is to avoid the use of potentially quite costly direct solvers replacing them by any well tested preconditioner, for the positive definite subproblems, such as an incomplete LU method, (algebraic) multigrid, etc. We are able to maintain the good features of the FETI method such as scalability and efficiency while decreasing the computer resources required.

The remainder of this chapter is organized as follows. In Section 2, we introduce our elliptic problems and the basic geometry of the decomposition; we have chosen to work only with the more interesting three dimensional case. In Section 3, we give a short introduction to the FETI method. In Section

4, we introduce our family of preconditioners and formulate our main result; our results could also be extended to certain other elliptic problems as in [15]. A connection between one element of our family of preconditioners and the method, recently developed by Rixen and Farhat [22] and by [1], is established in Section 5. In Section 6, we summarize the algorithm developed in [15] and give an improved condition number estimate using the new results proven in [16].

2 Elliptic model problem, finite elements, and geometry

Let $\Omega \subset \mathbf{R}^3$ be a bounded, polyhedral region, let $\partial\Omega_D \subset \partial\Omega$ be a closed set of positive measure, and let $\partial\Omega_N := \partial\Omega \setminus \partial\Omega_D$ be its complement. We impose Dirichlet and Neumann boundary conditions, respectively, on these two subsets and introduce the standard Sobolev space $H_0^1(\Omega, \partial\Omega_D) := \{v \in H^1(\Omega) : v = 0 \text{ on } \partial\Omega_D\}$.

For simplicity, we will only consider a first order, conforming finite element approximation of the following scalar, second order model problem:

Find $u \in H_0^1(\Omega, \partial\Omega_D)$, such that

$$a(u, v) = f(v) \quad \forall v \in H_0^1(\Omega, \partial\Omega_D), \tag{2.1}$$

where

$$a(u, v) := \int_\Omega \rho(x) \nabla u \cdot \nabla v dx, \quad f(v) := \int_\Omega f v dx. \tag{2.2}$$

Here $\rho(x) > 0$ for $x \in \Omega$. For simplicity, we choose zero Neumann boundary data on $\partial\Omega_N$.

We decompose Ω into non-overlapping subdomains $\Omega_i, i = 1, \ldots, N$, also known as substructures, and each of which is the union of shape-regular elements with the finite element nodes on the boundaries of neighboring subdomains matching across the interface $\Gamma := \left(\bigcup_{i=1}^N \partial\Omega_i\right) \setminus \partial\Omega$. We denote a standard finite element space of continuous, piecewise linear functions on Ω_i by $W^h(\Omega_i)$. For simplicity, we assume that the triangulation of each subdomain is quasi uniform. The diameter of Ω_i is H_i, or generically, H. We denote the corresponding finite element trace spaces by $W_i := W^h(\partial\Omega_i), i = 1, \ldots, N$, and by $W := \prod_{i=1}^N W_i$ the associated product space. We note that we will often consider elements of W which are discontinuous across the interface. The finite element approximation of the elliptic problem is continuous across the interface and we denote the corresponding subspace of W by $\widehat{W}$. We note that all the iterates of the Neumann–Neumann methods belong to $\widehat{W}$ while those of the FETI methods normally do not.

We assume that possible jumps of $\rho(x)$ are aligned with the subdomain boundaries and, for simplicity, that on each subregion Ω_i, $\rho(x)$ has the constant value ρ_i. Our bilinear form and load vector can then be written, in terms of contributions from individual subregions, as

$$a(u,v) = \sum_{i=1}^{N} \rho_i \int_{\Omega_i} \nabla u \cdot \nabla v dx, \quad f(v) = \sum_{i=1}^{N} \int_{\Omega_i} f v dx. \tag{2.3}$$

In our theoretical analysis, we assume that the subregions, Ω_i, are tetrahedra or parallelepipeds and that they are shape regular, i.e., not very thin. We also assume that if a face of a subdomain intersects $\partial\Omega_D$, then the measure of this set is comparable to that of the face. Similarly, if an edge of a subdomain intersects $\partial\Omega_D$, we assume that the length of this intersection is bounded from below in terms of the length of the edge as a whole. For the FETI methods and the case of arbitrary coefficients, ρ_i, we also have to make a further assumption on the geometry of the decomposition.

The sets of nodes in Ω_i, on $\partial\Omega_i$, and on Γ are denoted by $\Omega_{i,h}, \partial\Omega_{i,h}$, and Γ_h, respectively.

As in previous work on Neumann–Neumann algorithms, a crucial role is played by *the weighted counting functions* $\mu_i \in \widehat{W}$, which are associated with the individual $\partial\Omega_i$; cf. [2, 4, 17, 25]. They are defined, by a sum of contributions from Ω_i, and its relevant next neighbors, for $\gamma \in [1/2, \infty)$, and for $x \in \Gamma_h \cup \partial\Omega_h$ by

$$\mu_i(x) = \begin{cases} \sum_{j \in \mathcal{N}_x} \rho_j^{\gamma}(x) & x \in \partial\Omega_{i,h} \cap \partial\Omega_{j,h}, \\ \rho_i^{\gamma}(x) & x \in \partial\Omega_{i,h} \cap \partial\Omega_h, \\ 0 & x \in \Gamma_h \setminus \partial\Omega_{i,h}. \end{cases} \tag{2.4}$$

Here, $\mathcal{N}_x$ is the set of indices of the subregions which have x on its boundary.

The pseudo inverses $\mu_i^{\dagger} \in \widehat{W}$ are defined, for $x \in \Gamma_h \cup \partial\Omega_h$, by

$$\mu_i^{\dagger}(x) = \begin{cases} \mu_i^{-1}(x) & \text{if } \mu_i(x) \neq 0, \\ 0 & \text{if } \mu_i(x) = 0. \end{cases}$$

We note that these functions provide a partition of unity:

$$\sum_i \rho_i^{\gamma}(x) \mu_i^{\dagger}(x) \equiv 1 \quad \forall x \in \Gamma_h \cup \partial\Omega_h. \tag{2.5}$$

3 A review of the FETI method

In this section, we give a brief review of the original FETI method of Farhat and Roux [9, 8] and the variant with a Dirichlet preconditioner introduced in Farhat, Mandel, and Roux [7]. The more general projection operators, described in this section, were first introduced for heterogeneous problems in [8]; such methods have recently been tested in very large scale numerical experiments; see [1]. For a more detailed description and extensions beyond scalar elliptic problems, see [5, 6, 19, 21, 26]. Let us point out that there are also other variants of the FETI methods; see, e.g., Park, Justino, and Felippa [20]. The relation of one of them to the FETI method developed by Farhat and Roux is discussed in [23]. A convergence analysis of the method given in [20] can be found in Tezaur's dissertation [26].

For a chosen finite element method and for each subdomain Ω_i, we assemble the local stiffness matrix $K^{(i)}$ and the local load vector corresponding to a single, appropriate term in the sums of (2.3). Any nodal variable, not associated with Γ_h, is called interior and it only belongs to one substructure. The interior variables of any subdomain can be eliminated by a step of block Gaussian elimination; this work can clearly be parallelized across the subdomains. The resulting matrices are the Schur complements

$$S^{(i)} = K^{(i)}_{\Gamma\Gamma} - K^{(i)}_{\Gamma I}(K^{(i)}_{II})^{-1}K^{(i)}_{I\Gamma}, \quad i = 1, \ldots, N.$$

Here, Γ and I represent the interface and interior, respectively. We note that the $S^{(i)}$ are only needed in terms of matrix-vector products and that therefore the elements of these matrices need not be explicitly computed.

The elimination of the interior variables of a substructure can also be viewed in terms of an orthogonal projection, with respect to the bilinear form $\langle K^{(i)}\cdot, \cdot\rangle$, onto the subspace of vectors with components that vanish at all the nodes of $\partial\Omega_{i,h}$. Here $\langle\cdot,\cdot\rangle$ denotes the ℓ_2–inner product. We note that these vectors represent elements of $H_0^1(\Omega_i) \cap W^h(\Omega_i)$. These local subspaces are orthogonal, in this inner product, to the space of discrete harmonic vectors which represent discrete harmonic finite element functions. With v_Γ the vector of boundary values, they are defined, on the subdomain Ω_i, by

$$\langle K^{(i)}w, v\rangle = 0 \quad \forall v \text{ such that } v_\Gamma = 0, \tag{3.1}$$

or, equivalently, by

$$K^{(i)}_{II}w_I + K^{(i)}_{I\Gamma}w_\Gamma = 0. \tag{3.2}$$

Here (w_I, w_Γ) is the vector of nodal values of $w \in W^h(\Omega_i)$. We can regard w_Γ as a vector of Dirichlet data given on $\partial\Omega_{i,h}$ and note that a piecewise discrete harmonic function is completely defined by its values on the interface. In what follows, we will almost exclusively work with functions in the trace spaces W_i and, whenever convenient, consider such an element as representing a discrete harmonic function in Ω_i. We will denote the discrete harmonic extension of a function w_i in the trace space W_i to the interior of Ω_i by $\mathcal{H}_i(w_i)$. We then, obviously, have the identity $|w_i|^2_{S^{(i)}} = \rho_i |\mathcal{H}_i(w_i)|^2_{H^1(\Omega_i)}$. For $w \in W$, $\mathcal{H}(w)$ denotes the piecewise discrete harmonic extension into all the Ω_i. We also note that it is the piecewise discrete harmonic part of the solution, representing an element of $\widehat{W}$, that is determined by any iterative substructuring method; the other, interior, parts of the solution are computed locally as indicated above.

The values of the right hand vectors also change when the interior variables are eliminated. We denote the resulting vectors, representing the modified load originating in Ω_i, by f_i and the local vectors of interface nodal values by u_i.

We can now reformulate the finite element problem, reduced to the interface Γ, as a minimization problem with constraints given by the requirement of continuity across Γ :

Find $u \in W$, such that

$$\left.\begin{array}{c} J(u) := \frac{1}{2}\langle Su, u\rangle - \langle f, u\rangle \rightarrow \min \\ Bu = 0 \end{array}\right\} \tag{3.3}$$

where

$$u = \begin{bmatrix} u_1 \\ \vdots \\ u_N \end{bmatrix}, \quad f = \begin{bmatrix} f_1 \\ \vdots \\ f_N \end{bmatrix}, \quad \text{and } S = \begin{bmatrix} S^{(1)} & O & \cdots & O \\ O & S^{(2)} & \ddots & \vdots \\ \vdots & \ddots & \ddots & O \\ O & \cdots & O & S^{(N)} \end{bmatrix}.$$

The matrix $B = [B^{(1)}, \ldots, B^{(N)}]$ is constructed from $\{0, 1, -1\}$ such that the values of the solution u, associated with more than one subdomain, coincide when $Bu = 0$. We note that the choice of B is far from unique. The local Schur complements $S^{(i)}$ are positive semi–definite and they are singular for any subregion with a boundary which does not intersect $\partial\Omega_D$. The problem (3.3) is uniquely solvable if and only if $ker\,(S) \cap ker\,(B) = \{0\}$, i.e., if and only if S is invertible on $ker\,(B)$.

By introducing a vector of Lagrange multipliers λ, to enforce the constraints $Bu = 0$, we obtain a saddle point formulation of (3.3):

Find $(u, \lambda) \in W \times U$, such that

$$\left.\begin{array}{rcl} Su + B^t\lambda & = & f \\ Bu & = & 0 \end{array}\right\}. \tag{3.4}$$

We note that the solution λ of (3.4) is unique only up to an additive vector of $ker\,(B^t)$. The space of Lagrange multipliers U is therefore chosen as $range\,(B)$.

We will also use a full column rank matrix built from all of the null space elements of S; these elements are associated with individual subdomains (the rigid body motions in the case of elasticity),

$$R = \begin{bmatrix} R^{(1)} & O & \cdots & O \\ O & R^{(2)} & \ddots & \vdots \\ \vdots & \ddots & \ddots & O \\ O & \cdots & O & R^{(N)} \end{bmatrix}.$$

Thus, $range\,(R) = ker\,(S)$. We note that any subdomain of which the boundary intersects $\partial\Omega_D$ will not contribute to R.

The solution of the first equation in (3.4) exists if and only if $f - B^t\lambda \in range\,(S)$; this constraint will lead to the introduction of a projection P. We obtain,

$$u = S^\dagger(f - B^t\lambda) + R\alpha \text{ if } f - B^t\lambda \perp ker\,(S),$$

where $S^\dagger$ is a pseudoinverse of S. The value of α can be determined easily once λ has been found.

Substituting u into the second equation of (3.4) gives

$$BS^\dagger B^t\lambda = BS^\dagger f + BR\alpha. \tag{3.5}$$

We now introduce a symmetric, positive definite matrix Q which induces an inner product on U; it is defined by $\langle\lambda, \mu\rangle_Q := \langle\lambda, Q\mu\rangle$. By considering the component Q^{-1}–orthogonal to $G := BR$, we find that

$$\left.\begin{array}{rcl} P^tF\lambda & = & P^td \\ G^t\lambda & = & e \end{array}\right\} \tag{3.6}$$

with $F := BS^\dagger B^t, d := BS^\dagger f, P := I - QG(G^tQG)^{-1}G^t$, and $e := R^tf$. We note that P is an orthogonal projection, from U onto $ker\,(G^t)$, in the Q^{-1}–inner product, i.e., the inner product defined by $\langle\lambda, Q^{-1}\mu\rangle$.

There are different choices for Q. In the case of homogeneous coefficients, it is sufficient to use $Q = I$, while for problems with jumps in the coefficients, we have to make a more elaborate choice to make our proofs work satisfactorily. In our analysis, Q will be a diagonal scaling matrix or we will use the preconditioner; other alternatives are discussed in [1, 8].

Multiplying (3.5) by $(G^tQG)^{-1}G^tQ$, we find that $\alpha := (G^tQG)^{-1}G^tQ(F\lambda - d)$ which then fully determines the primal variables in terms of λ.

We introduce the spaces

$$V := \{\mu \in U \,:\, \langle \mu, Bz\rangle = 0 \quad \forall z \in ker\,(S)\} = ker\,(G^t) = range\,(P),$$

and

$$V' := \{\lambda \in U : \langle \lambda, Bz\rangle_Q = 0 \quad \forall z \in ker\,(S)\} = range\,(P^t).$$

It can be easily shown that V' is isomorphic to the dual space of V. Following Farhat, Chen, and Mandel [5], we call V the space of admissible increments. The original FETI method is a conjugate gradient method in the space V applied to

$$P^tF\lambda = P^td, \qquad \lambda \in \lambda_0 + V, \tag{3.7}$$

with an initial approximation λ_0 chosen such that $G^t\lambda_0 = e$. The most basic FETI preconditioner, as introduced in Farhat, Mandel, and Roux [7], is of the form

$$M^{-1} := BSB^t.$$

To apply M^{-1} to a vector, N independent Dirichlet problems have to be solved, one on each subregion; it is therefore called the Dirichlet preconditioner.

To keep the search directions of this preconditioned conjugate gradient method in the space V, the application of the preconditioner M^{-1} is followed by an application of the projection P. Hence, the Dirichlet variant of the FETI method is the conjugate gradient algorithm applied to the equation

$$PM^{-1}P^tF\,\lambda = PM^{-1}P^t\,d, \qquad \lambda \in \lambda_0 + V. \tag{3.8}$$

We note that for $\lambda \in V$, $PM^{-1}P^tF\lambda = PM^{-1}P^tP^tFP\lambda$, and we can therefore view the operator on the left hand side of (3.8) as the product of two symmetric matrices.

It is well known that an appropriate norm of the iteration error of the conjugate gradient method will decrease at least by a factor

$$2(\frac{\sqrt{\kappa}-1}{\sqrt{\kappa}+1})^k,$$

in k steps; cf., e.g., [10]. Here κ is the ratio of the largest and smallest eigenvalues of the iteration operator. The main task in the theory is therefore always to obtain a good bound for the condition number κ.

We note that several different possibilities of improving the FETI preconditioner M^{-1} already have been explored. Some interesting variants are discussed by Rixen and Farhat [22], in a framework of mechanically consistent preconditioners, in the case of redundant Lagrange multipliers; see also Section 5. A new family of improved FETI preconditioners, with non–redundant Lagrange multipliers, is introduced and analyzed in Section 4.

4 New FETI methods with non–redundant Lagrange multipliers

In this section, we present and outline some of our results on a family of new FETI preconditioners with an improved condition number estimate compared to that of Mandel and Tezaur [18]; the bound in their paper involves three powers of $(1 + \log(H/h))$, in the general case, ours only two. In addition, we obtain a uniform bound for arbitrary positive values of the ρ_i if the scaling matrix Q, which enters the definition of P, is chosen carefully. In our proofs, we use a number of arguments developed in [18], but our presentation also differs considerably in several respects. We remark that for the FETI method described in Park, Justino, and Felippa [20] and for the case of continuous coefficients, a bound involving only two powers of $(1 + \log(H/h))$ is given in Tezaur [26].

We now assume, for the rest of this section, that B has full row rank, i.e., the constraints are linearly independent and there are no redundant Lagrange multipliers.

Our new preconditioner is defined, for any diagonal matrix D with positive elements, as

$$\widehat{M}^{-1} := (BD^{-1}B^t)^{-1}BD^{-1}SD^{-1}B^t(BD^{-1}B^t)^{-1}.$$

To obtain a method, which converges at a rate that is independent of the coefficient jumps, we now choose a special family of matrices D; a careful choice of the scaling Q, introduced in the definition of the operator P, will also be required. As in previous work on Neumann–Neumann algorithms, a crucial role is played by the weighted counting functions μ_i, associated with the individual $\partial\Omega_i$, and already introduced in (2.4) in Section 2. The diagonal matrix $D^{(i)}$ has the diagonal entry $\rho_i^\gamma(x)\mu_i^\dagger(x)$ corresponding to the point $x \in \partial\Omega_{i,h}$. Finally, we set

$$D := \begin{bmatrix} D^{(1)} & O & \cdots & O \\ O & D^{(2)} & \ddots & \vdots \\ \vdots & \ddots & \ddots & O \\ O & \cdots & O & D^{(N)} \end{bmatrix}.$$

We note that this matrix operates on elements in the product space W.

An important role will be played by $P_D := D^{-1}B^t(BD^{-1}B^t)^{-1}B$; this is a projection which is orthogonal in the scaled ℓ_2–inner product x^tDy, where $x, y \in W$. In our analysis of the new preconditioner, we equip V' with the following norm

$$||\mu||^2_{V'} := |D^{-1}B^t(BD^{-1}B^t)^{-1}\mu|^2_S = \langle \widehat{M}^{-1}\mu, \mu\rangle,$$

where $|w|_S := \sqrt{\langle Sw, w\rangle}$ is the semi–norm on the space W induced by the Schur complement S. Since S is only semi–definite, we have to establish that $||\cdot||_{V'}$ in fact is a norm on V'; this is done in [16, Lemma 1].

We also equip the space of admissible increments V with the norm

$$||\lambda||_V := \sup_{\mu\in V'} \frac{\langle\lambda, \mu\rangle}{||\mu||_{V'}}.$$

We note that V' is isomorphic to the dual space of V. Since

$$||\mu||^2_{V'} = \langle \widehat{M}^{-1}\mu, \mu\rangle \quad \mu \in V', \tag{4.1}$$

we find by a simple computation that

$$\langle \widehat{M}\lambda, \lambda\rangle = ||\lambda||^2_V \quad \lambda \in V. \tag{4.2}$$

Throughout the rest of this section, we will add an extra assumption on the geometry of the subregions; this extra assumption is not necessary in the Neumann–Neumann theory; cf. Dryja and Widlund [4].

Assumption 1 *There are no subregions Ω_i that share only one or a few nodes with the set $\partial\Omega_{D,h}$.*

We now give a condition number estimate for the preconditioned FETI operator $P\widehat{M}^{-1}P^tF$; cf. [16]. The result holds for $Q = \widehat{M}^{-1}$ and also for a recipe which involves a special choice of B and a special diagonal Q.

Theorem 1 *The condition number of the FETI method, with the new preconditioner* $\widehat{M}$, *satisfies*

$$\kappa(P\widehat{M}^{-1}P^tF) \leq C\,(1+\log(H/h))^2.$$

Here, $\kappa(P\widehat{M}^{-1}P^tF)$ *is the spectral condition number of* $P\widehat{M}^{-1}P^tF$, *and* C *is independent of* h, H, γ, *and the values of the* ρ_i.

5 FETI with redundant Lagrange multipliers

In this section, we extend our discussion to the case of redundant Lagrange multipliers. For a detailed algorithmic description of FETI preconditioners in this case, with $\gamma = 1$, together with an analysis based on mechanics, see Rixen and Farhat [21, 22].

Following Rixen and Farhat, we consider the case where a maximum number of redundant Lagrange multipliers are introduced, i.e., when every possible pair of degrees of freedom of the primal variables u, that belongs to the same nodal point $x \in \Gamma_h$, is connected by a Lagrange multiplier. Any crosspoint, where at least three subregions meet, will then contribute at least one additional Lagrange multiplier in comparison with the non–redundant case. An illustration for a crosspoint with four subregions is given in Figure 5.1.

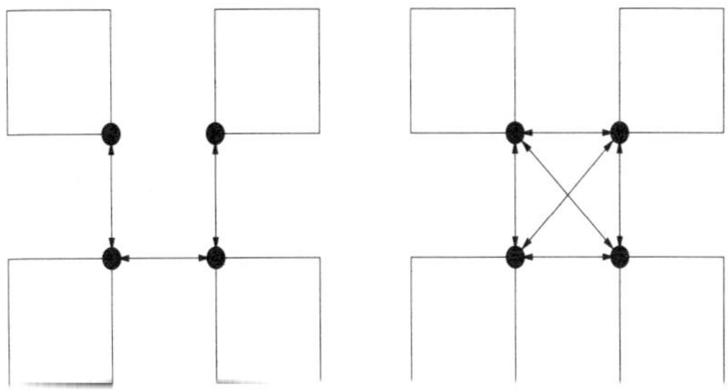

Figure 5.1: Left: U-shaped distribution of Lagrange multipliers at a crosspoint in a non–redundant case. Right: Distribution of Lagrange multipliers at a crosspoint in the redundant case.

We denote the new vector of Lagrange multipliers by λ_r. Similarly, we obtain a jump operator B_r with additional rows. We also introduce scaling matrices $D_r^{(i)}$, that operate on the Lagrange multiplier space, as follows: Consider, for a point $x \in \partial\Omega_{i,h} \cap \partial\Omega_{j,h}$, the Lagrange multiplier connecting the corresponding two degrees of freedom in W_i and W_j, respectively. Then the diagonal entry of $D_r^{(i)}$ for that point is chosen as $\rho_j^\gamma(x)\mu_j^\dagger(x)$.

We note that $D_r := diag_{i=1}^N(D_r^{(i)})$ is a mapping from the Lagrange multiplier space onto itself, in contrast to the matrix D of the non–redundant case, discussed in Section 4, which maps the space of primal variables W onto itself. We note that in the special case of continuous coefficients, we obtain the multiplicity scaling described in [22, Section 3]. Finally, we define a scaled jump operator by

$$B_{D_r} := [D_r^{(1)}B_r^{(1)}, \dots, D_r^{(N)}B_r^{(N)}],$$

and the FETI preconditioner by

$$\widehat{M}_r^{-1} := \sum_i D_r^{(i)}B_r^{(i)}S^{(i)}B_r^{(i)t}D_r^{(i)} = B_{D_r}SB_{D_r}^t.$$

This preconditioner, with $\gamma = 1$, was introduced in Rixen and Farhat [22, Section 5] in a framework of mechanically consistent preconditioners.

The matrix of the reduced linear system can be written as

$$F_r := B_r S^\dagger B_r^t.$$

Thus, we now have to solve the preconditioned system

$$P_r\widehat{M}_r^{-1}P_r^tF_r\lambda_r = P_r\widehat{M}_r^{-1}P_r^td_r, \tag{5.1}$$

with $P_r := I - Q_rG_r(G_r^tQ_rG_r)^{-1}G_r^t, G_r := B_rR$, and $d_r := B_rS^\dagger f$. Here, Q_r is again a properly chosen symmetric positive definite matrix. As before, we denote the inner product induced by Q_r by $\langle\lambda_r, \mu_r\rangle_{Q_r}$.

For the next theorem, see Klawonn and Widlund [16].

Theorem 2 *The condition number of the FETI method defined by $\widehat{M}_r$ satisfies*

$$\kappa(P_r\widehat{M}_r^{-1}P_r^tF_r) \le C\,(1 + \log(H/h))^2.$$

Here C is independent of h, H, γ, and the values of the ρ_i.

6 A FETI method with approximate subdomain solvers

In this section, we have to reinterpret some of our notations since, in particular, we are going to work with the entire stiffness matrix K and not just the Schur complement S; this is a routine matter and no details are provided.

We first note that using the decomposition $\lambda = \lambda_0 + \mu$, with $\mu \in V$, we can rewrite (3.7) as

$$PBK^{\dagger}B^t\mu = PBK^{\dagger}(f - B^t\lambda_0). \tag{6.1}$$

Since $u = K^{\dagger}(f - B^t\lambda) + R\underline{\alpha}$, we see immediately that the solution of (6.1) can also be obtained by solving

$$\begin{bmatrix} K & B^t \\ PB & O \end{bmatrix}\begin{bmatrix} u \\ \mu \end{bmatrix} = \begin{bmatrix} f - B^t\lambda_0 \\ 0 \end{bmatrix}.$$

Using that $\mu \in V$, i.e., $P\mu = \mu$, we can make the system matrix symmetric

$$\begin{bmatrix} K & (PB)^t \\ PB & O \end{bmatrix}\begin{bmatrix} u \\ \mu \end{bmatrix} = \begin{bmatrix} f - B^t\lambda_0 \\ 0 \end{bmatrix}. \tag{6.2}$$

We note that in this formulation we are not enforcing $Bu = 0$ but only its projected version $PBu = 0$. The addition of an element of $ker\,(K)$ does not change the solution u of the first equation in (6.2). Since $B^t\mu \perp ker\,(K)$ for $\mu \in V$, this is also true for the second equation in (6.2). We use this fact to post-process u, such that $Bu = 0$ is finally satisfied. This can be done by setting $u_{cor} := u - R(G^tG)^{-1}G^tBu$; it easily follows that $Bu_{cor} = PBu = 0$. Thus, we first compute the component of the solution in *range* (K) and then add the correct null space component, such that the solution has no jumps across the interface.

For the solution of the saddle point problem (6.2), we propose a preconditioned conjugate residual method with a block–diagonal preconditioner. For a detailed description of this algorithm, see Hackbusch [11] or Klawonn [12, 13]. We note that this algorithm will be designed such that the first component of the iterates belong to *range* (K).

Our preconditioner has the form

$$\mathcal{B} = \begin{bmatrix} \bar{K} & O \\ O & \bar{M} \end{bmatrix}$$

Here $\bar{K}$ is assumed to be symmetric and to be a good preconditioner for $K + D_H\,Q$, where

$$Q = \begin{bmatrix} Q_1 & O & \cdots & O \\ O & Q_2 & \ddots & \vdots \\ \vdots & \ddots & \ddots & O \\ O & \cdots & O & Q_N \end{bmatrix},$$

with Q_i being the mass matrices associated with the mesh on Ω_i and $D_H = diag_{i=1}^{N}(H_i^{-2} I_i)$ is a diagonal matrix. Here H_i denotes the diameter of the subdomain Ω_i. We further assume that $\bar{M}$ is symmetric and a good preconditioner for $\widehat{M}$, where $\widehat{M}$ can be either the new preconditioner from Section 4 or Section 5, depending on the implementation. More precisely, we assume there exists constants k_0, k_1, m_0, m_1 with $0 < c \leq m_0 \leq m_1 \leq C < \infty$ and $0 < c \leq k_0 \leq k_1 \leq C < \infty$, such that

$$\begin{array}{lclcll} k_0 \langle (K + D_H Q)u, u\rangle & \leq & \langle \bar{K}u, u\rangle & \leq & k_1 \langle (K + D_H Q)u, u\rangle & \forall u \in W, \\ m_0 \langle \widehat{M}\lambda, \lambda\rangle & \leq & \langle \bar{M}\lambda, \lambda\rangle & \leq & m_1 \langle \widehat{M}\lambda, \lambda\rangle & \forall \lambda \in V. \end{array} \tag{6.3}$$

Here c and C are generic constants independent of, or only weakly dependent on, the mesh size. Because of the block-diagonal structure of K and M and the preconditioners, c and C do not depend on the number of subdomains and the values of the ρ_i. We note that all that is required here are preconditioners for quite benign positive definite problems on individual subregions and that the bounds are independent of the number of subdomains.

From these assumptions it is clear that our preconditioner $\mathcal{B}$ is symmetric positive definite and thus it can be used with the preconditioned conjugate residual method. In order to have a computationally efficient preconditioner, we must also assume that $\bar{K}^{-1}$ and $\bar{M}^{-1}$ can be applied to a vector at a low cost.

To guarantee that the iterates belong to *range* (K), we introduce the projection P_R onto *range* (K) by

$$P_R := I - R(R^t R)^{-1} R^t.$$

We recall that *range* (R) = *ker* (K) and note that $R^t R$ is a diagonal matrix with an entry for each interior subdomain; the expense of applying P_R to a vector is therefore very modest.

The resulting domain decomposition method is the conjugate residual algorithm using the $\mathcal{B}$–inner product applied to the preconditioned system

$$\mathcal{B}^{-1}\mathcal{A}x = \mathcal{B}^{-1}F$$

with

$$\mathcal{A} = \left[\begin{array}{cc} K & (PB)^t \\ PB & O \end{array} \right], \mathcal{B}^{-1} = \left[\begin{array}{cc} P_R \widehat{K}^{-1} P_R^t & O \\ O & P\widehat{M}^{-1}P^t \end{array} \right], \tag{6.4}$$

$$x = \begin{bmatrix} u \\ \mu \end{bmatrix}, F = \begin{bmatrix} f - B^t\lambda_0 \\ 0 \end{bmatrix}.$$

We note that it is easy to see that only two matrix-vector products with the projection P and one with the projection P_R are required in each step. We note that the iterates of the conjugate residual method belong to $W_R \times V$ with $W_R := range\,(K)$.

We now give a condition number estimate for the block–diagonal preconditioner resulting in a convergence estimate for the preconditioned conjugate residual method.

An upper bound for the convergence rate of the conjugate residual method can be given in terms of the condition number $\kappa(\mathcal{B}^{-1}\mathcal{A})$ of the preconditioned system. The estimate follows directly by combining the arguments given in [15] and [16].

Theorem 3

$$\kappa(\mathcal{B}^{-1}\mathcal{A}) \leq C\,(1 + \log(H/h)),$$

with a constant C independent of H, h, γ, and the values of the ρ_i.

Remark 1 *Note that the convergence bound for the conjugate residual method depends linearly on the condition number $\kappa(\mathcal{B}^{-1}\mathcal{A})$ in contrast to the conjugate gradient method in the symmetric positive definite case where it depends on $\sqrt{\kappa(P\widehat{M}^{-1}P^tF)}$. Thus, all of the methods discussed in this chapter have an asymptotic convergence rate on the order of $1 + \log(H/h)$.*

We refer to [15] for a report on numerical experiments with the algorithm, described in this section, for an elasticity problem in two dimensions.

References

[1] Manoj Bhardwaj, David Day, Charbel Farhat, Michel Lesoinne, Kendall Pierson, and Daniel Rixen. Application of the FETI method to ASCI problems: Scalability results on one thousand processors and discussion of highly heterogeneous problems. Technical Report CU-CAS-99-05, University of Colorado at Boulder, Department of Aerospace Engineering, March 1999.

[2] Maksymilian Dryja, Marcus V. Sarkis, and Olof B. Widlund. Multilevel Schwarz methods for elliptic problems with discontinuous coefficients in three dimensions. *Numer. Math.*, 72(3):313–348, 1996.

[3] Maksymilian Dryja, Barry F. Smith, and Olof B. Widlund. Schwarz analysis of iterative substructuring algorithms for elliptic problems in three dimensions. *SIAM J. Numer. Anal.*, 31(6):1662–1694, December 1994.

[4] Maksymilian Dryja and Olof B. Widlund. Schwarz methods of Neumann-Neumann type for three-dimensional elliptic finite element problems. *Comm. Pure Appl. Math.*, 48(2):121–155, February 1995.

[5] Charbel Farhat, Po-Shu Chen, and Jan Mandel. A scalable Lagrange multiplier based domain decomposition method for time-dependent problems. *Int. J. Numer. Meth. Engrg.*, 38(22):3831–3853, 1995.

[6] Charbel Farhat, Po-Shu Chen, Jan Mandel, and François-Xavier Roux. The two-level FETI method - part II: extensions to shell problems, parallel implementation, and performance results. *Comp. Meth. Appl. Mech. Engrg.*, 155:153–179, 1998.

[7] Charbel Farhat, Jan Mandel, and François-Xavier Roux. Optimal convergence properties of the FETI domain decomposition method. *Comp. Meth. Appl. Mech. Engrg.*, 115:367–388, 1994.

[8] Charbel Farhat and François-Xavier Roux. Implicit parallel processing in structural mechanics. In J. Tinsley Oden, editor, *Computational Mechanics Advances*, volume 2 (1), pages 1–124. North-Holland, 1994.

[9] Charbel Farhat and François-Xavier Roux. A method of finite element tearing and interconnecting and its parallel solution algorithm. *Int. J. Numer. Meth. Engrg.*, 32:1205–1227, 1991.

[10] Gene H. Golub and Charles F. Van Loan. *Matrix Computations.* Second Edition. Johns Hopkins University Press, Baltimore, MD, 1989.

[11] Wolfgang Hackbusch. *Iterative Solution of Large Sparse Systems of Equations.* Springer, New York, 1994.

[12] Axel Klawonn. *Preconditioners for Indefinite Problems.* Ph.D. thesis, Westfälische Wilhelms-Universität Münster, 1995. TR 716, Courant Institute of Mathematical Sciences, New York University, New York, February 1996.

[13] Axel Klawonn. An optimal preconditioner for a class of saddle point problems with a penalty term. *SIAM J. Sci. Comp.*, 19(2):540–552, March 1998.

[14] Axel Klawonn and Olof B. Widlund. A domain decomposition method with Lagrange multipliers for linear elasticity. In Choi-Hon Lai, Petter Bjørstad, Mark Cross, and Olof B. Widlund, editors, *Proceedings of the 11th International Conference on Domain Decomposition Methods, Greenwich, UK, July 20–24, 1998*, pages 49–56, URL: `http://www.ddm.org/`.

[15] Axel Klawonn and Olof B. Widlund. A domain decomposition method with Lagrange multipliers for linear elasticity. Technical Report TR 780, Courant Insitute of Mathematical Sciences, New York University, New York, February 1999. URL: `file://cs.nyu.edu/pub/tech-reports/tr780.ps.gz`.

[16] Axel Klawonn and Olof B. Widlund. FETI and Neumann–Neumann Iterative Substructuring Methods: Connections and New Results. Technical Report, Computer Science Department, Courant Institute of Mathematical Sciences, New York University, New York, 1999. To appear.

[17] Jan Mandel and Marian Brezina. Balancing domain decomposition for problems with large jumps in coefficients. *Math. Comp.*, 65:1387–1401, 1996.

[18] Jan Mandel and Radek Tezaur. Convergence of a substructuring method with Lagrange multipliers. *Numer. Math.*, 73:473–487, 1996.

[19] Jan Mandel, Radek Tezaur, and Charbel Farhat. A scalable substructuring method by Lagrange multipliers for plate bending problems. *SIAM J. Numer. Anal.*, 36(5):1370–1391, 1999.

[20] K.C. Park, M.R. Justino, and C.A. Felippa. An algebraically partitioned FETI method for parallel structural analysis: algorithm description. *Int. J. Numer. Meth. Engrg.*, 40:2717–2737, 1997.

[21] Daniel Rixen and Charbel Farhat. Preconditioning the FETI and balancing domain decomposition methods for problems with intra- and inter-subdomain coefficient jumps. In Petter Bjørstad, Magne Espedal, and David Keyes, editors, *Proceedings of the Ninth International Conference on Domain Decomposition Methods in Science and Engineering, Bergen, Norway, June 1996*, pages 472–479, 1998. URL: `http://www.ddm.org/DD9/Rixen.ps.gz`.

[22] Daniel Rixen and Charbel Farhat. A simple and efficient extension of a class of substructure based preconditioners to heterogeneous structural mechanics problems. *Int. J. Numer. Meth. Engrg.*, 44:489–516, 1999.

[23] Daniel Rixen, Charbel Farhat, Radek Tezaur, and Jan Mandel. Theoretical Comparison of the FETI and Algebraically Partitioned FETI Methods, and Performance Comparisons with a Direct Sparse Solver. Technical Report CU-CAS-98-17, University of Colorado at Boulder, Department of Aerospace Engineering, October 1998.

[24] Marcus V. Sarkis. Two-level Schwarz methods for nonconforming finite elements and discontinuous coefficients. In N. Duane Melson, Thomas A. Manteuffel, and Steve F. McCormick, editors, *Proceedings of the Sixth Copper Mountain Conference on Multigrid Methods, Volume 2*, number 3224, pages 543–566, Hampton VA, 1993. NASA.

[25] Marcus V. Sarkis. *Schwarz Preconditioners for Elliptic Problems with Discontinuous Coefficients Using Conforming and Non-Conforming Elements.* Ph.D. thesis, Courant Institute of Mathematical Sciences, New York University, New York, September 1994.

[26] Radek Tezaur. *Analysis of Lagrange Multiplier Based Domain Decomposition.* Ph.D. thesis, University of Colorado at Denver, 1998.

[27] Andrea Toselli and Axel Klawonn. A FETI domain decomposition method for Maxwell's equations with discontinuous coefficients in two dimensions. Technical Report 788, Department of Computer Science, Courant Institute of Mathematical Sciences, New York University, New York, September 1998.

Acknowledgements

The work of the first author was supported in part by the National Science Foundation under Grants NSF-CCR-9732208. The work of the second author was supported in part by the National Science Foundation under Grants NSF-CCR-9732208 and in part by the U.S. Department of Energy under Contract DE-FG02-92ER25127.

Axel Klawonn
SCAI - Institute for Scientific Computing and Algorithms
GMD - German National Research Center for Information Technology
Schlosss Birlinghoven, D-53754 Sankt Augustin
Germany
klawonn@gmd.de
http://www.gmd.de/SCAI/people/klawonn

Olof B. Widlund
Courant Institute of Mathematical Sciences
New York University
251 Mercer St
New York, NY 10012
USA
widlund@cs.nyu.edu
http://www.cs.nyu.edu/cs/faculty/widlund

I. H. SLOAN AND R. S. WOMERSLEY

The search for good polynomial interpolation points on the sphere

Abstract This chapter considers the search for good points for interpolation by polynomials of degree at most n on the sphere $\mathbb{S}^2 \subset \mathbb{R}^3$. Points which minimize the potential energy have a regular geometric structure, but they are not very good for polynomial interpolation. On the other hand the eigenvalue points, chosen to minimize a bound on the norm $\|\Lambda_n\|$ of the interpolation operator as a map from $C(\mathbb{S}^2)$ to $C(\mathbb{S}^2)$, are much better for polynomial interpolation, but have a more irregular clustered structure. For these points and for $n < 30$ the empirical growth in $\|\Lambda_n\|$ is very close to $n+1$. However it is still an open question as to how small the order of the growth of $\|\Lambda_n\|$ can be made by a good choice of the interpolation points. An example of what can be achieved with a non-interpolatory polynomial approximation is the hyperinterpolation approximation, for which the norm of the operator grows as $n^{1/2}$, which is optimal among all linear projections on $\mathbb{S}^2$.

1 Introduction

The distribution of a set $X_m = \{x_1, \ldots, x_m\}$ of m points on the unit sphere $\mathbb{S}^{r-1}$ in $\mathbb{R}^r$ is of widespread interest in coding theory, approximation, quadrature, complexity theory, chemistry, physics, and astronomy (see [19, 2, 7] for general reviews). Key questions are how to generate "good" sets of points and the asymptotic properties of these distributions. This chapter concentrates on the $r = 3$ case.

Of particular interest here is interpolation by polynomials of degree at most n, which has been extensively studied by Reimer [15]. For *extremal fundamental systems*, which maximize the determinant of the interpolation matrix (see Section 3), Reimer has shown that the norm $\|\Lambda_n\|$ of the interpolation operator, considered as a map from $C(\mathbb{S}^2)$ to $C(\mathbb{S}^2)$, is bounded by $(n+1)^2$. Given a set of points for polynomial interpolation, it is then straightforward to use these in a numerical quadrature rule, (see for example Reimer [16] and Fliege and Maier [6]).

Applications such as a boundary integral equation approach to the scattering of sound in $\mathbb{R}^3$ (Graham and Sloan [8]) require a polynomial approximation in which the norm of the operator grows no faster than $O(n^{1-\epsilon})$ for some $\epsilon > 0$. This can be achieved by hyperinterpolation, which is a non-interpolatory polynomial approximation introduced by Sloan [20]. It was shown recently [23] that the norm of the hyperinterpolation operator grows as $O(n^{1/2})$, which is known to be optimal for a linear projection of the sphere $\mathbb{S}^2$.

It is still an open question as to how good polynomial interpolation, which uses the number of points equal to the dimension of the space of polynomials of degree at most n on $\mathbb{S}^2$, can be made by a good choice of the interpolation points. This question is further explored experimentally in [26], but even the experiments leave open the question of whether the order of growth can be slower than $O(n)$.

2 Points on the sphere

The "equidistribution" of a set of points on the sphere $\mathbb{S}^{r-1}$ can be measured by many different criteria. For $X_m = \{x_1, \dots, x_m\}$ the *mesh norm* is

$$h(X_m) = \sup_{x \in \mathbb{S}^{r-1}} \operatorname{dist}(x, X_m), \tag{2.1}$$

where

$$\operatorname{dist}(x, X_m) = \min_{j=1,\dots,m} \operatorname{dist}(x, x_j) \quad \text{and} \quad \operatorname{dist}(x, y) = \cos^{-1}(x \cdot y), \tag{2.2}$$

the maximum great circle distance from the closest point of $\{x_1, \dots, x_m\}$. Here $x \cdot y$ is the standard inner product in $\mathbb{R}^r$. Another geometrical measure is the *spherical cap discrepancy*

$$D(x_1, \dots, x_m) = \sup_S \left| \frac{|\{j : x_j \in S\}|}{m} - |S| \right|, \tag{2.3}$$

where $|S|$ is the area of the spherical cap S, and the supremum is taken over all spherical caps S. The spherical cap with opening α, $0 < \alpha < \pi$, centred at $a \in \mathbb{S}^{r-1}$ is $S_\alpha^{r-1}(a) = \{x \in \mathbb{S}^{r-1} : x \cdot a \geq \cos\alpha\}$.

Lubotzky, Phillips and Sarnak [11] produce equidistributed sets of $m = (3 \times 5^L - 1)/2$ points in $\mathbb{S}^2$, (where $L \geq 2$ is a positive integer giving the 'level') as a sequence of rotations. They measure equidistribution by making

$$\delta f = \frac{|\mathbb{S}^{r-1}|}{m} \sum_{j=1}^{m} f(x_j) - \int_{\mathbb{S}^{r-1}} f(x)dx, \tag{2.4}$$

where dx is the surface measure on $\mathbb{S}^{r-1}$, small in the L^2 norm for all functions f with L^2 norm 1. A *spherical n–design* is a set of points which makes $\delta f = 0$ for all polynomials of degree less than or equal to n. There is considerable interest [5, 27, 2] in the minimal number of points required for a spherical n–design. Equation (2.4) is the error for the special case of the quadrature rule

$$\int_{\mathbb{S}^{r-1}} f(x)dx \approx \sum_{j=1}^{m} w_j f(x_j) \tag{2.5}$$

with equal weights $w_j = |\mathbb{S}^{r-1}|/m$ for $j = 1, \ldots, m$.

Rahkmanov, Saff and Zhou [13] studied the α-energy

$$\widehat{E}(\alpha, m) = \begin{cases} \inf_{X_m \subset \mathbb{S}^2} E(\alpha, X_m) & \alpha \le 0 \\ \sup_{X_m \subset \mathbb{S}^2} E(\alpha, X_m) & \alpha > 0, \end{cases}$$

where

$$E(\alpha, X_m) = \begin{cases} \displaystyle\sum_{1 \le i < j \le m} \|x_i - x_j\|^\alpha & \alpha \ne 0 \\ \displaystyle\sum_{1 \le i < j \le m} \log \frac{1}{\|x_i - x_j\|} & \alpha = 0 \end{cases} \tag{2.6}$$

and $\|\cdot\|$ is the Euclidean norm in $\mathbb{R}^r$. As an initial approximation in a search for minimal energy points they propose the set of *generalized spiral points*, which are easily generated by

$$x_j = \begin{bmatrix} \cos(\phi_j)\sin(\theta_j) \\ \sin(\phi_j)\sin(\theta_j) \\ \cos(\theta_j) \end{bmatrix} \quad \text{for } j = 1, \ldots, m, \tag{2.7}$$

where the spherical polar coordinates $\theta_j \in [0, \pi]$ and $\phi_j \in [0, 2\pi)$ are generated by

$$\begin{aligned} z_j &= 1 - \frac{2(j-1)}{m-1}, \quad j = 1, \ldots, m \\ \theta_j &= \cos^{-1} z_j, \quad j = 1, \ldots, m \\ \phi_2 = 0, \quad \phi_j &= \left(\phi_{j-1} + \frac{C}{\sqrt{m}} \frac{1}{\sqrt{1 - z_j^2}} \right) \mod 2\pi, \quad j = 3, \ldots, m-1. \end{aligned}$$

The points have been rotated, so that the second point lies on the prime meridian ($\phi_2 = 0$). The values of ϕ_1 and ϕ_m can be arbitrarily set to zero since

$\theta_1 = 0$ and $\theta_m = \pi$. The constant C is chosen so that successive points will have approximately the same Euclidean distance apart on $\mathbb{S}^2$. They propose a value of $C = (8\pi/\sqrt{3})^{1/2}$ for points with low logarithmic energy [13].

Fliege and Maier [6] use the points obtained by minimizing

$$E(-1, X_m) = \sum_{1\le i<j\le m} \frac{1}{||x_i - x_j||}. \tag{2.8}$$

We shall call the points found by [6] the *minimum energy* points. A linear system (see Section 5) can then be solved to get the weights w_j for use in the quadrature formula (2.5).

Freeden, Gervens and Schreiner [7] give several examples of points which are approximately uniformly distributed over the sphere. Example 7.1.9 of [7] gives the points with spherical coordinates $(\theta_i, \phi_{i,j})$ for $i = 1, \dots, \ell$, $j = 1, \dots, n_i$, where

$$\begin{aligned}
&\theta_i = i\pi/\ell \quad i = 0, \dots, \ell \\
&n_0 = 1, \phi_{0,1} = 0, \quad n_\ell = 1, \phi_{\ell,1} = 0 \\
&n_i = \lfloor 2\pi/\cos^{-1}\left((\cos(\pi/\ell) - \cos^2\theta_i)/\sin^2\theta_i\right)\rfloor \quad i = 1, \dots, \ell-1 \\
&\phi_{i,j} = (j - 1/2)(2\pi/n_i) \quad j = 1, \dots, n_i \quad i = 1, \dots, \ell-1.
\end{aligned} \tag{2.9}$$

We call these the FGS points. For each θ_i, $i = 1, \dots, \ell-1$, there are $n_i \approx 2\ell\sin\theta_i$ points.

The above criteria are primarily concerned with the geometric distribution of the points $\{x_1, \dots, x_m\}$ on the sphere, and are not directly related to the problem of polynomial interpolation. The mesh norm function $\mathrm{dist}(x, X_m)$ for $x \in \mathbb{S}^2$ is plotted in Figure 2.1 for the generalized spiral and FGS points. The mesh norm function provides a convenient way of visualizing the points in the set $\{x_1, \dots, x_m\}$, in that each dark region has a point at its centre. Many other criteria, such as the Green's function potential [11], have also been used.

3 Polynomial interpolation

Of primary interest here is the problem of finding good polynomial interpolation points on the unit sphere $\mathbb{S}^2$ in $\mathbb{R}^3$. Let $\mathbb{P}_n$ denote the space of all spherical polynomials of degree at most n on $\mathbb{S}^2$ (i.e., the space of all polynomials in three variables restricted to $\mathbb{S}^2$). Using the usual spherical polar coordinates $\theta \in [0, \pi]$ and $\phi \in [0, 2\pi)$, a common basis for $\mathbb{P}_n$ is provided by the spherical harmonics [12]

$$Y_{\ell k}(\theta, \phi) = c_\ell P_\ell^{|k|}(\cos\theta)\cos(|k|\phi), \quad \ell = 0, \dots, n \text{ and } k = -\ell, \dots, \ell,$$

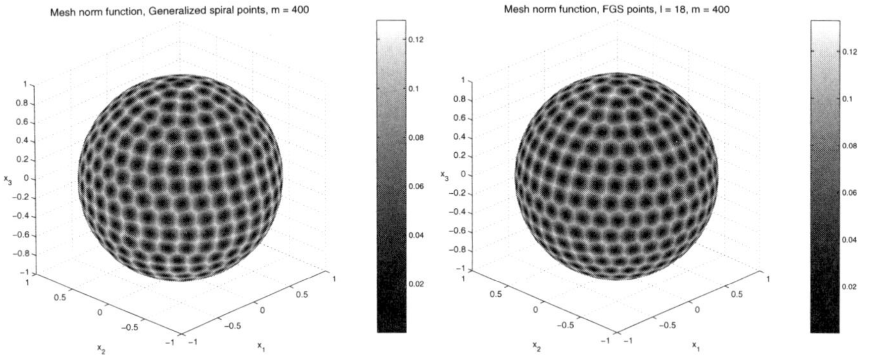

Figure 2.1: Mesh norm function for generalized spiral points for $m = 400$ and FGS points for $\ell = 18, m = 404$

where c_ℓ are appropriate normalization constants and $P_\ell^{|k|}$ are the associated Legendre functions. The dimension of the space $\mathbb{P}_n$ is $d_n = (n+1)^2$. We shall assume the spherical harmonics are normalized by

$$\int_{\mathbb{S}^{r-1}} Y_{\ell k}(x) Y_{\ell' k'}(x) dx = \delta_{\ell k} \delta_{\ell' k'}.$$

The polynomial interpolant $\Lambda_n f$ coincides with a given continuous function f at a prescribed set of points $\{x_1, \ldots, x_{d_n}\} \subseteq \mathbb{S}^2$. A set of $m = d_n$ points on $\mathbb{S}^2$ is a possible set of interpolation points for the space $\mathbb{P}_n$ if and only if it is a *fundamental system*, which is to say that the zero polynomial is the only member of $\mathbb{P}_n$ that vanishes at each point x_j, $j = 1, \ldots, d_n$.

The question is, for fixed n, to find $d_n = (n+1)^2$ points on $\mathbb{S}^2$ so that polynomial interpolation is a good approximation. Figure 3.1 plots the mesh norm function for two point sets with 400 points. The first set is (for $n = 19$) the minimum energy point set found by [6]. The second set (of *eigenvalue points*) will be explained later in this section. Clearly, the first set is much more geometrically regular, yet we will see that this set is far worse for interpolation.

First we must decide the criterion to measure a good approximation. Let $\mathbb{P}_n^r$ denote the space of all spherical harmonics of degree at most n on $\mathbb{S}^{r-1}$, and let $d_n^r = \dim \mathbb{P}_n^r$ (so $d_n \equiv d_n^3$). Using the uniform norm

$$||f||_\infty = \max_{x \in \mathbb{S}^{r-1}} |f(x)|$$

to measure the error, the norm of the polynomial interpolation operator, as a

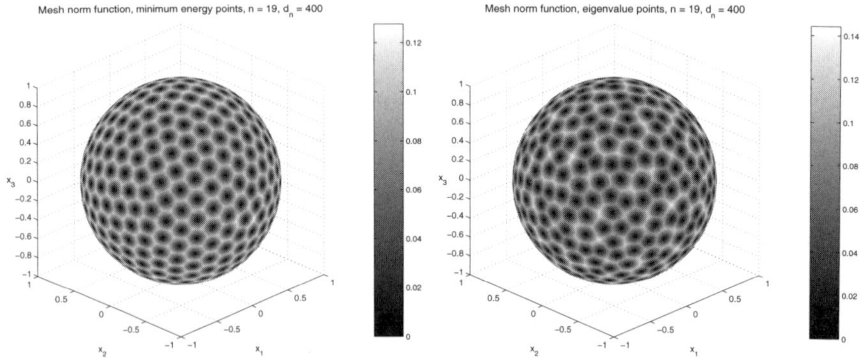

Figure 3.1: Mesh norm function for minimum energy and eigenvalue points for $n = 19$, $d_n = 400$

map from $C(\mathbb{S}^{r-1})$ to $C(\mathbb{S}^{r-1})$ is

$$||\Lambda_n|| = \sup_{f \in C(\mathbb{S}^{r-1}), f \neq 0} \frac{||\Lambda_n f||_\infty}{||f||_\infty}.$$

As Λ_n is a projection onto $\mathbb{P}_n^r$, i.e., Λ_n is linear and $\Lambda_n^2 = \Lambda_n$, for any polynomial p in $\mathbb{P}_n^r$ we have

$$\begin{aligned} ||\Lambda_n f - f||_\infty &= ||\Lambda_n(f-p) - (f-p)||_\infty \\ &\leq ||\Lambda_n(f-p)||_\infty + ||f-p||_\infty \\ &\leq (||\Lambda_n|| + 1)||f-p||_\infty. \end{aligned}$$

Hence

$$||\Lambda_n f - f||_\infty \leq (1 + ||\Lambda_n||) E_n(f), \tag{3.1}$$

where $E_n(f) = \inf_{p \in \mathbb{P}_n^r} ||f-p||_\infty$ is the error of best uniform approximation. Because of its role in the upper bound (3.1) we use the *Lebesgue constant* $||\Lambda_n||$ as the criterion for evaluating the interpolation points.

Let $b_i \in \mathbb{P}_n^r$ for $i = 1, \ldots, d_n^r$ be a basis for $\mathbb{P}_n^r$, and define the vector valued function $\mathbf{b} : \mathbb{S}^{r-1} \to \mathbb{R}^{d_n^r}$ by $\mathbf{b}(x) = [b_1(x) \cdots b_{d_n^r}(x)]^T$ and the interpolation matrix B by

$$B = [\mathbf{b}(x_1) \cdots \mathbf{b}(x_{d_n^r})], \quad \text{i.e., } B_{ij} = b_i(x_j) \quad i,j = 1, \ldots, d_n^r. \tag{3.2}$$

The matrix B is non-singular if and only if the set of points $\{x_1, \ldots, x_{d_n^r}\}$ is a fundamental system. Given a vector $\mathbf{f} = [f(x_1) \cdots f(x_{d_n^r})]^T \in \mathbb{R}^{d_n^r}$ of function values, the interpolant is

$$(\Lambda_n f)(x) = \sum_{i=1}^{d_n^r} v_i b_i(x), \quad \text{where} \quad B\mathbf{v} = \mathbf{f}. \tag{3.3}$$

The norm of the interpolation operator is then

$$||\Lambda_n|| = \max_{x \in \mathbb{S}^{r-1}} ||B^{-1}\mathbf{b}(x)||_1. \tag{3.4}$$

This follows because, as is easily seen, the 'fundamental Lagrange polynomial' associated with the point x_j is $(B^{-1}\mathbf{b})_j$.

The value of $||\Lambda_n||$ depends on the fundamental system $\{x_1, \ldots, x_{d_n^r}\}$. The norm $||\Lambda_n||$ can be made arbitrarily large if the fundamental system is badly chosen. The interesting question is how small $||\Lambda_n||$ can be made by a good choice of fundamental system, especially for $r = 3$.

Figure 3.2 plots the interpolation norms for three sets of interpolation points. The minimum energy points of [6] are chosen to minimize the potential energy (2.8). The maximum determinant (or extremal fundamental systems) were obtained by maximizing det G, starting from the minimum energy points, where G is a symmetric positive definite matrix which depends on the fundamental system of points (see (5.4)). The eigenvalue points were obtained by maximizing λ_{min}, the smallest eigenvalue of G, starting from the minimum energy points. Note the very large values of $||\Lambda_n||$ obtained with the minimum energy points. The generalized spiral points (not shown) produce even larger values of $||\Lambda_n||$.

Figure 3.3 visualizes the interpolation norm function $\psi(x) = ||B^{-1}\mathbf{b}(x)||_1$ over $x \in \mathbb{S}^2$ for the minimum energy points and the eigenvalue points. The function plotted is $(1 + \xi/(\psi_u - \psi_l)(\psi(x) - \psi_l))x$, for $x \in S^2$, where $\psi_l = \min_{x \in S^2} \psi(x)$, $\psi_u = \max_{x \in S^2} \psi(x)$ and $\xi = 0.3$ is a scaling parameter. The completely different scale and character of the two functions is apparent; in the latter case there are many peaks, all of comparable height, and all of them orders of magnitude smaller than the peaks of the mountain range in the dramatic landscape in the first picture.

4 What is known theoretically?

Reimer and Sündermann [18] and Reimer [14] have shown that an *extremal fundamental system*, i.e., a fundamental system which maximizes $|\det(B)|$, has,

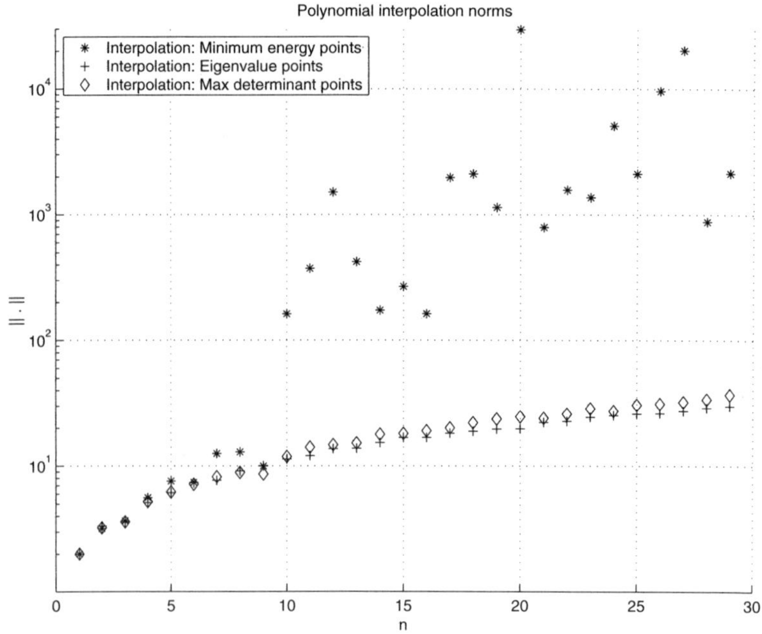

Figure 3.2: Interpolation norms for minimum energy, maximum determinant and eigenvalue points

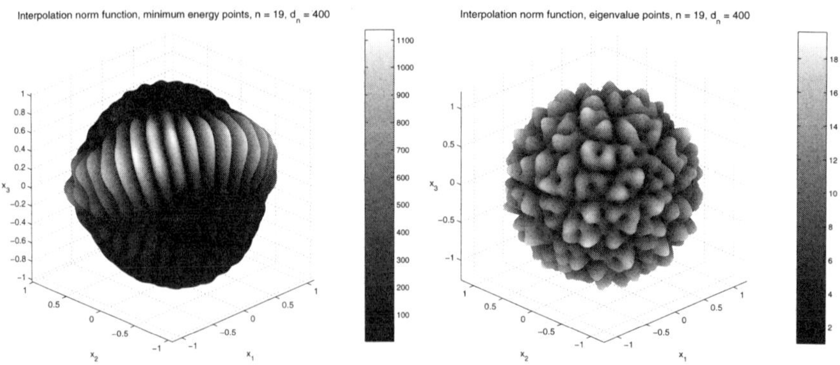

Figure 3.3: Interpolation norm function for minimum energy and eigenvalue points for $n = 19$, $d_n = 400$

for $r = 3$,

$$||\Lambda_n|| \leq d_n = (n+1)^2. \tag{4.1}$$

It can be seen from Figure 3.2 that this bound is quite pessimistic.

Another bound given by Reimer [15] is

$$||\Lambda_n|| \leq (n+1)\left(\frac{\lambda_{\rm avg}}{\lambda_{\rm min}}\right)^{1/2}, \tag{4.2}$$

where $\lambda_{\rm avg}$ and $\lambda_{\rm min}$ are the average and minimum eigenvalues of the positive-definite matrix G determined by the fundamental system (see Section 5). The ratio $\lambda_{\rm avg}/\lambda_{\rm min}$ is certainly not less that 1, and depends on the choice of points $\{x_1, \ldots, x_{d_n}\}$. Less obviously (Reimer [15]), $\lambda_{\rm avg}/\lambda_{\rm min} > 1$ for $n \geq 3$.

It is known that the optimal order of growth for the norm of a linear projection onto $\mathbb{S}^2$ is $O(n^{1/2})$, which is achieved by both the L_2-orthogonal projection and hyperinterpolation [23].

5 Reproducing kernel basis

The *reproducing kernel* (Reimer [15]) on the space $\mathbb{P}_n^r$ is

$$G_n(x,y) := \sum_{\ell=0}^{n} \sum_{k=1}^{N(r,\ell)} Y_{\ell k}^r(x) Y_{\ell k}^r(y), \quad x, y \in \mathbb{S}^{r-1}, \tag{5.1}$$

where

$$N(r,0) = 1, \quad N(r,\ell) = \frac{2\ell + r - 2}{\ell} \binom{\ell + r - 3}{\ell - 1} \quad \text{for} \quad \ell \geq 1.$$

The Addition Theorem [12] for spherical harmonics shows that $G_n(x,y)$ is bizonal, depending only on the angle between x and y, so

$$G_n(x,y) = \widetilde{G}_n(x \cdot y).$$

In particular, for $r = 3$

$$\widetilde{G}_n(z) = \frac{1}{4\pi} \sum_{\ell=0}^{n} (2\ell+1) P_\ell(z), \tag{5.2}$$

where $P_\ell(\,\cdot\,)$ is the usual Legendre polynomial.

For each point x_j of the fundamental system $\{x_1, \ldots, x_{d_n}\}$ we define the *kernel polynomials* $g_j \in \mathbb{P}_n^r$ with axis x_j, by

$$g_j(x) = G_n(x, x_j) = \widetilde{G}_n(x \cdot x_j), \quad j = 1, \ldots, d_n. \tag{5.3}$$

Define the vector valued function $\mathbf{g} : \mathbb{S}^{r-1} \to \mathbb{R}^{d_n}$ by $\mathbf{g}(x) = [g_1(x) \cdots g_{d_n}(x)]^T$. The corresponding basis matrix is

$$G = [\mathbf{g}(x_1) \cdots \mathbf{g}(x_{d_n})], \quad \text{i.e., } G_{ij} = \widetilde{G}_n(x_i \cdot x_j) \text{ for } i, j = 1, \ldots, d_n. \tag{5.4}$$

Then the weights $\mathbf{w}$ such that $(\Lambda_n f)(x) = \mathbf{w} \cdot \mathbf{g}(x)$ are given by the linear system $G\mathbf{w} = \mathbf{f}$, where $\mathbf{f} = [f(x_1) \cdots f(x_{d_n})]^T$. Using this reproducing kernel basis the norm of the interpolation operator is

$$||\Lambda_n|| = \max_{x \in \mathbb{S}^{r-1}} ||G^{-1}\mathbf{g}(x)||_1. \tag{5.5}$$

The value of $\widetilde{G}_n(1)$ is [15]

$$G_n(x, x) = \widetilde{G}_n(1) = \frac{d_n}{|\mathbb{S}^2|} \quad \forall\, x \in \mathbb{S}^2, \tag{5.6}$$

so the matrix G in (5.4) has diagonal elements

$$G_{ii} = \frac{(n+1)^2}{4\pi} \quad i = 1, \ldots, d_n.$$

For any fundamental system of points X_{d_n} the gram matrix $G(X_{d_n})$ is symmetric positive definite. Given the matrix G the weights w_j in the quadrature rule (2.5) are given by the linear system

$$Gw = e, \tag{5.7}$$

where $e \in \mathbb{R}^{d_n}$ is the vector of all ones.

6 Best possible projection

The best possible projection on $\mathbb{P}_n^r$ is the L_2-orthogonal projection, given by

$$P_n f = \sum_{\ell=0}^{n} \sum_{k=1}^{N(r,\ell)} (f, Y_{lk}^r)\, Y_{lk}^r\,, \tag{6.1}$$

where $(\cdot, \cdot)$ is the L_2 inner product on $\mathbb{S}^{r-1}$,

$$(u, v) := \int_{\mathbb{S}^{r-1}} u(x)v(x)dx.$$

For $r = 3$, the orthogonal projection can be written, using the addition theorem, as

$$P_n f(x) = \sum_{\ell=0}^{n} \frac{2\ell+1}{4\pi} \int_{\mathbb{S}^{r-1}} f(t)P_\ell(x \cdot t)dt = \int_{\mathbb{S}^{r-1}} f(t)\widetilde{G}_n(x \cdot t)dt,$$

from which it follows that

$$||P_n|| = 2\pi \int_{-1}^{1} |\widetilde{G}_n(z)|dz. \tag{6.2}$$

Berman [1] (for the circle $r = 2$) and Daugavet [3] for general r have proved that if T_n is an arbitrary linear projection onto $\mathbb{P}_n^r$, then

$$||P_n|| \leq ||T_n||.$$

Moreover, it is known (see [4]) that $||P_n|| \asymp \log n$ for $r = 2$, where $a \asymp b$ means that there exist constants c_1 and c_2 such that $c_1 a \leq b \leq c_2 a$, while for $r = 3$ (see Gronwall [9])

$$\frac{||P_n||}{\sqrt{n}} \to 2\sqrt{\frac{2}{\pi}} \quad \text{as } n \to \infty. \tag{6.3}$$

For arbitrary $r \geq 3$, see Reimer [15, Section 11], the result is

$$||P_n|| \asymp n^{(r-2)/2}.$$

7 Hyperinterpolation

The hyperinterpolation approximation $L_n f$ introduced in [20] is obtained by approximating the inner product in the definition (6.1) of $P_n f$ by a positive-weight quadrature rule with the property of integrating all spherical polynomials of degree at most $2n$ exactly. Thus

$$L_n f = \sum_{\ell=0}^{n} \sum_{k=1}^{N(r,\ell)} (f, Y_{lk}^r)_m \, Y_{lk}^r, \tag{7.1}$$

where $(\cdot, \cdot)_m$ is a discrete version of the inner product obtained by application of an m-point quadrature formula,

$$(u, v)_m := \sum_{j=1}^{m} w_j u(x_j) v(x_j),$$

and where the weights w_j and points x_j in the quadrature rule Q_m,

$$Q_m g := \sum_{j=1}^{m} w_j g(x_j) \approx \int_{\mathbb{S}^{r-1}} g(x) dx, \tag{7.2}$$

must satisfy

$$w_j > 0, \quad x_j \in \mathbb{S}^{r-1}, \quad j = 1, \dots, m, \tag{7.3}$$

and

$$Q_m p = \int_{\mathbb{S}^{r-1}} p(x) dx, \quad \forall\, p \in \mathbb{P}_{2n}. \tag{7.4}$$

From this it follows easily that L_n is a projection.

Several choices for the quadrature rule Q_m are possible. Three possible rules are the following tensor product rules (Stroud [24]), using the $2(n+1)$-point rectangle rule with respect to ϕ and

1. the $(n+1)$–point Gauss rule with respect to $\cos\theta$ (see the first plot in Figure 7.1 for an example);
2. the $(2n+1)$–point Clenshaw-Curtis rule (see the second plot in Figure 7.1 for an example);
3. the $(2n+1)$–point Fejér rule. Hyperinterpolation with this rule was studied by Kushpel and Levesley [10].

Non-product rules also exist, but are harder to use. For approximations exact for all spherical polynomials of degree at most n these tensor product rules use between 2 and 4 times the number of points needed for polynomial interpolation.

Some basic properties of hyperinterpolation when using an m-point quadrature rule Q_m are (Sloan [20])

- $m \geq d_n$,
- $m = d_n \Longrightarrow L_n = \Lambda_n$,

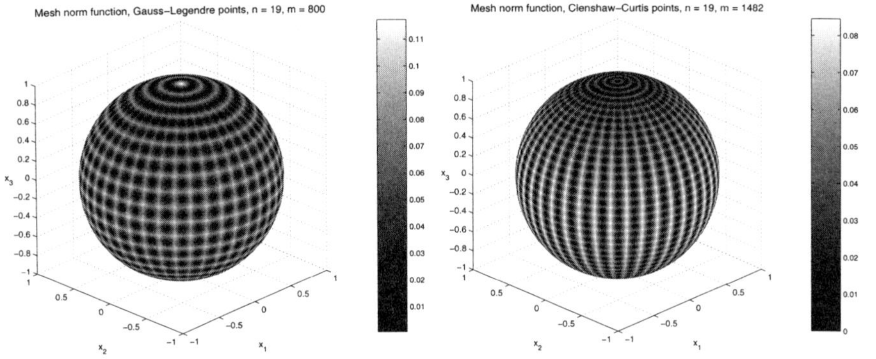

Figure 7.1: Mesh norm function for Gauss-Legendre ($n = 19, m = 800$) and Clenshaw-Curtis ($n = 19, m = 1482$) tensor product rules

- $m > d_n$ for $n \geq 3$. For example, in the product Gauss case we have $m = 2(n+1)^2 > (n+1)^2 = d_n$.
- $||L_n||_{C\to L_2} = (4\pi)^{1/2}$, where

$$||L_n||_{C\to L_2} = \sup\left\{\frac{||L_n f||_2}{||f||_\infty} : f \in C(\mathbb{S}^2), f \neq 0\right\}$$

- (Sloan [21])

$$||\Lambda_n||_{C\to L_2} > (4\pi)^{1/2} \quad \text{for } n \geq 3.$$

Sloan and Womersley [23] established that the norm of the hyperinterpolation operator L_n in the setting C to C is given by (cf. (6.2))

$$||L_n|| = \max_{x\in\mathbb{S}^{r-1}} \sum_{j=1}^{m} w_j |\widetilde{G}_n(x \cdot x_j)|, \tag{7.5}$$

and that $||L_n||$ is bounded above by $d_n^{1/2}$. For $r = 3$ this yields

$$||L_n|| \leq n + 1.$$

For $r = 3$, and under a certain *quadrature regularity* assumption, they also show [23] that

$$||L_n|| \asymp n^{1/2}, \tag{7.6}$$

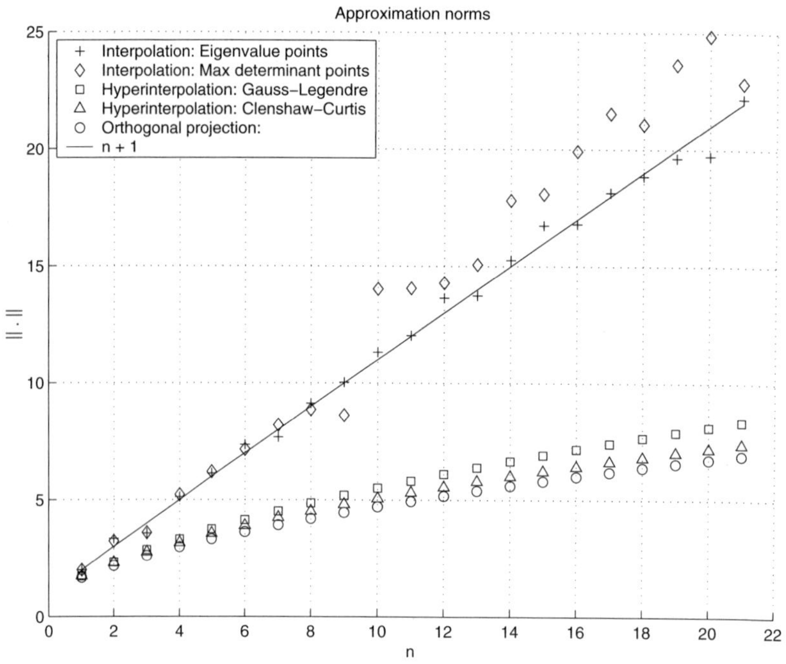

Figure 7.2: Interpolation and hyperinterpolation norms

which is optimal with respect to order. Moreover empirical evidence from [22] and Figure 7.2 suggests that the constant is close to that for P_n. Recently Reimer [17] has shown that the quadrature regularity assumption is automatically satisfied for any positive weight quadrature rule of precision $2n$, provided $m \leq cd_n$. Note that the interpolation norms $||\Lambda_n||$ in Figure 7.2 are upper bounds on what can be achieved by minimizing $\det G$ or maximizing $\lambda_{\min}$, in that the point sets are only approximate local optimizers.

An expression analogous to (3.3) for the hyperinterpolation approximation $L_n f$ is obtained by interchanging the order of summation in (7.1) giving

$$L_n f = \sum_{j=1}^{m} w_j f(x_j) g_j, \tag{7.7}$$

where g_j denotes the kernel polynomial with axis x_j defined by (5.3). Equation (7.7) has a very simple structure, similar to the formula (3.3) for $\Lambda_n f$, but not requiring the solution of a linear system. It is also easy to use for the numerical evaluation of the hyperinterpolant.

8 Uniform error on a cosine cap

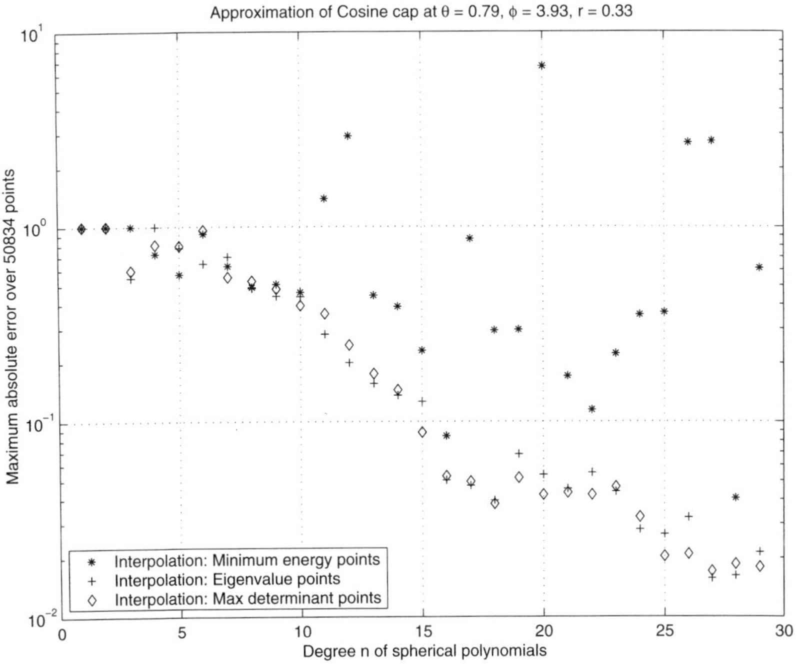

Figure 8.1: Uniform errors in approximation of a cosine cap

Figure 7.2 gives, for $n < 30$, the norms of the interpolation operators $||\Lambda_n||$ for the maximum determinant and eigenvalue points, plus the norms $||L_n||$ for the hyperinterpolation with the Gauss-Legendre and Clenshaw-Curtis tensor product rules. Even for these small values of n, the difference in the growth of the interpolation and hyperinterpolation norms is apparent. For the eigenvalue points, the interpolation norm is approximately $n + 1$. The growth of the interpolation norm for the maximum determinant points is larger, but much less than the bound $||\Lambda_n|| \leq d_n = (n + 1)^2$.

To verify that the norms $||\Lambda_n||$ give a realistic feeling for the quality of the

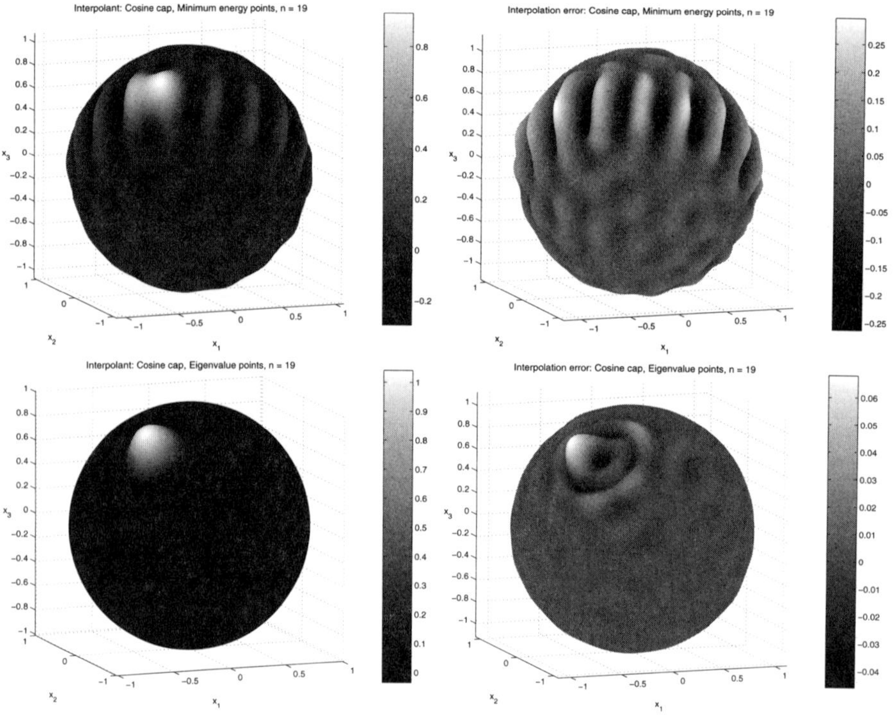

Figure 8.2: Cosine cap: Interpolant and error, minimum energy and eigenvalue points, $n = 19$, $m = 400$

interpolant, the uniform error is estimated for the specific example of a cosine cap. Further numerical experiments can be found in [26]. The cosine cap with centre $x_0 \in \mathbb{S}^2$, radius R, $0 < R < \pi$, and height h_0 is

$$f(x) = \begin{cases} h_0 \cos^2(\frac{\pi}{2}\mathrm{dist}(x, x_0)/R) & \text{if } \mathrm{dist}(x, x_0) < R \\ 0 & \text{if } \mathrm{dist}(x, x_0) \geq R. \end{cases} \tag{8.1}$$

The cosine cap is part of a standard test set [25] for numerical approximations to the shallow water equations in spherical geometry. The results in Figure 8.1 used the centre x_0 with polar coordinates $(\theta_0, \phi_0) = (\pi/4, 5\pi/4)$, radius $R = 1/3$ and amplitude $h_0 = 1$. Other parameters can easily be used. The cosine cap has local support, and on the boundary of the support is only once continuously differentiable, so it is difficult for polynomials to approximate.

Figure 8.2 gives the interpolants and their error when using the minimum energy and eigenvalue points for $n = 19$, corresponding to $m = d_n = 400$. For the minimum energy points we have $||\Lambda_n f - f||_\infty \approx 0.3$, while for the eigenvalue points $||\Lambda_n f - f||_\infty \approx 0.07$. This difference in the error is reflected in the different scales on the error plots. More striking still is the globally different nature of the two approximations: the interpolant with the eigenvalue points is qualitatively the true cosine cap, whereas the interpolant with the minimum energy points completely fails to reflect the locally supported nature of the cosine cap function.

The eigenvalue points and associated weights w_j in (2.5) are available from the web site `http://www.maths.unsw.edu.au/~rsw/Sphere/`.

References

[1] D. L. Berman, *On a class of linear operators*, Dokl. Akad. Nauk SSSR, 85 (1952), pp. 13–16. (Russian), Math. Reviews **14**, 57.

[2] J. H. Conway and N. J. A. Sloane, *Sphere Packings: Lattices and Groups*, Springer-Verlag, New York, Berlin, third ed., 1998.

[3] I. K. Daugavet, *Some applications of the Marcinkiewicz-Berman identity*, Vestnik Leningrad Univ., Math., 1 (1974), pp. 321–327.

[4] P. J. Davis, *Interpolation and Approximation*, Blaisdell, New York, 1963.

[5] P. Delsarte, J. M. Goethals, and J. J. Seidel, *Spherical codes and designs*, Geom. Dedicata, 6 (1977), pp. 363–388.

[6] J. Fliege and U. Maier, *The distribution of points on the sphere and corresponding cubature formulae*, IMA J. Num. Anal., 19 (1999), pp. 317–334. `http://www.mathematik.uni-dortmund.de/lsx/fliege/nodes.html`

[7] W. Freeden, T. Gervens, and M. Schreiner, *Constructive Approximation on the Sphere*, Clarendon Press, Oxford, U.K., 1998.

[8] I. Graham and I. H. Sloan, *Analysis of fully discrete boundary integral methods on smooth closed surfaces in $\mathbb{R}^3$*, tech. rep., Department of Mathematics, University of Bath, U.K., 1999.

[9] T. H. Gronwall, *On the degree of convergence of Laplace's series*, Trans. Amer. Math. Soc., 15 (1914), pp. 1–30.

[10] A. K. Kushpel and J. Levesley, *Radial quasi-interpolation on S^2*, tech. rep., Mathematics and Computer Science, University of Leicester, U.K., 1998.

[11] A. Lubotzky, R. Phillips, and P. Sarnak, *Hecke operators and distributing points on $\mathbb{S}^2$ I*, Comm. Pur Appl. Math., 39 (1986), pp. S149–S186.

[12] C. Müller, *Spherical Harmonics*, vol. 17 of Lecture Notes in Mathematics, Springer-Verlag, Berlin, 1966.

[13] E. A. Rahkmanov, E. B. Saff, and Y. M. Zhou, *Minimal discrete energy on the sphere*, Mathematical Research Letters, 1 (1994), pp. 647–662.

[14] M. Reimer, *Interpolation on the sphere and bounds for the Lagrangian square sums*, Resultate Math., 11 (1987), pp. 144–166.

[15] ——, *Constructive Theory of Multivariate Functions*, BI Wissenschaftsverlag, Mannheim, Wien, Zürich, 1990.

[16] ——, *Quadrature rules for the surface integral of the unit sphere based on extremal fundamental systems*, Mathematische Nachrichten, 169 (1994), pp. 235–241.

[17] ——, *Hyperinterpolation on the sphere at the minimal projection order*, tech. rep., Department of Mathematics, University of Dortmund, Germany, 1999.

[18] M. Reimer and B. Sündermann, *A Remez-type algorithm for the calculation of extremal fundamental systems on the sphere*, Computing, 37 (1986), pp. 43–58.

[19] E. B. Saff and A. B. J. Kuijlaars, *Distributing many points on a sphere*, Mathematical Intelligencer, 19 (1997), pp. 5–11.

[20] I. H. Sloan, *Polynomial interpolation and hyperinterpolation over general regions*, J. Approx. Theory, 83 (1995), pp. 238–254.

[21] ——, *Interpolation and hyperinterpolation on the sphere*, in Multivariate Approximation: Recent Trends and Results, W. Haussmann, K. Jetter, and M. Reimer, eds., Akademie Verlag GmbH, Berlin (Wiley-VCH), 1997, pp. 255–268.

[22] I. H. SLOAN AND R. S. WOMERSLEY, *The uniform error of hyperinterpolation on the sphere*, in Advances in Multivariate Approximation, W. Haussmann, K. Jetter, and M. Reimer, eds., Wiley–VCH, 1999.

[23] ——, *Constructive polynomial approximation on the sphere*, J. Approx. Theory, (to appear).

[24] A. H. STROUD, *Approximate Calculation of Multiple Integrals*, Prentice-Hall, Englewood Cliffs, N.J., 1971.

[25] D. L. WILLIAMSON, J. B. BRAKE, J. J. HACK, R. JAKOB, AND P. N. SWARZTRAUBER, *A standard test set for numerical approximations to the shallow water equations in spherical geometry*, J. Comp. Phys., 102 (1992), pp. 211–224.

[26] R. S. WOMERSLEY AND I. H. SLOAN, *How good can polynomial interpolation on the sphere be?*, tech. rep., School of Mathematics, University of New South Wales, Sydney, Australia, in preparation.

[27] V. A. YUDIN, *Lower bounds for spherical designs*, Isvestiya RAN: Ser. Mat., 61 (1997), pp. 213–223.

Acknowledgement

This research was supported by the Australian Research Council, with programming support by Dave Dowsett.

Ian H. Sloan
School of Mathematics
University of New South Wales
Sydney, NSW 2052
Australia
`I.Sloan@unsw.edu.au`
`http://www.maths.unsw.edu.au/~sloan/`

Robert S. Womersley
School of Mathematics
University of New South Wales
Sydney, NSW 2052
Australia
`R.Womersley@unsw.edu.au`
`http://www.maths.unsw.edu.au/~rsw/`

P. M. VAN DOOREN

Gramian based model reduction of large-scale dynamical systems

Abstract We describe model reduction techniques for large-scale dynamical systems, modeled via systems of equations of the type

$$\begin{cases} F(\dot{x}(t), x(t), u(t)) = 0 \\ y(t) = H(x(t), u(t)), \end{cases}$$

as encountered in the study of control systems with input $u(t) \in \Re^m$, state $x(t) \in \Re^N$ and output $y(t) \in \Re^p$. These models arise from the discretization of continuum problems and correspond to sparse systems of equations $F(.\,,.\,,.)$ and $H(.\,,.)$. The state dimension N is typically very large, while m and p are usually reasonably small. Although the numerical simulation of such systems may still be viable for large state dimensions N, most control problems of such systems are of such high complexity that they require model reduction techniques, i.e., techniques that construct a lower order model via a projection on a state space of lower dimension. We survey such techniques and put emphasis on the case where $F(.\,,.\,,.)$ and $H(.\,,.)$ are linear time-invariant or linear time-varying.

1 Linear, time-varying and nonlinear models

Model reduction can have a lot of meanings. In this chapter we refer to the problem of representing a complex dynamical system by a much simpler one. More precisely, we focus on dynamical systems describing a relation between a vector $u(.) \in \Re^m$ of input functions of time and a vector $y(.) \in \Re^p$ of output functions of time. The relation between inputs and outputs is a "system" $\mathcal{S}$ which is often represented as a "black box":

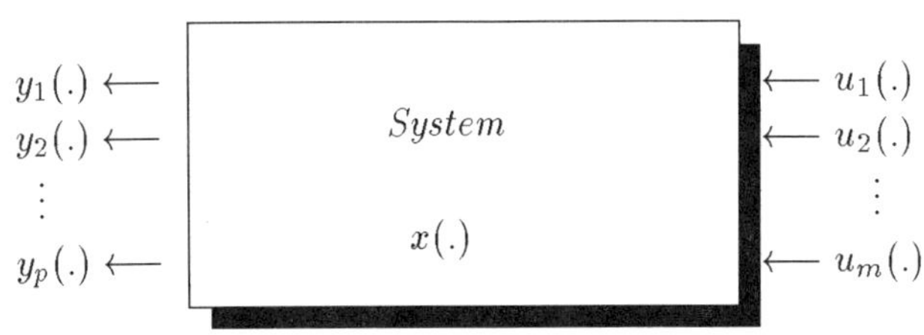

and uses an internal "state" $x(.) \in \Re^N$ to describe the relation between inputs and outputs.

Even though the original physical model typically has an infinite dimensional state-space to start with we assume that the model has already been discretized in space by, e.g., using a finite element model. This however leads to a state-space dimension N that is typically very large if one wants to obtain a sufficient approximation of the true physical phenomenon. On the other hand, the resulting models are then often sparse and when simulating such systems using ODE techniques, this can be exploited to yield a reasonable complexity.

For control applications one often has to solve associated problems that are not sparse anymore and hence have a much higher complexity. Typical examples of this are optimal control or Kalman filtering over a finite horizon. These both require the solution of an adjoint problem without sparsity.

The models that are used in control systems design can be both continuous-time and discrete-time and are nonlinear in their most general form. For computational reasons these are typically linearized around their trajectory. The resulting linear time-varying systems can then subsequently be approximated by a time invariant one over a period of time where the model does not change too much. So we differentiate between six types of models that are used in this area. Notice that a discrete-time model will, e.g., result from using an ODE solver for the simulation of a corresponding continuous-time model. Moreover, we simplified the models by choosing *explicit* state equations, although the ideas presented in later sections generalize to a large extent to *implicit* state equations.

continuous-time	$\Longrightarrow$	discrete-time
$\begin{cases} \dot{x}(t) = G(x(t), u(t)) \\ y(t) = H(x(t), u(t)) \end{cases}$		$\begin{cases} x(k+1) = G(x(k), u(k)) \\ y(k) = H(x(k), u(k)) \end{cases}$
$\Downarrow$	linearize	$\Downarrow$
$\begin{cases} \dot{x}(t) = A(t)x(t) + B(t)u(t) \\ y(t) = C(t)x(t) + D(t)u(t) \end{cases}$		$\begin{cases} x(k+1) = A(k)x(k) + B(k)u(k) \\ y(k) = C(k)x(k) + D(k)u(k) \end{cases}$
$\Downarrow$	freeze time	$\Downarrow$
$\begin{cases} \dot{x}(t) = Ax(t) + Bu(t) \\ y(t) = Cx(t) + Du(t) \end{cases}$		$\begin{cases} x(k+1) = Ax(k) + Bu(k) \\ y(k) = Cx(k) + Du(k) \end{cases}$

In order to reduce the complexity of the solution of a control problem for any of these models one can use reduced order models. These are essentially projectors $P(.)$ of the state-space vector $x(.)$ to a vector $\hat{x}(.) \doteq P(.)x(.)$ of much lower dimension. The control problem is then solved in the lower dimensional setting, after which the solution can be "lifted back" to the original coordinate system. This of course does not yield the correct solution but an approximation whose quality depends on several factors. We now describe the different types of projection techniques that can be used for each of these models, starting with the simplest case, the linear time-invariant case. Throughout this chapter we assume all signals and models are real. When we need an infinite dimensional setting we assume that we are working in a Hilbert space with the usual inner product leading to the ℓ_2 norm. We also assume that vectors (i.e., signals) have bounded ℓ_2 norm.

2 Linear time-invariant systems

This is the simplest case of model reduction and it is also the one that has been studied the most. We formulate the approximation problem in continuous-time but everything extends to discrete-time systems as well. The original system is given by the state-space equations

$$\left\{ \begin{array}{l} \dot{x}(t) = Ax(t) + Bu(t) \\ y(t) = Cx(t) + Du(t), \end{array} \right. \tag{2.1}$$

where $u(t) \in \Re^m$, $y(t) \in \Re^p$ and $x(t) \in \Re^N$ are the vectors of input variables, output variables and state variables, respectively. The input dimension m and output dimension p are assumed much smaller than the state dimension N. A reduced-order approximation of (2.1) takes the corresponding form

$$\left\{ \begin{array}{l} \dot{\hat{x}}(t) = \hat{A}\hat{x}(t) + \hat{B}u(t) \\ \hat{y}(t) = \hat{C}\hat{x}(t) + \hat{D}u(t), \end{array} \right. \tag{2.2}$$

where $\hat{y}(t) \in \Re^p$ and $\hat{x}(t) \in \Re^n$. Notice that both systems are driven by the same input $u(t)$ and that their outputs $y(t)$ and $\hat{y}(t)$ are of the same dimension, and hence can directly be compared (this does not hold for $x(t)$ and $\hat{x}(t)$). The idea of model reduction is to find a smaller model (2.2) whose output $\hat{y}(t)$ is close to the original output $y(t)$. In other words, we are trying to construct a dynamical system of much lower complexity that nevertheless closely approximates the behavior of the original system. How can we assess the quality of the reduced order model and how can we compute good approximations at low cost? These are the questions to be addressed here.

2.1 Transfer functions and norms

Linear time-invariant systems have a transfer function which (for continuous-time systems) is obtained under Laplace transform of the differential equations

[8]. This yields

$$T(s) = C(sI_N - A)^{-1}B + D, \quad \hat{T}(s) = \hat{C}(sI_n - \hat{A})^{-1}\hat{B} + \hat{D},$$

respectively, for the full and reduced order models. These transfer functions play an important role since the approximation problem can now be phrased in the frequency domain. Let

$$u_f(\omega) = \mathcal{F}u(t), \quad y_f(\omega) = \mathcal{F}y(t), \quad \hat{y}_f(\omega) = \mathcal{F}\hat{y}(t)$$

be the Fourier transforms of the corresponding time domain signals defined for $t \in (-\infty, +\infty)$ (we assume these signals to be of bounded ℓ_2 norm). If both transfer functions $T(s)$ and $\hat{T}(s)$ correspond to *stable* dynamical systems (i.e., their poles must lie in the open left half plane), then the inputs and outputs of these systems are related at each frequency ω by the so-called frequency response of the systems defined as the transfer function evaluated at $s = j\omega$ [8] :

$$y_f(\omega) = T(j\omega)u_f(\omega), \quad \hat{y}_f(\omega) = \hat{T}(j\omega)u_f(\omega).$$

As a consequence we also have that the error $e(t) \doteq [y(t) - \hat{y}(t)]$ has a Fourier transform

$$\mathcal{F}e(t) = e_f(\omega) = [T(j\omega) - \hat{T}(j\omega)]u_f(\omega).$$

Since the Fourier transform is a linear transformation preserving the energy of a signal (namely $||x_f(\omega)||_2 = \sqrt{2\pi}||x(t)||_2$), the energy $||e(t)||_2$ of the error signal $e(t)$ is minimized for all unit norm inputs $u(t)$ by minimizing $||e_f(\omega)||_2$ for all $||u_f(\omega)||_2 = 1$. The so-called H_∞-norm, denoted by $||\cdot||_\infty$, is defined as follows [14] :

$$||T(j\omega)||_\infty = \sup_{u_f \neq 0} ||T(j\omega)u_f(\omega)||_2 / ||u_f(\omega)||_2.$$

This norm is thus defined as the largest possible energy increase between inputs and outputs of a dynamical system and since Fourier transforms are linear, this holds as well in the time domain as in the frequency domain. For a stable transfer function, one shows that [14] :

$$||T(.)||_\infty \doteq \max_\omega ||T(j\omega)||_2.$$

Applying this to the difference between the original and approximate system, we obtain

$$||T(.) - \hat{T}(.)||_\infty \doteq \max_\omega ||T(j\omega) - \hat{T}(j\omega)||_2,$$

so that we will minimize the worst case error $||e_f(\omega)||_2$ by minimizing $||T(.) - \hat{T}(.)||_\infty$.

So the problem of model reduction of a stable linear time-invariant system can be stated as follows : *find the best stable approximation $\hat{T}(.)$ of a given degree for $T(.)$ where the error is measured in the H_∞ norm.*

Unfortunately, it is not easy to find good (stable) approximations $\hat{T}(.)$ of given (stable) transfer functions $T(.)$ using this norm. Another norm that is quite close to the H_∞-norm is the Hankel norm, which is still the largest possible energy increase between inputs and outputs, but where we restrict inputs to be nonzero in the interval $(-\infty, 0)$ (i.e., the "past") and outputs to be nonzero in the interval $[0, \infty)$ (i.e., the "future"). It turns out that the input/output map is then easy to describe and its norm easy to compute. For the continuous-time case we have

$$y(t) = \int_{-\infty}^{0} Ce^{A(t-\tau)}Bu(\tau)d\tau = Ce^{At} \cdot \int_{0}^{\infty} e^{A\tau}Bu(-\tau)d\tau, \quad t \in [0, \infty).$$

This map obviously factorizes into two sub-maps :

$$y(t) = Ce^{At}x(0), \quad x(0) = \int_{0}^{\infty} e^{A\tau}Bu(-\tau)d\tau.$$

The infinite dimensional "Hankel" operator $\mathcal{H}$ mapping $u(t)$, $t \in (-\infty, 0)$ to $y(t)$, $t \in [0, \infty)$ is then of rank at most N since $x(0) \in \Re^N$.

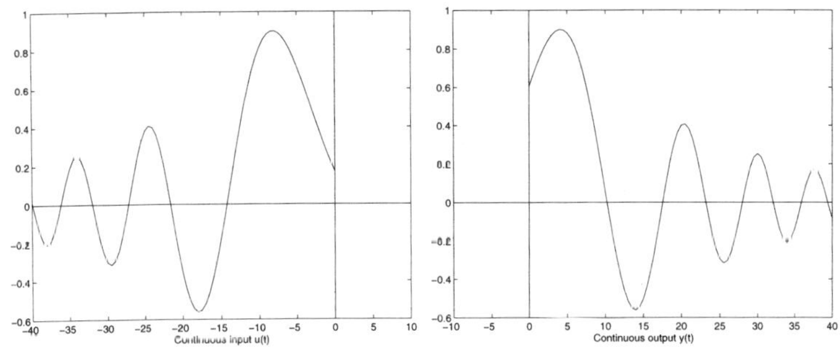

$$u(t),\ t \in (-\infty, 0) \Longrightarrow x(0) \Longrightarrow y(t),\ t \in [0, \infty)$$

The same holds for discrete-time systems where the input/output relation becomes

$$y(k) = \sum_{-\infty}^{0} CA^{(k-j)}Bu(j) = CA^k \cdot \sum_{0}^{\infty} A^j Bu(-j), \quad k \in [0, \infty).$$

This map obviously factorizes into two sub-maps since

$$y(k) = CA^k x(0), \quad x(0) = \sum_0^\infty A^j Bu(-j).$$

The infinite dimensional "Hankel" operator $\mathcal{H}$ mapping $u(k),\ k \in (-\infty, 0)$ to $y(k),\ k \in [0, \infty)$ is then of rank at most N since $x(0) \in \Re^N$.

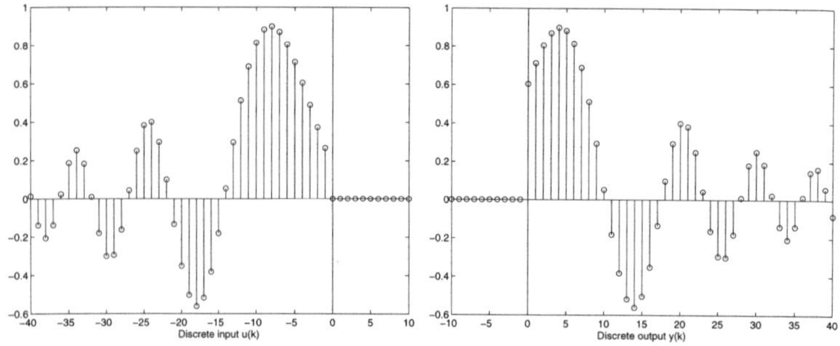

$$u(k),\ k \in (-\infty, 0) \Longrightarrow x(0) \Longrightarrow y(k),\ k \in [0, \infty)$$

Since the Hankel operator has finite rank N, it has only a finite number of nonzero singular values and these can easily be computed. For this, we represent the two linear maps as follows :

$$y([0, \infty)) = \mathcal{O}x(0), \quad x(0) = \mathcal{C}u((-\infty, 0))$$

and define the dual maps as

$$\mathcal{O}^* : y([0, \infty)) \mapsto x(0), \quad \mathcal{C}^* : x(0) \mapsto u((-\infty, 0)).$$

2.2 Gramians

The products $G_o \doteq \mathcal{O}^*\mathcal{O}$ and $G_c \doteq \mathcal{C}\mathcal{C}^*$ are $N \times N$ matrices and are called the Gramians of the system.

For a continuous-time system (2.1) the observability and controllability Gramians are equal to [8] :

$$G_o = \int_0^{+\infty} (Ce^{At})^T (Ce^{At}) dt, \tag{2.3}$$

$$G_c = \int_0^{+\infty} (e^{At}B)(e^{At}B)^T dt, \tag{2.4}$$

which by Parseval's theorem are also equal to

$$G_o = \frac{1}{2\pi}\int_{-\infty}^{+\infty} (-j\omega I - A^T)^{-1} C^T C (j\omega I - A)^{-1} d\omega, \tag{2.5}$$

$$G_c = \frac{1}{2\pi}\int_{-\infty}^{+\infty} (j\omega I - A)^{-1} B B^T (-j\omega I - A^T)^{-1} d\omega. \tag{2.6}$$

These Gramians can be computed as the solution of the Lyapunov equations

$$A^T G_o + G_o A + C^T C = 0 \quad \text{and} \quad A G_c + G_c A^T + B B^T = 0. \tag{2.7}$$

For a discrete-time system the corresponding time domain definitions are :

$$G_o = \sum_0^{+\infty} (CA^k)^T (CA^k), \tag{2.8}$$

$$G_c = \sum_0^{+\infty} (A^k B)(A^k B)^T, \tag{2.9}$$

which by Parseval's theorem can be transformed to the frequency domain :

$$G_o = \frac{1}{2\pi}\sum_{-\infty}^{+\infty} (e^{-j\omega} I - A^T)^{-1} C^T C (e^{j\omega} I - A)^{-1} \tag{2.10}$$

$$G_c = \frac{1}{2\pi}\sum_{-\infty}^{+\infty} (e^{j\omega} I - A)^{-1} B B^T (e^{-j\omega} I - A^T)^{-1}. \tag{2.11}$$

These Gramians can again be computed as the solution of the Stein equations

$$A^T G_o A - G_o + C^T C = 0 \quad \text{and} \quad A G_c A^T - G_c + B B^T = 0. \tag{2.12}$$

How can one compute the norm of the Hankel map

$$\mathcal{H} = \mathcal{O}\mathcal{C}\ ?$$

One shows that for any two positive definite matrices there always exists a (so-called contragradient) transformation

$$\hat{G}_o \doteq T^T G_o T, \quad \hat{G}_c \doteq T^{-1} G_c T^{-T},$$

such that both new Gramians are equal and diagonal [9] :

$$\hat{G}_o = \hat{G}_c = \Lambda \doteq \text{diag}\ \{\lambda_1, \ldots, \lambda_N\}. \tag{2.13}$$

The transformed Gramians $\hat{G}_o$ and $\hat{G}_c$ are easily seen to be the new Gramians of the transformed system

$$\left\{\hat{A},\hat{B},\hat{C},\hat{D}\right\} \doteq \{T^{-1}AT, T^{-1}B, CT, D\} \stackrel{T}{\Longleftarrow} \{A,B,C,D\}.$$

Note that the corresponding state-to-outputs and inputs-to-state maps of the transformed system equal

$$\hat{\mathcal{O}} = \mathcal{O}T, \quad \hat{\mathcal{C}} = T^{-1}\mathcal{C}.$$

In this new coordinate system we thus have

$$\hat{\mathcal{O}}^*\hat{\mathcal{O}} = \Lambda = \hat{\mathcal{C}}\hat{\mathcal{C}}^*. \tag{2.14}$$

The transformation T is in fact a similarity that diagonalizes the product G_cG_o since

$$T^{-1}(G_cG_o)T = \hat{G}_c\hat{G}_o = \Lambda^2.$$

Moreover, the λ_i's are the singular values of the Hankel map since the maps

$$\mathcal{H}^*\mathcal{H} = \mathcal{C}^*\mathcal{O}^*\mathcal{O}\mathcal{C} \quad \text{and} \quad \mathcal{C}\mathcal{C}^*\mathcal{O}^*\mathcal{O} = T\Lambda^2T^{-1}$$

have the same nonzero eigenvalues. Finally (2.14) also implies that Λ contains the square of the singular values of the transformed maps $\hat{\mathcal{O}}$ and $\hat{\mathcal{C}}$.

2.3 Approximation via projection methods

How is this now used for model reduction? It is well known that the best approximation M_n of a given rank $n \ll N$ to a linear map M of rank N is obtained from the singular value decomposition of M. Notice that this decomposition always exists if M is bounded and has finite rank N to start with. Let us partition the eigenvalue matrix Λ and the system matrices $\{\hat{A},\hat{B},\hat{C}\}$ conformably such that the $(1,1)$ blocks are $n \times n$:

$$\Lambda \doteq \begin{bmatrix} \Lambda_{11} & 0 \\ 0 & \Lambda_{22} \end{bmatrix}, \hat{A} \doteq \begin{bmatrix} \hat{A}_{11} & \hat{A}_{12} \\ \hat{A}_{21} & \hat{A}_{22} \end{bmatrix}, \hat{B} \doteq \begin{bmatrix} \hat{B}_1 \\ \hat{B}_2 \end{bmatrix}, \hat{C} \doteq \begin{bmatrix} \hat{C}_1 & \hat{C}_2 \end{bmatrix}. \tag{2.15}$$

Define also

$$\hat{Y}^T \doteq \begin{bmatrix} I_n & 0 \end{bmatrix}, \quad \hat{X} \doteq \begin{bmatrix} I_n \\ 0 \end{bmatrix}, \quad \hat{P} \doteq \hat{X}\hat{Y}^T, \tag{2.16}$$

then $\hat{P}$ is a projector since $\hat{Y}^T\hat{X} = I_n$ and $\hat{P}^2 = \hat{P}$. It follows from (2.14) that there exist orthogonal transformations U and V which we partition conformably, such that

$$\hat{\mathcal{O}} = \begin{bmatrix} U_1\Lambda_{11}^{\frac{1}{2}} & U_2\Lambda_{22}^{\frac{1}{2}} \end{bmatrix}, \quad \hat{\mathcal{C}} = \begin{bmatrix} \Lambda_{11}^{\frac{1}{2}}V_1^T \\ \Lambda_{22}^{\frac{1}{2}}V_2^T \end{bmatrix},$$

$$\|\mathcal{H} - \hat{\mathcal{O}}\hat{P}\hat{\mathcal{C}}\|_2 = \|U_2\Lambda_{22}V_2^T\|_2 = \lambda_{n+1}.$$

The mapping $\hat{\mathcal{O}}\hat{P}\hat{\mathcal{C}}$ is therefore an optimal rank n approximation to $\mathcal{H}$ but it does not necessarily have a Hankel structure, in which case it does not correspond to a time-invariant system [14]. It turns out that if the gap $\lambda_n - \lambda_{n+1}$ is small then the projected system

$$\left\{\hat{A}_{11}, \hat{B}_1, \hat{C}_1, D\right\} = \left\{\hat{Y}^T\hat{A}\hat{X}, \hat{Y}^T\hat{B}, \hat{C}\hat{X}, D\right\}$$

has a corresponding Hankel map that is very close to $\hat{\mathcal{O}}\hat{P}\hat{\mathcal{C}}$. For obvious reasons, this technique of producing reduced order models is called "balanced truncation". One shows [9] that balanced truncation of stable systems always produces stable reduced order models as well. This technique has the additional advantage to yield simultaneously *all* reduced order models of degree 1 up to $N-1$ and one can choose, e.g., the order n of the approximate model based on the error estimate λ_n between both Hankel maps. A more involved model reduction technique that is derived from the balanced coordinate system is the so-called optimal Hankel norm approximation [4]. Its construction involves more work but yields a stable and optimal Hankel norm approximation of a given order n. When the gap $\lambda_n - \lambda_{n-1}$ is small, both techniques yield very close models [14].

2.4 Large-scale models

We pointed out that balanced truncation is based on the construction of two restrictions X and Y satisfying $Y^TX = I_n$ and hence of a projector $P = XY^T$. This projector $\hat{P}$ is orthogonal in the coordinate system of the balanced realization (2.16), but the corresponding projector P is not orthogonal in the original coordinate system. Defining

$$X = T\hat{X}, \quad Y^T = \hat{Y}^TT^{-1},$$

it follows that $Y^TX = \hat{Y}^TT^{-1}T\hat{X} = I_n$ and that the balanced truncation amounts to

$$\left\{\hat{A}_{11}, \hat{B}_1, \hat{C}_1, D\right\} = \left\{Y^TAX, Y^TB, CX, D\right\}$$

in the original coordinate system. The construction of the projector in the previous section was based on an eigenvalue decomposition of the product of two Gramians, which is a dense matrix even if the original system $\{A, B, C, D\}$ is sparse.

So how can one produce an "approximation" of this projector P without having to perform eigenvalue or singular value decompositions of dense $N \times N$ matrices? For this it is informative to look at the Hankel map of the discrete-time case :

$$\begin{bmatrix} y_0 \\ y_1 \\ y_2 \\ \vdots \end{bmatrix} = \mathcal{O}\mathcal{C} \begin{bmatrix} u_{-1} \\ u_{-2} \\ u_{-3} \\ \vdots \end{bmatrix} = \begin{bmatrix} C \\ CA \\ CA^2 \\ \vdots \end{bmatrix} \begin{bmatrix} B & AB & A^2B & \dots \end{bmatrix} \begin{bmatrix} u_{-1} \\ u_{-2} \\ u_{-3} \\ \vdots \end{bmatrix}.$$

From this one expects Krylov sequences to play an important role in approximating both factors $\mathcal{O}$ and $\mathcal{C}$. A Krylov subspace is defined as follows :

$$\mathcal{K}_j(M, R) = Im\left\{R, MR, M^2R, \dots, M^{j-1}R\right\} \tag{2.17}$$

where the matrices M and R are $N \times N$ and $N \times k$, respectively. The construction of bases of Krylov spaces of a particular order have been studied for many years in the numerical linear algebra literature (especially for sparse M and R). The link with the present problem is obviously the eigenvalue problem since we would like to approximate the dominant eigenspace of the product of the Gramians G_oG_c. One wishes in fact to construct an n dimensional basis with $n \ll N$ and yet capture the dominant features of both Gramians.

For linear time invariant systems, one can exploit the frequency domain identities (2.6, 2.11) very efficiently. They suggest that the Gramians are well approximated if the transfer function is interpolated in points where the expressions (2.6, 2.11) are large. The transfer function $T(s)$ can be expanded in a Taylor series around any point σ that is not a pole of $T(s)$:

$$T(s) = T_0 + T_1(s-\sigma)^1 + T_2(s-\sigma)^2 + \cdots, \tag{2.18}$$

where the coefficients T_i – called moments – are equal to :

$$T_0 \doteq T(\sigma) = D - C(A - \sigma I)^{-1}B,$$
$$T_i \doteq \frac{1}{i!}\frac{\partial^i T(s)}{\partial s^i}\bigg|_{s=\sigma} = -C(A-\sigma I)^{-(i+1)}B, \; i > 0.$$

In general, one can produce a reduced-order model that interpolates the frequency response and its derivatives at multiple points $\{\sigma^{(1)}, \sigma^{(2)}, \dots, \sigma^{(K)}\}$.

A model meeting these constraints is denoted by a multipoint Padé approximation or a rational interpolant. Recently, a general theory for such interpolations, based on Krylov subspace computations was presented [5]. It is shown in [6] that these approximations also satisfy certain Galerkin conditions for the Lyapunov equations (2.7, 2.12). Moreover, it is shown there that these techniques also apply to implicit systems of differential equations. Another approach is to use interpolation ideas to accelerate particular iterative methods for computing the solution of Lyapunov and Stein equations directly [10].

3 Time-varying systems

Time-varying systems are described by systems of differential or difference equations. Assume that we are given a large-scale time-varying system

$$\left\{ \begin{array}{l} \dot{x}(t) = A(t)x(t) + B(t)u(t) \\ y(t) = C(t)x(t) + D(t)u(t) \end{array} \right. \quad \left\{ \begin{array}{l} x(k+1) = A(k)x(k) + B(k)u(k) \\ y(k) = C(k)x(k) + D(k)u(k) \end{array} \right. \tag{3.1}$$

which we would like to approximate by a lower order model

$$\left\{ \begin{array}{l} \dot{\hat{x}}(t) = \hat{A}(t)\hat{x}(t) + \hat{B}(t)u(t) \\ \hat{y}(t) = \hat{C}(t)\hat{x}(t) + \hat{D}(t)u(t) \end{array} \right. \quad \left\{ \begin{array}{l} \hat{x}(k+1) = \hat{A}(k)\hat{x}(k) + \hat{B}(k)u(k) \\ \hat{y}(k) = \hat{C}(k)\hat{x}(k) + \hat{D}(k)u(k). \end{array} \right. \tag{3.2}$$

It is obvious that projectors have to be time-varying as well in order to capture the dynamics of the system at each time instant. Since there is no transfer function anymore, one should use instead the Gramians of the state-to-outputs and inputs-to-state maps.

Gramians of time-varying systems are based on the fundamental solution matrix $\Phi(.,.)$ of the corresponding homogeneous system, which yields the state at a particular time in terms of the state at a previous time [8] :

$$\begin{array}{ll} \dot{x}(t) = A(t)x(t), & x(k+1) = A(k)x(k), \\ x(T) = \Phi(T,t)x(t), & x(K) = \Phi(K,k)x(k). \end{array}$$

For a continuous-time system the observability and controllability Gramians can be defined as follows for a finite time window $t \in [t_i, t_f]$:

$$G_o(t) = \int_t^{t_f} (C(\tau)\Phi(\tau,t))^T (C(\tau)\Phi(\tau,t))d\tau, \tag{3.3}$$

$$G_c(t) = \int_{t_i}^{t} (\Phi(t,\tau)B(\tau))(\Phi(t,\tau)B(\tau))^T d\tau. \tag{3.4}$$

Using $\frac{d}{d(-t)}\Phi(T,t) = \Phi(T,t)A(t)$ and $\frac{d}{dT}\Phi(T,t) = A(T)\Phi(T,t)$ we observe that the Gramians can be computed as the solution of the Lyapunov differential equations

$$\begin{aligned} \tfrac{d}{d(-t)}G_o(t) &= A(t)^T G_o(t) + G_o(t)A(t) + C(t)^T C(t), \quad G_o(t_f) = 0, \\ \tfrac{d}{dt}G_c(t) &= A(t)G_c(t) + G_c(t)A(t)^T + B(t)B(t)^T, \quad G_c(t_i) = 0. \end{aligned}$$

Notice that the first equation goes "backward" in time, while the second goes forward in time. For a discrete-time system the corresponding Gramians are defined for $k \in [k_i, k_f]$ as follows :

$$G_o(k) = \sum_{j=k}^{k_f} (C(j)\Phi(j,k))^T (C(j)\Phi(j,k)), \tag{3.5}$$

$$G_c(k) = \sum_{j=k_i}^{k} (\Phi(k,j)B(j))(\Phi(k,j)B(j))^T. \tag{3.6}$$

Using $\Phi(K,k) = \Phi(K,k+1)A(k)$ and $\Phi(K+1,k) = A(K)\Phi(K,k)$ we observe that the Gramians can be computed as the solution of the Lyapunov difference equations

$$\begin{aligned} G_o(k) &= A(k)^T G_o(k+1)A(k) + C(k)^T C(k), \quad G_o(k_f+1) = 0, \\ G_c(k+1) &= A(k)G_c(k)A(k)^T + B(k)B(k)^T, \quad G_c(k_i-1) = 0. \end{aligned}$$

Again we point out that both recurrences evolve differently with time. In the literature (see, e.g., [11, 12]) one typically assumes $t_i = k_i = -\infty$ and $t_f = k_f = \infty$ but in order to be able to compute the Gramians we consider here a finite interval of time.

The continuous-time and discrete-time problems are very similar but since continuous-time systems need to be discretized anyway, we focus here on discrete-time systems. The state-to-outputs and inputs-to-state maps of the discrete-time case are :

$$\begin{bmatrix} y(k) \\ y(k+1) \\ y(k+2) \\ \vdots \end{bmatrix} = \begin{bmatrix} C(k) \\ C(k+1)A(k) \\ C(k+2)A(k+1)A(k) \\ \vdots \end{bmatrix} x(k),$$

$$x(k) = \begin{bmatrix} B(k-1) & A(k-1)B(k-2) & A(k-1)A(k-2)B(k-3) & \dots \end{bmatrix} \begin{bmatrix} u(k-1) \\ u(k-2) \\ u(k-3) \\ \vdots \end{bmatrix}.$$

The Gramians clearly reflect the *energy* of both these maps and hence play an important role in their approximation. Provided we compute only a finite window of the above Gramians then one can easily come up with a square root version for the Gramians. Let $R_o^T(k)R_o(k) = G_o(k)$ and $R_c^T(k)R_c(k) = G_c(k)$ be Cholesky factorizations of these Gramians then $R_o(k)$ and $R_c(k+1)$ are the upper triangular factors of the QR factorizations

$$Q_o(k)R_o(k) = \begin{bmatrix} R_o(k+1)A(k) \\ C(k) \end{bmatrix}, \quad Q_c(k+1)R_c(k+1) = \begin{bmatrix} R_c(k)A(k)^T \\ B(k)^T \end{bmatrix},$$

respectively. Rather than computing the exact factors, one can keep a low rank approximation of both Gramians

$$\hat{R}_o^T(k)\hat{R}_o(k) \approx G_o(k), \quad \hat{R}_c^T(k)\hat{R}_c(k) \approx G_c(k). \tag{3.7}$$

The matrices $\hat{R}_o(k+1)$ and $\hat{R}_c(k)$ will have, e.g., $n \ll N$ rows, implying that at each step of the QR factorization only the n "dominant" rows of the triangular factors should be kept. The basic idea for this is to keep at each step the leading n row vectors of the singular value decomposition rather than performing the above QR decompositions. How to do this is described in a different context in [13], [7].

It is important to point out here that one can still define the eigenvalues of the product of the Gramians :

$$T^{-1}(k)G_c(k)G_o(k)T(k) = \Lambda(k)^2$$

and these will be positive real if the system is completely controllable and completely observable over the considered time interval [11, 12]. The contragradient transformation then exists and reduced order models can be constructed provided there is a nonzero gap $\lambda_n(k) - \lambda_{n+1}(k) > 0$ at each time instant. The optimal projector would be $P(k) = Y^T(k)X(k)$ where $X(k)$ contains the first n columns of $T(k)$ and $Y^T(k)$ contains the n first rows of $T^{-1}(k)$. It follows from (3.7) that a good approximation is given by

$$\hat{P}(k) \doteq \hat{R}_o^T(k)[\hat{R}_c(k)\hat{R}_o^T(k)]^{-1}\hat{R}_c(k)$$

and its quality will depend on the gap $\lambda_n(k) - \lambda_{n+1}(k) > 0$ at each time step.

For continuous-time problems with Gramians defined over infinite horizon intervals $(-\infty, t]$ and $[t, +\infty)$, it is shown in [11, 12] that if the original system is stable, then so will the reduced order model provided the original system is uniformly completely controllable and observable. If finite intervals are considered, the issue of stability is less crucial of course.

Notice that since we are defining projectors for finite time windows, this could also be applied to linear time-invariant systems that are unstable. One can then not show any property of stability for the reduced order system, but the finite horizon Hankel map will at least be well approximated.

3.1 Time-varying linearized problems

The nonlinear problems in dynamical systems are much harder to handle

$$\begin{cases} \dot{x}(t) = G(x(t), u(t)) \\ y(t) = H(x(t), u(t)), \end{cases} \qquad \begin{cases} x(k+1) = G(x(k), u(k)) \\ y(k) = H(x(k), u(k)). \end{cases}$$

One typically linearizes such models along a "nominal" trajectory $(x(t), u(t))$ by computing $A(.), B(.), C(.)$ and $D(.)$ from the Taylor expansion of $G(.,.)$ and $H(.,.)$ around that trajectory. Model reduction techniques for time-varying models could then be applied to this, but the construction of the Gramians becomes too complex, since they evolve in two different time directions.

A simpler idea that has gained popularity in nonlinear systems is to just compute a trajectory $x(.)$ and consider

$$\int_{t_i}^{t_f} x(\tau)x(\tau)^T d\tau \quad \text{and} \quad \sum_{k=k_i}^{k_f} x(k)x(k)^T,$$

as approximations of Gramians (or "energy functions") for constructing an appropriate projector. This is known as the Proper Orthogonal Decomposition (POD) technique [1].

How does this relate to the time-varying schemes described in the previous section? If we consider a linear time-varying system :

$$\dot{x}(t) = A(t)x(t) \quad \text{and} \quad x(k+1) = A(k)x(k),$$

with initial conditions $x(t_i)$ and $x(k_i)$, respectively, then

$$x(t) = \Phi(t, t_i)x(t_i) \quad \text{and} \quad x(k) = \Phi(k, k_i)x(k_i).$$

The above expressions then become

$$\int_{t_i}^{t_f} (\Phi(\tau, t_i)x(t_i))(\Phi(\tau, t_i)x(t_i))^T d\tau \quad \text{and} \sum_{k=k_i}^{k_f} (\Phi(k, k_i)x(k_i))(\Phi(k, k_i)x(k_i))^T,$$

which shows the link with Gramians over a finite time interval. The difference here is that no input matrix $B(.)$ is involved.

4 Concluding remarks

Model reduction of dynamical systems has its roots in many different fields of applied mathematics. The earlier occurrence of such techniques is the approximation of rational functions of high degree by one of lower degree. The first results in that area were formulated in a mathematical setting and included techniques such as Padé approximations, continued fraction expansions and so on [3]. Such results were also used in the context of model reduction techniques or approximation techniques in application areas such as signals and systems, and lead to the synthesis of approximating systems by one of a prespecified degree. Examples of algorithmic developments in this area are the Remez algorithm in filter design and the Massey Berlekamp algorithm in convolutional codes. More recent developments in linear systems theory are nicely synthesized in [2].

But these developments are referring mainly to the area of linear time-invariant dynamical systems. In this chapter we have tried to establish connections between different projection techniques used in the area of systems and control and in particular showed how to extend this to the time-varying case. We also tried to show the connection with a popular technique for nonlinear dynamical systems.

References

[1] P. Holmes, J. Lumley and G. Berkooz, *Turbulence, Coherent Structures, Dynamical Systems and Symmetry*, Cambridge University Press, Cambridge, U.K., 1996.

[2] J. Ball, I. Gohberg and L. Rodman, *Interpolation of Rational Matrix Functions*, Operator Theory : Advances and Applications **45**, Birkhäuser Verlag, Basel, 1990.

[3] C. Brezinski, *Padé-Type Approximation and General Orthogonal Polynomials*, **ISNM 50**, Birkhäuser, Basel, 1980.

[4] K. Glover, All optimal Hankel norm approximations of linear time multivariable systems and their L_∞-error bounds, *Int. J. Contr.* **39**, pp. 1115-1193, 1984.

[5] E. Grimme, K. Gallivan and P. Van Dooren, On some recent developments in projection-based model reduction, in *ENUMATH 97, 2nd European Conference on Numerical Mathematics and Advanced Applications*,

H.G. Bock, F. Brezzi, R. Glowinski, G. Kanschat, Yu.A. Kuznetsov, J. Périaux, R. Rannacher (eds.), World Scientific Publishing, Singapore, pp. 98-113, 1998.

[6] E. Grimme, K. Gallivan and P. Van Dooren, Model reduction of large-scale systems. Rational Krylov versus balancing techniques, in *Error Control and Adaptivity in Scientific Computing*, H. Bulgak and C. Zenger (eds.), Kluwer Academic Publishers, Dordrecht, The Netherlands, pp. 177-190, 1999.

[7] K. Gallivan and P. Van Dooren, Recursive calculation of dominant singular subspaces, Internal Report, CESAME, Univ. Catholique de Louvain, Belgium, 1999.

[8] T. Kailath, *Linear Systems*, Prentice Hall, Englewood Cliffs, N.J., 1980.

[9] B.C. Moore, Principal component analysis in linear systems : controllability, observability and model reduction, *IEEE Trans. Aut. Contr.* **26**, pp. 17-32. 1981.

[10] T. Penzl, *A Cyclic Low Rank Smith Method for Large Sparse Lyapunov Equations with Applications in Model Reduction and Optimal Control*, T.U. Chemnitz, Dept. Mathematics, Preprint SFB393/98-6, 1998.

[11] S. Shokoohi, L. Silverman and P. Van Dooren, Linear time-variable systems: balancing and model reduction, *IEEE Trans. Aut. Contr.*, **28**, pp. 810-822, 1983.

[12] S. Shokoohi, L. Silverman and P. Van Dooren, Stable approximation of time-variable systems, *Automatica*, **20**, pp. 59-67, 1984.

[13] M. Verlaan and A. Heemink, Tidal flow forecasting using reduced rank square root filters, *Stochastic Hydrology and Hydraulics*, **11**, pp. 349-368, 1997.

[14] K. Zhou, J.C. Doyle and K. Glover, *Robust and Optimal Control*, Upper Saddle River, Prentice Hall, Englewood Cliffs, N.J., 1996.

Acknowledgements

This work was supported by the National Science Foundation under grant CCR-9796315. Parts of this chapter present research results of the Belgian Programme on Inter-university Poles of Attraction, initiated by the Belgian State,

Prime Minister's Office for Science, Technology and Culture. The scientific responsibility rests with its authors.

Discussions with K. Gallivan and N. Nichols regarding this chapter are also gratefully acknowledged.

Paul M. Van Dooren
CESAME
Université Catholique de Louvain
B-1348 Louvain-la-Neuve, Belgium
`vdooren@csam.ucl.ac.be`
`http://www.auto.ucl.ac.be/~vdooren/`

G. A. WATSON

Solving data fitting problems in l_p norms with bounded uncertainties in the data

Abstract We consider a range of robust data fitting problems which have attracted interest so far in the special case when least squares norms are involved. We extend the analysis of these problems to l_p norms, and consider some numerical methods for obtaining solutions.

1 Introduction

A central problem in data analysis is to estimate a set of parameters in a model from inexact data. If the model is linear in the parameters, then one formulation of the problem is that of determining $x \in R^n$ to solve the overdetermined system of linear equations

$$Ax \approx b,$$

where $A \in R^{m \times n}$, $b \in R^m$ arise from the data. For given $x \in R^n$, define

$$r = Ax - b.$$

Then perhaps the most common way of determining x is to minimize $||r||$ over $x \in R^n$, where the norm is some norm on R^m. This involves an underlying assumption that A is exact, and all the errors are in b, and this may not be the case in many practical situations. The effect of errors in A as well as b has been recognized and studied for many years, mainly in the statistics literature. One way to take the more general case into account is to solve the problem

$$\text{minimize } ||E : d|| \text{ subject to } (A + E)x = b + d, \tag{1.1}$$

where the matrix norm is one on $(m \times (n + 1))$ matrices. This problem, when the matrix norm is the Frobenius norm, was first analyzed by Golub and Van Loan [7], who used the term total least squares, and developed an algorithm based on the singular value decomposition of $[A : b]$. However, for problems for which A is nearly exact, this formulation may exaggerate the errors in A, and this has lead to consideration of a new formulation in which the quantities E and d are bounded, having known bounds. This idea gives rise to a number of different, but closely related problems. Algorithms and analysis for problems of this type based on least squares norms are given in [1], [2], [5], [6].

In Section 2, we consider problems where separate bounds on norms of E and d are assumed known. In Section 3, we consider the case when a bound on the matrix $[E : d]$ is known. The problems which arise can be interpreted as permitting more robust solutions to be obtained for the underlying data fitting problems: for an explanation of the significance of the term robustness in this context, see for example [6]. A simpler interpretation of the problems being solved is that they guarantee that the effect of the uncertainties in the data will never be overestimated, beyond the assumptions made by knowledge of the bounds.

Of interest here are problems which are based on the use of l_p norms, $1 < p < \infty$, where the matrix norm (on matrices of any size) is defined by

$$||M|| = (\sum_{ij} |M_{ij}|^p)^{1/p}, \tag{1.2}$$

or

$$||M|| = \max_{||x||_p=1} ||Mx||_p, \tag{1.3}$$

with the vector norm the usual l_p norm. Let p' be dual to p so that

$$1/p + 1/p' = 1.$$

Then useful relations which are easily verified are that

$$||Mx||_p \leq ||M|| ||x||_{p'}, \text{ for } ||M|| \text{ defined by (1.2)},$$

$$||Mx||_p \leq ||M|| ||x||_p, \text{ for } ||M|| \text{ defined by (1.3)}.$$

The above definitions and inequalities extend in an obvious way to the limiting values $p = 1$ and $p = \infty$. Indeed much of what follows extends also. However, since the norms involved would no longer be smooth, the algorithmic development would be significantly different.

2 Known bounds on $||E||$ and $||d||_p$

In this section, it is assumed that separate bounds on $||E||$ and $||d||_p$ are known. Then following [2], we consider the problem of determining

$$\min_{x \in R^n} \max_{||E|| \leq \rho, \ ||d||_p \leq \rho_d} ||(A + E)x - (b + d)||_p, \tag{2.1}$$

for matrix norms from (1.2) or (1.3). For both cases (2.1) can be replaced by a more tractable problem, as shown by the following two theorems.

Theorem 2.1 *Let the norm on E be defined by (1.2). For any x, the maximum in (2.1) is attained when*

$$E = \rho u w^T,$$

$$d = -\rho_d u,$$

where

$$w_i = x_i |x_i|^{p'-2} ||x||_{p'}^{1-p'}, \quad i = 1, \ldots, n,$$

where $u = \frac{r}{||r||_p}$ if $r \neq 0$, otherwise u is arbitrary but $||u||_p = 1$. The maximum value is

$$||r||_p + \rho ||x||_{p'} + \rho_d.$$

Proof We have for any E, d such that $||E|| \leq \rho$, $||d||_p \leq \rho_d$,

$$\begin{aligned} ||(A+E)x - (b+d)||_p &= ||r + Ex - d||_p \\ &\leq ||r||_p + \rho ||x||_{p'} + \rho_d. \end{aligned}$$

Now let E and d be as in the statement of the theorem. Then since $||w||_p = 1$, and $||u||_p = 1$,

$$||E|| = \rho, \qquad ||d||_p = \rho_d,$$

and further since $w^T x = ||x||_{p'}$,

$$\begin{aligned} ||(A+E)x - (b+d)||_p &= ||r + \rho ||x||_{p'} u + \rho_d u||_p \\ &= ||r||_p + \rho ||x||_{p'} + \rho_d. \end{aligned}$$

The result follows. □

A consequence of this result is that the problem (2.1) with the norm on E given by (1.2) is solved by minimizing with respect to x

$$f(x) = ||Ax - b||_p + \rho ||x||_{p'}. \tag{2.2}$$

The next theorem can be proved in a similar way.

Theorem 2.2 *Let the norm on E be defined by (1.3). For any x, the maximum in (2.1) is attained when*

$$E = \rho u w^T,$$

$$d = -\rho_d u,$$

where

$$w_i = x_i |x_i|^{p-2} ||x||_p^{1-p}, \quad i = 1, \ldots, n,$$

where $u = \frac{r}{||r||_p}$ if $r \neq 0$, otherwise u is arbitrary but $||u||_p = 1$. The maximum value is

$$||r||_p + \rho ||x||_p + \rho_d.$$

A consequence of this result is that the problem (2.1) with the norm on E given by (1.3) is solved by minimizing with respect to x

$$f(x) = ||Ax - b||_p + \rho ||x||_p. \tag{2.3}$$

Both (2.2) and (2.3) are special cases of the general problem of minimizing

$$f(x) = ||Ax - b||_p + \rho ||x||_q, \tag{2.4}$$

with

$$1 < p, \ q < \infty.$$

In addition to the cases considered above, other combinations of norms in (2.4) correspond to other norms from generalizations of the class (1.3). The expression (2.4) is not differentiable if either x or $r = Ax - b$ is identically zero. However, we can characterize a minimum using standard convex analysis as follows. We introduce the following notation, that for any vector $v \in R^t$ we will write

$$D_v = \text{diag}\{|v_1|, \ldots, |v_t|\},$$

and

$$\theta_v = (\text{sign}(v_1), \ldots, \text{sign}(v_t))^T.$$

Just as p' is dual to p, let q' be dual to q.

Theorem 2.3 *The function (2.4) is minimized at x if and only if*

$$A^T v + \rho w = 0, \tag{2.5}$$

where

$$v = ||r||_p^{1-p} D_r^{p-1} \theta_r, \quad \text{if } r \neq 0,$$

otherwise v is arbitrary except $||v||_{p'} \leq 1$, and

$$w = ||x||_q^{1-q} D_x^{q-1} \theta_x, \quad \text{if } x \neq 0,$$

otherwise w is arbitrary except $||w||_{q'} \leq 1$.

Theorem 2.4 *Let $r = b$ and let v as in Theorem 2.3 satisfy*

$$||A^T v||_{q'} \leq \rho.$$

Then $x = 0$ minimizes (2.2).

Proof For $x = 0$ to give a minimum we must have v as in Theorem 2.2 with $r = b$ so that (2.5) is satisfied with $||w||_{q'} \leq 1$. The result follows. □

Let us make the assumption (quite reasonable in practice) that there is no x which makes $||Ax - b||_p = 0$, so that $||Ax - b||_p$ is differentiable for all x. Then the condition (2.5) of Theorem 2.3 can be written

$$||r||_p^{1-p} A^T D_r^{p-2} r + \rho w = 0, \tag{2.6}$$

where w is as defined there.

At points where $x \neq 0$, then $||x||_q$ is also differentiable; in particular if $x = 0$ is not a solution then $f(x)$ is differentiable at all x in a neighbourhood of the minimum. If this is assumed, then (2.6) can be written

$$||r||_p^{1-p} A^T D_r^{p-2} r + \rho ||x||_q^{1-q} D_x^{q-2} x = 0,$$

or

$$(A^T D_r^{p-2} A + \alpha D_x^{q-2}) x = A^T D_r^{p-2} b, \tag{2.7}$$

where

$$\alpha = \frac{\rho ||Ax - b||_p^{p-1}}{||x||_q^{q-1}}. \tag{2.8}$$

There is a connection with the problem of minimizing the l_p norm of r. Defining

$$A(x) = A + \rho \|x\|_q^{1-q} \|r\|_p^{-1} r x^T D_x^{q-2},$$

we can write (2.7) as

$$A(x)^T D_r^{p-2} r = 0, \tag{2.9}$$

Thus the optimality conditions can be represented as an orthogonality relationship which is related to the condition for optimality for the minimization of the l_p norm of r (replace $A(x)$ by A).

When the norm on E is given by (1.2), and $q = p'$, then clearly for any x, we have from Theorem 2.1 that a maximizing perturbation E is given by $A(x) - A$, or equivalently $A(x)$ is the perturbed value of A. While evidently E is rank one, $A(x)$ is normally full rank.

Theorem 2.5 *Let A have rank n, and let $b \notin \text{range}(A)$. For any non-zero x, $A(x)$ has rank n.*

Proof Let $x \neq 0$, let $r \neq 0$, and let rank $(A(x)) < n$. Thus there exists $z \neq 0$ such that

$$A(x)z = 0.$$

If A has rank n, it follows that

$$x^T D_x^{q-2} z \neq 0.$$

Define the non-zero scalar

$$s = \rho \|x\|_q^{1-q} \|r\|_p^{-1} x^T D_x^{q-2} z.$$

Then it follows that

$$A(x + \frac{z}{s}) = b.$$

This gives a contradiction if $b \notin \text{range}(A)$, and the result is proved. □

The special case of (2.4) when both norms are the least squares norm corresponds to the use of the Frobenius norm or spectral norm on E. Note that the presence of least squares norms on both terms of (2.4) for the Frobenius norm reflects the fact that the 2-norm is self dual. This problem has been considered by Chandrasekaran et al. [1], [2] who give an efficient method of solution.

Least squares orthogonality results, and the special case of Theorem 2.5, are considered in [8]. When $p = q = 2$ and (2.4) is differentiable, (2.7) becomes

$$(A^T A + \alpha I)x = A^T b,$$

where from (2.8),

$$\alpha = \frac{\rho ||r||_2}{||x||_2}.$$

Let the singular value decomposition of A be

$$A = U \begin{bmatrix} \Sigma \\ 0 \end{bmatrix} V^T,$$

where $U \in R^{m\times m}$ and $V \in R^{n\times n}$ are orthogonal and $\Sigma = \text{diag}\{\sigma_1, \ldots, \sigma_n\}$ is the matrix of singular values in descending order of magnitude. Let

$$\begin{bmatrix} b_1 \\ b_2 \end{bmatrix} = U^T b,$$

where $b_1 \in R^n$ and $b_2 \in R^{m-n}$. It will be assumed in what follows that A has rank n, and further that $x = 0$ is not a solution (which means, in particular, that $b_1 \neq 0$) and $b \notin \text{range}A$ (which means that $b_2 \neq 0$). From Theorem 2.3, we require that

$$\rho < \frac{||A^T b||_2}{||b||_2} = \frac{||\Sigma b_1||_2}{||b||_2}. \tag{2.10}$$

Then it is shown in [2] that α satisfies the equation

$$\alpha = g(\alpha), \tag{2.11}$$

where

$$g(\alpha) = \frac{\rho\sqrt{||b_2||_2^2 + \alpha^2||(\Sigma^2 + \alpha I)^{-1} b_1||_2^2}}{||\Sigma(\Sigma^2 + \alpha I)^{-1} b_1||_2}.$$

For the general (p, q) problem, x cannot be eliminated between (2.7) and (2.8) to give an equation in α alone. Of course, even if that were possible, x is then not immediately available from (2.7).

It may be shown that if (2.10) holds, then simple iteration based on (2.11) is locally convergent to the unique positive root of the equation.

Theorem 2.6 *Let ρ satisfy (2.10). Then (2.11) has exactly one positive root α^* and the iteration*

$$\alpha_k = g(\alpha_{k-1}), \quad k = 1, 2, \ldots$$

is locally convergent to α^*.

Proof Let ρ satisfy (2.10). Then as in [2], (2.11) has a unique positive root α^*. Differentiating $G(\alpha)$ gives

$$G'(\alpha) = -2b_1^T(\Sigma^2 - \rho^2 I)(\Sigma^2 + \alpha I)^{-3} b_1 + 2\frac{\rho^2 \|b_2\|_2^2}{\alpha^3},$$

and so

$$\alpha^* G'(\alpha^*) = 2b_1^T \Sigma^2 (\Sigma^2 - \rho^2 I)(\Sigma^2 + \alpha^* I)^{-3} b_1, \tag{2.12}$$

using $G(\alpha^*) = 0$. Now $g(\alpha)$ and $G(\alpha)$ are related by

$$G(\alpha) = (1 - (\frac{g(\alpha)}{\alpha})^2)\|\Sigma(\Sigma^2 + \alpha I)^{-1} b_1\|_2^2,$$

and so

$$G'(\alpha^*) = \frac{2(1 - g'(\alpha^*))}{\alpha^*}\|\Sigma(\Sigma^2 + \alpha^* I)^{-1} b_1\|_2^2, \tag{2.13}$$

using $g(\alpha^*) = \alpha^*$. Thus

$$g'(\alpha^*) = 1 - \frac{\alpha^* G'(\alpha^*)}{2\|\Sigma(\Sigma^2 + \alpha^* I)^{-1} b_1\|_2^2}. \tag{2.14}$$

Substituting from (2.12) gives

$$\begin{aligned} g'(\alpha^*) &= 1 - \frac{b_1^T \Sigma^2 (\Sigma^2 - \rho^2 I)(\Sigma^2 + \alpha^* I)^{-3} b_1}{b_1^T \Sigma^2 (\Sigma^2 + \alpha^* I)^{-2} b_1} \\ &= 1 - \frac{b_1^T \Sigma^2 (\Sigma^2 + \alpha^* I)^{-2} b_1 - (\alpha^* + \rho^2) b_1^T \Sigma^2 (\Sigma^2 + \alpha^* I)^{-3} b_1}{b_1^T \Sigma^2 (\Sigma^2 + \alpha^* I)^{-2} b_1} \\ &= \frac{(\alpha^* + \rho^2) b_1^T \Sigma^2 (\Sigma^2 + \alpha^* I)^{-3} b_1}{b_1^T \Sigma^2 (\Sigma^2 + \alpha^* I)^{-2} b_1} \\ &> 0. \end{aligned}$$

It follows using (2.14) and $G'(\alpha^*) > 0$ that

$$0 < g'(\alpha^*) < 1,$$

and the result is proved. □

Numerical evidence suggests that simple iteration is an effective method, although clearly there are other possibilities. It is also clearly possible to use (2.7) and (2.8) to provide a simple iteration scheme in more general cases. Let an initial value of α be chosen, say $\alpha = 0$. Let an initial value of x be chosen, say the least squares solution of $Ax = b$. Then (2.8) can be used to update α, and then x can be updated from (2.7) by calculating

$$x = (A^T D_r^{p-2} A + \alpha D_x^{q-2})^{-1} A^T D_r^{p-2} b,$$

with the current values used in the right hand side. This iteration scheme can work; however, it requires p and q to be close to 2, and even if convergence is obtained as these values move away from 2, convergence may be very slow. Therefore this does not represent a generally useful solution method, and we consider next an efficient and practical method.

If f is differentiable, and also if both $p, q \geq 2$, then f is twice differentiable and this means that Newton's method, for example, can be used. This extends to $1 < p < 2$, $1 < q < 2$, provided that no component of r or x is zero in a neighbourhood of the solution. Because f is convex, then the direction defined by the Newton step is a descent direction, and so a line search can be introduced to force convergence. A possible starting point to use is $x = 0$. If that is not a solution and we descend from there, differentiability of f is subsequently assured. Therefore the next theorem is a very useful one.

Theorem 2.7 *Let*

$$c = -||b||_p^{1-p} A^T D_b^{p-2} b.$$

If $||c||_{q'} \leq \rho$, *then* $x = 0$ *minimizes* f. *Otherwise,* $d \in R^n$ *defined by*

$$d_i = -||c||_{q'}^{1-q'} c_i |c_i|^{q'-2}, \quad i = 1, \ldots, n, \tag{2.15}$$

is a descent direction for f *at* $x = 0$.

Proof For $x = 0$ to be a solution, putting $x = 0$ in (2.6) shows that there must exist w, $||w||_{q'} \leq 1$ such that

$$c + \rho w = 0.$$

This is possible if and only if $||c||_{q'} \leq \rho$.

Now assume that $||c||_{q'} > \rho$, and let d be defined by (2.15). Then

$$\begin{aligned} c^T d &= -||c||_{q'}^{1-q'} \sum_{i=1}^{n} |c_i|^{q'} \\ &= -||c||_{q'}^{1-q'} \, ||c||_{q'}^{q'} \\ &= -||c||_{q'}. \end{aligned}$$

Further

$$\begin{aligned} ||d||_q^q &= \sum_{i=1}^{n} |d_i|^q \\ &= ||c||_{q'}^{q(1-q')} \sum_{i=1}^{n} |c_i|^q |c_i|^{q(q'-2)} \\ &= ||c||_{q'}^{-q'} \sum_{i=1}^{n} |c_i|^{q(q'-1)} \\ &= ||c||_{q'}^{-q'} \, ||c||_{q'}^{q'} \\ &= 1. \end{aligned}$$

Now the directional derivative of f at $x = 0$ in the direction d is given by

$$\begin{aligned} \max_{||w||_{q'} \le 1} \{c^T d + \rho w^T d\} &= c^T d + \rho ||d||_q \\ &= -||c||_{q'} + \rho \\ &< -||c||_{q'} + ||c||_{q'} \\ &= 0, \end{aligned}$$

and the result is proved. □

Suppose therefore that x is not a solution and we wish to implement a damped Newton method to find the solution, following a step away from $x = 0$ along the direction d as defined by (2.15). Then we can use

$$g = \nabla_x f = ||r||_p^{1-p} A^T D_r^{p-2} r + \rho ||x||_q^{1-q} D_x^{q-2} x,$$

$$\begin{aligned} H = \nabla_x^2 f =& (p-1)||r||_p^{1-2p} (||r||_p^p A^T D_r^{p-2} A - A^T D_r^{p-2} r r^T D_r^{p-2} A) \\ &+ \rho(q-1)||x||_q^{1-2q} (||x||_q^q D_x^{q-2} - D_x^{q-2} x x^T D_x^{q-2}). \end{aligned}$$

Example 1 Let

$$A = [1,\ 2,\ 3,\ 4]^T, \quad b = [3, 7, 1, 3]^T,$$

let $p = 3$, $q = 1.5$ and $\rho = 0.5$.

Table 2.1 shows the progress of the method to satisfy $||d||_2 < 10^{-8}$ starting from $x = 0$. For this example, $x = 0$ is a solution for any $\rho \geq 2.69834499$.

x	f	γ
0.0	7.35576237	1
1.00000000	5.71710345	1
1.26609635	5.63445637	1
1.20106175	5.62094082	1
1.19800434	5.62091432	1
1.19799620	5.62091432	–

Table 2.1: Example 1

Example 2 Consider now the stack loss data set of Daniel and Wood [3], which has $m = 21$, $n = 4$. The performance of the method when $p = 3$, $q = 1.5$, $\rho = 0.5$ is shown in Table 2.2, again starting with $x = 0$ and a descent step as in Theorem 2.6, followed by the (damped) Newton's method, and asking for the same accuracy as before. For this example, for $x = 0$ to be a solution, we would require $\rho \geq 219.169$.

x_1	x_2	x_3	x_4	f	γ
−0.00114712	1.52176679	0.50396952	−1.03489907	23.58373889	.25
−0.00823909	0.74591275	1.08176297	−0.57861415	12.70857672	1
−0.04939979	0.81570802	1.17694568	−0.65587841	12.50016328	1
−0.09309212	0.80578363	1.14864140	−0.64038610	12.48356020	1
−0.10181497	0.80558564	1.14784481	−0.63992317	12.48353668	1
−0.10202577	0.80558432	1.14784872	−0.63992072	12.48353667	1
−0.10202589	0.80558432	1.14784872	−0.63992072	12.48353667	–

Table 2.2: Example 2

We can offer some sort of comparison with the work involved in solving a least squares problem using simple iteration, by comparing output for the data set in Example 2 from MATLAB programmes. If this method is used with $\rho = 0.5$, and $p = 3$, $q = 1.5$, then for the norm of the Newton step to be smaller than 10^{-6} it takes 46635 flops. A sample flop count for a programme to produce a solution for the least squares case with $\rho = 0.5$ by using simple iteration, starting with $\alpha = 0$, and terminating after 12 iterations when successive values of α differ by less than 10^{-6}, is 12783 flops.

Example 3 Consider now the Iowa wheat data from Draper and Smith [4], which has $m = 33$, $n = 9$. The performance of the method for $p = q = 3$, and some different values of ρ is summarized in Table 2.3. Here $f(0) = 170.336820$, k denotes the number of iterations, and f^* denotes the minimum value of f, for the same accuracy as before.

ρ	f^*	k
1.0	26.47610742	9
0.5	25.49994806	8
0.1	24.60983096	10
0.05	24.48871752	9
0.005	24.37742667	9

Table 2.3: Example 3; $p = q = 3$

3 A known bound on $||E : d||$

Now suppose that the underlying problem is such that we know upper bounds on the uncertainties in A and b, in the form

$$||E : d|| \leq \rho,$$

where ρ is given, and the matrix norm is as defined by (1.2) or (1.3). Consider the problem of determining

$$\min_x \max_{||E:d|| \leq \rho} ||(A + E)x - (b + d)||_p. \tag{3.1}$$

This problem and variants have been considered for example by El Ghaoui and Lebret [5], [6], where the matrix norm is the Frobenius norm, and the vector norm is the 2-norm. Arguing as in Theorems 2.1 and 2.2 gives the following results.

Theorem 3.1 *Let the matrix norm be given by (1.2). For any x, the maximum in (3.1) is attained when*

$$[E : d] = \rho u w^T, \tag{3.2}$$

where $u = \frac{r}{||r||_p}$ if $r \neq 0$, otherwise any vector with $||u||_p = 1$, and where $w \in R^{n+1}$ with

$$w_i = x_i |x_i|^{p'-2} ||[x^T, -1]^T||_{p'}^{1-p'}, \quad i = 1, \ldots, n,$$

$$w_{n+1} = -||[x^T, -1]^T||_{p'}^{1-p'}.$$

The maximum value is

$$||r||_p + \rho ||[x^T : -1]^T||_{p'}. \tag{3.3}$$

Theorem 3.2 *Let the matrix norm be given by (1.3). For any x, the maximum in (3.1) is attained when*

$$[E : d] = \rho u w^T,$$

where $u = \frac{r}{||r||_p}$ if $r \neq 0$, otherwise any vector with $||u||_p = 1$, and where $w \in R^{n+1}$ with

$$w_i = x_i |x_i|^{p-2} ||[x^T, -1]^T||_p^{1-p}, \quad i = 1, \ldots, n,$$

$$w_{n+1} = -||[x^T, -1]^T||_p^{1-p}.$$

The maximum value is

$$||r||_p + \rho ||[x^T : -1]^T||_p. \tag{3.4}$$

These examples are both special cases of the problem of minimizing

$$f(x) = ||Ax - b||_p + \rho ||[x^T : -1]^T||_q, \tag{3.5}$$

where

$$1 < p,\ q < \infty.$$

Again this allows the treatment of a wider class of problems, including more general norms of the form (1.3). The essential difference between this problem and the previous one is that the second term in this expression can never be zero. Since the first term is unlikely to be zero, this means that (3.5) is normally differentiable. However, standard convex analysis gives the following general characterization result.

Theorem 3.3 *The function (3.5) is minimized at x if and only if*

$$A^T v + \rho w_1 = 0, \tag{3.6}$$

where v is as in Theorem 2.2, and w_1 denotes the first n components of w as defined in Theorem 3.1.

Assuming that f is differentiable for all x, we have

$$\nabla_x \|Ax - b\|_p = \|r\|_p^{1-p} A^T D_r^{p-2} r,$$

$$\nabla_x \|[x^T : -1]^T\|_q = \|[x^T : -1]^T\|_q^{1-q} D_x^{q-2} x.$$

Thus

$$g = \nabla_x f = \|r\|_p^{1-p} A^T D_r^{p-2} r + \rho \|[x^T : -1]^T\|_q^{1-q} D_x^{q-2} x,$$

and when $p, q \geq 2$ or when no component of r or x is zero,

$$\begin{aligned} H = \nabla_x^2 f = & (p-1)\|r\|_p^{1-2p}(\|r\|_p^p A^T D_r^{p-2} A - A^T D_r^{p-2} r r^T D_r^{p-2} A) \\ & + \rho(q-1)\|[x^T : -1]^T\|_q^{1-2q}(\|[x^T : -1]^T\|_q^q D_x^{q-2} - D_x^{q-2} x x^T D_x^{q-2}), \end{aligned}$$

and Newton's method can be used. The Newton step d is defined by

$$Hd = -g,$$

and a line search in the direction d will always guarantee descent, because f is convex. Eventually we must be able to take full steps and get a second order convergence rate.

Consider now the special case of (3.5) when $p = q = 2$. An analysis similar to that given in Section 2.1 can be given in this case, leading to an efficient numerical method. An alternative approach is considered by El Ghaoui and Lebret [5], [6], but that seems likely to be less efficient. Let A have singular value decomposition as before, and have full rank. Assume also that $b \notin \text{range} A$. Then optimality conditions are

$$\|r\|_2^{-1} A^T r + \rho \|[x^T : -1]^T\|_2^{-1} x = 0,$$

or

$$(A^T A + \alpha I)x = A^T b,$$

where

$$\alpha = \frac{\rho||r||_2}{||[x^T : -1]^T||_2}.$$

It can be shown as before that x can be eliminated and α satisfies the equation

$$\alpha = h(\alpha),$$

where

$$h(\alpha) = \frac{\rho\sqrt{||b_2||_2^2 + \alpha^2||(\Sigma^2 + \alpha I)^{-1}b_1||_2^2}}{\sqrt{||\Sigma(\Sigma^2 + \alpha I)^{-1}b_1||_2^2 + 1}}.$$

Note that here there is no restriction on ρ except that it should be positive, and in that case it may be shown as in Theorem 2.6 that the simple iteration scheme

$$\alpha_k = h(\alpha_{k-1}), \quad k = 1, 2, \ldots$$

is locally convergent to the unique positive root.

Consider again Example 1 for the values of p, q, ρ given there. Table 3.1 shows the progress of the method to satisfy $||d||_2 < 10^{-8}$, starting from $x = 1.0$. Because $q < 2$, we require that components of x are not zero, and this is indeed the case.

x	f	γ
1.00000000	6.01080397	1
1.28823425	5.91251707	1
1.21611254	5.89534717	1
1.21290902	5.89531697	1
1.21290093	5.89531697	–

Table 3.1: Example 1

For Example 2, again starting with $x = (1, 1, 1, 1)^T$, we obtain the results shown in Table 3.2.

4 Conclusions

We have given an analysis in an l_p setting of a range of robust data fitting problems. These have attracted interest so far in the special case when least squares

x_1	x_2	x_3	x_4	f	γ
−0.57752515	0.80194065	1.17619011	−0.63983467	12.75192244	1
−0.29921499	0.80384192	1.15678981	−0.63858592	12.72046535	1
−0.08762724	0.80468901	1.15496729	−0.64120056	12.71528568	1
−0.12424185	0.80448346	1.15571823	--0.64081319	12.71489091	1
−0.12817450	0.80446270	1.15578949	−0.64077021	12.71488732	1
−0.12820707	0.80446253	1.15579010	−0.64076986	12.71488732	−

Table 3.2: Example 2

norms are involved, and there has been considerable algorithmic development. We have shown how the analysis extends to l_p problems, $1 < p < \infty$, and how solutions can be obtained for these other l_p cases. While special properties of the $p = 2$ case are not available for other values, nevertheless straightforward methods are available. The algorithmic development given here can readily be extended to deal with polyhedral norms (typified by l_1 and l_∞ norms), although significantly different methods are required. For example the case $p = 1$ of (1.2) leads, in the setting of Section 2, to the problem of minimizing

$$f(x) = ||Ax - b||_1 + \rho||x||_\infty.$$

Descent methods based on reduced or projected gradients may be applied to the minimization of this function, which is a finite problem. However, we will not consider this further here.

References

[1] Chandrasekaran, S., Golub, G. H., Gu, M. and A. H. Sayed, Efficient algorithms for least squares type problems with bounded uncertainties, in *Recent Advances in Total Least Squares Techniques and Errors-in-Variables Modeling,* ed S. Van Huffel, SIAM, Philadelphia, pp. 171–180 (1997).

[2] Chandrasekaran, S., Golub, G. H., Gu, M. and A. H. Sayed, Parameter estimation in the presence of bounded data uncertainties, *SIAM J. Matrix Anal. Appl.* **19**, pp. 235–252 (1998).

[3] Daniel, C. and F. S. Wood, *Fitting Equations to Data*, Wiley, New York (1971).

[4] Draper, N. R. and H. Smith, *Applied Regression Analysis*, Wiley, New York (1966).

[5] El Ghaoui, L. and H. Lebret, Robust solutions to least squares problems with uncertain data, in *Recent Advances in Total Least Squares Techniques and Errors-in-Variables Modeling,* ed S. Van Huffel, SIAM, Philadelphia, pp. 161–170 (1997).

[6] El Ghaoui, L. and H. Lebret, Robust solutions to least squares problems with uncertain data, *SIAM J. Matrix Anal. Appl.* **18**, pp. 1035–1064 (1997).

[7] Golub, G. H. and C. F. Van Loan, An analysis of the total least squares problem, *SIAM J. Num. Anal.* **17**, pp. 883–893 (1980).

[8] Sayed, A. H., Nascimento, V. H. and S. Chandrasekaran, Estimation and control with bounded data uncertainties, *Lin. Alg. Appl.* **284**, pp. 259–306 (1998).

G. Alistair Watson
Department of Mathematics
University of Dundee
Dundee DD1 4HN
Scotland
gawatson@maths.dundee.ac.uk

Contributed Papers[1]

Accelerating the limited memory BFGS method with inexact line searches
M. Al-Baali (Sultan Qaboos University, Dept. of Mathematics & Statistics, P.O. Box 36, Al-Khod 123, Muscat, Sultanate of Oman).

On the solution of singular perturbation problems using non-monotone methods on Shishkin meshes
Ali Ansari (University of Limerick, Department of Mathematics & Statistics, Limerick, Ireland) & Alan Hegarty.

Error estimates for finite element methods with anisotropic meshes
Thomas Apel (TU Chemnitz, Facultät für Mathematik, D-09107 Chemnitz, Germany).

Error propagation in the numerical integration of solitary waves. The regularized long wave equation
A. Aráujo (University of Coimbra, Department of Mathematics, 3000 Coimbra, Portugal) & A. Durán.

A Nyström method for integral equations on the real line and an application to scattering by diffraction gratings
T. Arens (Brunel University, Dept. of Mathematics and Statistics, Kingston Lane, Uxbridge UB8 3PH, UK), A. Meier, S. N. Chandler-Wilde & A. Kirsch.

Dynamic iteration for coupled differential-algebraic systems
Martin Arnold (DLR German Aerospace Center, Dept. Vehicle System Dynamics, P.O. Box 1116, D - 82230 Wessling, Germany).

Computation of a few of the largest eigenvalues of symmetric matrices
Ran Baik (Honam University, Mathematics Department, 59 1 Seobongdong, KwangsanGu KwangJu, 506-090, Korea), Karabi Datta & Yoopyo Hong.

The linear rational collocation method for PDEs
R. Baltensperger (University of Fribourg, Mathematical Institute, Perolles, 1700 Fribourg, Switzerland) & J.-P. Berrut.

Dynamics of the theta method
G. J. Barclay (University of Strathclyde, Department of Mathematics, Livingstone Tower, 26 Richmond Street, Glasgow G1 1XH, UK), D. F. Griffiths & D. J. Higham.

[1] The addresses given are those of the first named author, the presenter of the paper.

More accurate bidiagonal reduction for computing the SVD
Jesse L. Barlow (The Pennsylvania State University, Department of Computer Science and Engineering, 312 Pond Laboratory, University Park, PA 16802-6106, USA).

A study of moving mesh methods as applied to a thin flame propagating in a detonator delay element
T. Basebi (UMIST, Mathematics Department, P.O. Box 88, Manchester M60 1QD, England) & R. M. Thomas.

The adaptive solution of singularly perturbed boundary value problems using a posteriori error estimation
G. Beckett (University of Strathclyde, Department of Mathematics, Livingstone Tower, Richmond Street, Glasgow G1 1XH, UK) & J. A. Mackenzie.

A matrix for determining lower complexity barycentric representations of rational interpolants
J.-P. Berrut (University of Fribourg, Mathematical Institute, Perolles, CH - 1700 Fribourg, Switzerland).

Iterative multiscale methods for state estimation
Luise Blank (RWTH Aachen, Institut für Geometrie und Praktische Mathematik, Templergraben 55, 52056 Aachen, Germany), Thomas Binder, Wolfgang Dahmen & Wolfgang Marquardt.

Discrete compactness for edge elements
Daniele Boffi (Università di Pavia and Penn. State University, 413 McAllister Bldg., University Park, PA 16802, USA).

Introduction to the numerical treatment of stochastic delay differential equations
Evelyn Buckwar (University of Manchester, Dept. of Mathematics, Oxford Road, Manchester M13 9PL, UK) & Christopher T. H. Baker.

Modified Newton's method: convergence issues and properties
M. Isabel R. F. Caiado (Universidade do Minho, Departamento de Matematica Campus de Gualtar 4710 Braga, Portugal), M. Teresa Monteiro & Edite M. G. P. Fernandes.

Numerical solution of 2-phase Stefan problems
Jim Caldwell (City University of Hong Kong, 83 Tat Chee Avenue, Kowloon Tong, Hong Kong).

Avoiding the order reduction of Runge-Kutta methods when integrating boundary value problems for autonomous linear partial differential equations with nonhomogeneous terms
M. P. Calvo (Universidad de Valladolid, Departamento de Matematica Aplicada y Computacion Facultad de Ciencias Prado de la Magdalena s/n 47005 - Valladolid, Spain) & C. Palencia.

Approximating the logarithm of a matrix with variable accuracy
Sheung Hun Cheng (Centre for Novel Computing, University of Manchester, Dept. of Computer Science, Manchester M13 9PL, UK), Nicholas J. Higham, Charles S. Kenney & Alan J. Laub.

Towards highly accurate numerical approximation through defect correction and domain decomposition
A. S. Chibi (University of Annaba, Mathematics Department, B.P 12 Annaba, Algeria).

On the accuracy of difference schemes on nonuniform grids
Raimondas Ciegis (Vilnius Gediminas Technical University, Institute of Mathematics and Informatics, Akademijos 4, LT-2600, Lithuania).

Factorized Quasi-Newton structured techniques for nonlinear least squares problems
M. Fernanda P. Costa (Universidade do Minho, Departamento de Matematica, Campus de Azurim, 4810 Guimares, Portugal) & Edite M. G. P. Fernandes.

Up-wind based schemes in the simulation of free surface flows
J. A. Cuminato (University of Sao Paulo, ICMC-USP, Caixa Postal 668, 13560-970-Sao Carlos-SP, Brazil), A. Castelo Filho, V. G. Ferreira & M. F. Tome.

Generating test matrices for the one- and two-sided Jacobi methods
Philip I. Davies (Manchester University, Oxford Road, Manchester, UK) & N. J. Higham.

Frequency domain electromagnetic scattering from a thin wirc
P. J. Davies (University of Strathclyde, Department of Mathematics, 26 Richmond Street, Glasgow G1 1XH, UK), D. B. Duncan & S. A. Funken.

On a multistep method approximating a linear sectorial evolution equation
Nicoletta Del Buono (Università di Bari, Dipartimento Interuniversitario di Matematica, via Orabona 4, I-70125, Bari, Italy) & A. T. Hill.

Least squares data smoothing by non-negative divided differences
I. C. Demetriou (University of Athens, Department of Economics, 8 Pesmazoglou Street, Athens 105 59, Greece).

An efficient spectral method for nonlinear wave equations
Tobin A. Driscoll (University of Colorado, Department of Applied Mathematics, Campus Box 526, Boulder, CO 80309, USA) & Bengt Fornberg.

Sparse preconditioning of dense problems from electromagnetic applications
Iain Duff (RAL and CERFACS, Atlas Centre, Chilton, Didcot Oxon OX11 0QX, England), Bruno Carpentieri & Luc Giraud.

Approximating a convolution model of phase separation
D. B. Duncan (Heriot-Watt University, Department of Mathematics, Edinburgh EH14 4AS, UK), M. Grinfeld & I. Stoleriu.

GMRES convergence bounds, estimates, and restarts
Mark Embree (Oxford University Computing Laboratory, Wolfson Building, Parks Road, Oxford OX1 3QD, UK).

Multiple shooting for index-one differential-algebraic problems using a dichotomically stable integrator
Roland England (The Open University, Dept. of Applied Mathematics, Milton Keynes MK7 6AA, UK), René Lamour & Jesús López Estrada.

Feasible weight functions for generalised quadrature methods
Gwynne A. Evans (De Montfort University, Department of Mathematics, James Went Building, The Gateway, Leicester LE1 9BH, England).

Reliable and accurate software for evaluating the complex arctangent function
Thomas F. Fairgrieve (Ryerson Polytechnic University, Department of Mathematics, Physics and Computer Science, 350 Victoria Street, Toronto, Ontario M5B 2K3, Canada), T. E. Hull & Ping Tak Peter Tang.

Optimizing distillation columns
Roger Fletcher (University of Dundee, Department of Mathematics, Dundee DD1 4HN, Scotland).

Analysis of fractional differential equations
Neville J. Ford (Chester College, Department of Mathematics, Cheyney Road, Chester CH1 4BJ, UK) & Kai Diethelm.

Staggered time integrators for wave equations
Bengt Fornberg (University of Colorado, Department of Applied Mathematics, CB-526 Boulder, CO 80309, USA), Michelle Ghrist & Tobin Driscoll.

A scalable algorithm for heat integrated process synthesis
E. S. Fraga (University College London, Department of Chemical Engineering, Torrington Place, London WC1E 7JE, UK) & K. I. M. McKinnon.

Linearly implicit Runge-Kutta methods for advection-reaction-diffusion equations
Javier de Frutos (Universidad de Valladolid, Dpto. Matematica Aplicada y Computacion, Facultad de Ciencias , c/ Prado de la Magdalena s/n, 47011-Valladolid, Spain), M. P. Calvo & J. Novo.

Numerical simulation of reaction diffusion systems
M. Ganesh (University of New South Wales, School of Mathematics, Sydney, NSW 2052, Australia), M. A. J. Chaplain & I. G. Graham.

Stability and stabilization of some mixed finite element methods
Izaskun Garrido (University of Dundee, Department of Mathematics, Dundee DD1 4HN, UK) & D. F. Griffiths.

Numerical simulation of mixing in weakly nonhomogeneous viscous fluids
Dimitri V. Georgievskii (The Moscow State University, 119899 Vorobyovy Gory, Mechanical & Mathematical Faculty, Moscow, Russia).

Splitting methods for mixed hyperbolic-parabolic systems
A. Gerisch (Martin-Luther Universität Halle-Wittenberg, Institut für Numerische Mathematik, Fachbereich Mathematik und Informatik, Postfach, 06099 Halle (Saale), Germany) & R. Weiner.

On the numerical solution of Volterra integral equations of the second kind by a spline based method
Laura Gori (University of Rome "La Sapienza", Dipartimento Metodi e Modelli Matematici per le Scienze Applicate, Italy) & Elisabetta Santi.

An iterative active-set method for large-scale non-convex quadratic programming
Nick Gould (Rutherford Appleton Laboratory, Computational Science and Engineering Dept., Chilton, Oxon OX11 0QX, UK) & Philippe Toint.

Analysis of iterative, collocation and projection methods
Lechoslaw Hacia (Poznan University of Technology, Institute of Mathematics, Piotrowo 3a, 60-965 Poznan, Poland).

Exploiting hyper-sparsity in the revised simplex method
Julian Hall (University of Edinburgh, Department of Maths and Stats, JCMB, King's Buildings, Edinburgh EH9 3JZ, UK) & Ken McKinnon.

Adaptive finite element solution of electrochemical problems at microelectrodes
Kathryn Harriman (University of Oxford Computing Laboratory, Wolfson Building, Parks Road, Oxford OX1 3QD, UK), David Gavaghan, Paul Houston & Endre Süli.

Preconditioning eigenproblems arising from discretizations of stiff parameter dependent semilinear PDEs
Stuart C. Hawkins (University of Bath, Claverton Down, Bath BA2 7AY, UK) & Alastair Spence.

Uniformly accurate numerical solution of singular perturbation problems with parabolic boundary layers.
Alan Hegarty (University of Limerick, Plassey, Limerick, Ireland), Paul A. Farrell, John J. H. Miller, Eugene O'Riordan & G. I. Shishkin.

Adaptive error control in solving ordinary differential equations by the discontinuous Galerkin method
V. Heuveline (University of Heidelberg, Numerical Analysis Group, INF 293, D-69120 Heidelberg, Germany) & R. Rannacher.

Numerical stability for stochastic ODEs
D. J. Higham (University of Strathclyde, Department of Mathematics, Glasgow G1 1XH, UK).

A block algorithm for matrix 1-norm estimation, with an application to 1-norm pseudospectra
Nicholas J. Higham (University of Manchester, Department of Mathematics, Manchester M13 9PL, UK) & Françoise Tisseur.

Higher order approximation of inertial manifolds
Adrian Hill (University of Bath, Department of Mathematical Sciences, Claverton Down, Bath BA2 7AY, UK) & M. A. Falcon.

The 'enriched' hybrid Galerkin method
Robert A. Horrocks (University of Bath, Department of Mathematical Sciences, Bath BA2 7AY, UK), Ivan G. Graham & Stefan A. Sauter.

hp-adaptive finite element methods for first-order hyperbolic problems
Paul Houston (Oxford University Computing Laboratory, Wolfson Building, Parks Road, Oxford OX1 3QD, UK), E. Süli & Ch. Schwab.

Dynamics of variable time-stepping numerical methods
A. R. Humphries (Sussex University, School of Mathematical Sciences, Falmer, Brighton BN1 9QH, England).

BDF2 blends for convection dominated flows
Willem Hundsdorfer (CWI, P.O. Box 94079, Amsterdam, The Netherlands).

Numerical solution of the critical flow on a rotating cylinder
Roland Hunt (Strathclyde University, Mathematics Department, Livingstone Tower, 26 Richmond Street, Glasgow G1 1XH, UK).

Updating a correlation matrix by making use of hyperbolic transformations
D. Janovska (Prague Institute of Chemical Technology, Department of Mathematics, Technicka 1905/5, 166 28 Prague 6, Czech Republic) & Gerhard Opfer.

An optimal domain decomposition preconditioner
P. K. Jimack (University of Leeds, School of Computer Studies, Leeds LS2 9JT, UK), R. E. Bank & S. V. Nepomnyaschikh.

The box scheme for transcritical flow
Thomas C. Johnson (University of Reading, Department of Mathematics, Whiteknights P.O. Box 220, Reading RG6 6AH, UK), P. K. Sweby & M. J. Baines.

Accuracy in solving multivariate polynomials using Macaulay resultants
Gudbjorn F. Jonsson (Cornell University, Center for Applied Mathematics, 657 Rhodes Hall, Ithaca, NY 14853-3801, USA) & Stephen Vavasis.

Green's tensors for preconditioning steady-state Navier-Stokes equations: Part II - Application
David Kay (Oxford University, O.U.C.L., Wolfson Building, Parks Road, Oxford OX1 4QD, UK) & Daniel Loghin.

Constraint preconditioning for indefinite linear systems
Carsten Keller (University of Oxford, Numerical Analysis Group, Computing Laboratory, Wolfson Building, Parks Road, Oxford, OX1 3QD, UK), Andy Wathen & Nick Gould.

A shooting method for the generation of best L_1 piecewise linear interpolation
R. Ketzscher (Cranfield University, Applied Mathematics and Operational Research Group, Department of Informatics and Simulation, RMCS, Shrivenham SN6 8LA, UK) & S. A. Forth.

Newton's method for solving a quadratic matrix equation
Hyun-Min Kim (University of Manchester, Department of Mathematics, Oxford Road, Manchester M13 9PL, UK) & Nicholas J. Higham.

Homoclinic shadowing
Huseyin Kocak (University of Miami, Department of Mathematics and Computer Science, Coral Gables, FL 33124, USA), B. Coomes & K. Palmer.

The convergence of shooting methods for singular boundary value problems
Othmar Koch (Vienna University of Technology, Institut für Angewandte und Numerische Mathematik (E115), Wiedner Hauptstrasse 8-10, A-1040 Wien, Austria) & Ewa B. Weinmueller.

Structural optimization with stability constraints: solving nonconvex semidefinite programs
M. Kocvara (University of Erlangen, Institute for Applied Maths., Martensstr. 3, 91058 Erlangen, Germany) & F. Jarre.

The application of multiple shooting to singular boundary value problems
Peter Kofler (Vienna University of Technology, Department for Applied Mathematics and Numerical Analysis (E115/2), Wiedner Hauptstrasse 8-10, A-1040 Wien, Austria) & Winfried Auzinger.

The dynamics of adaptive timestepping ODE solvers
Harbir Lamba (George Mason University, Department of Mathematical Sciences, 4400 University Drive, Fairfax, VA 22030, USA).

On certain bivariate local spline quasi-interpolating operators
Paola Lamberti (University of Turin, Department of Mathematics, Via Carlo Alberto 10, 10123 Torino, Italy) & Catterina Dagnino.

Boundary integral methods for singularly perturbed boundary value problems
S. Langdon (Durham University, Department of Mathematics, Science Laboratories, South Road, Durham DH1 3LE, UK) & I. G. Graham.

Discrete least squares for hyperbolic equations and systems incorporating mesh movement
S. J. Leary (University of Reading, Department of Mathematics, Reading RG6 6AX, UK), M. J. Baines & M .E. Hubbard.

Singular differential equations in fluid flow
A. C. Lemos (University of Reading, Department of Mathematics, P.O. Box 220 Whiteknights, Reading, Berkshire RG6 6AX, UK), M. J. Baines & N. K. Nichols.

A quadratic-bundle-based decomposition for stochastic QP
Sven Leyffer (University of Dundee, Department of Mathematics, Dundee, DD1 4HN, UK), Jeff Linderot & Steve Wright.

Uniform superconvergence of a Galerkin finite element method on layer-adapted meshes
Torsten Linß (Technische Universität Dresden, Institut für Numerische Mathematik, D-01062 Dresden, Germany).

Green's tensors for preconditioning the steady-state Navier-Stokes equations: Part I - theory
Daniel Loghin (Oxford University Computing Laboratory, Parks Road, Oxford OX1 3QD, UK) & David Kay.

Quadrature rules for the hypercube of optimum trigonometric degree
James Lyness (Argonne National Laboratory, Mathematics and Computer Science Division, Argonne, IL 60439, USA).

A moving finite element method for two-dimensional Stefan problems
J. A. Mackenzie (Strathclyde University, Livingstone Tower, 26 Richmond Street, Glasgow G1 1XH, UK) & M. L. Robertson.

On the numerical recovery of a holomorphic mapping from a finite set of approximate values
J. M. Marban (University of Valladolid, Spain, Dpto. Matematica Aplicada y Computacion, Facultad de Ciencias, Paseo del Prado de la Magdalena s/n, 47005 Valladolid, Spain) & C. Palencia.

Using data assimilation to estimate systematic model error
M. J. Martin (University of Reading, Department of Mathematics, Reading RG6 6AX, UK), N. K. Nichols & M. J. Bell.

Convergence analysis of the discrete QR method for computing Lyapunov exponents
Edward McDonald (University of Strathclyde, Livingstone Tower, Richmond Street, Glasgow G1 1XH, UK) & Des Higham.

The rational Lanczos method for the Hermitian eigenvalue problem
Karl Meerbergen (Rutherford Appleton Laboratory, Chilton, Didcot OX11 0QX, UK).

Integral equation methods for scattering by rough surfaces
Anja Meier (Department of Mathematics and Statistics, Brunel University, Kingston Lane, Uxbridge UB8 3PH, UK) & Simon N. Chandler-Wilde.

A spectral method for computing viscoelastic flows
Sha Meng (De Montfort University, Department of Mathematics, James Went Building, Leicester LE1 9BH, UK).

On finite-difference methods for elliptic problems with discontinuous coefficients and jumps
Mathias Moog (ITWM, University of Kaiserslautern, ITWM, Erwin - Schroedinger - Str. 49, D-67663 Kaiserslautern, Germany).

Laguerre approximation of stable manifolds
Gerald Moore (Imperial College, Department of Mathematics, Huxley Building, 180, Queen's Gate, London SW7 2BZ, UK).

Cheap global error estimation of explicit Runge-Kutta methods
Ander Murua (University of the Basque Country, Informatika Fakultatea, K.Z.A.A. saila, 649 P.K., 20080 Donostia, Spain) & Joseba Makazaga.

The effect of roundoff error on the numerical solution of a problem with boundary layers
A. Musgrave (Trinity College, Department of Mathematics, Dublin 2, Ireland), J. J. H. Miller, E. O'Riordan & G. I. Shishkin.

Structured perturbations and robust eigenstructure assignment in quadratic matrix pencils
N. K. Nichols (University of Reading, Department of Mathematics, Reading RG6 6AX, UK) & J. Kautsky.

A new fitting for a spectral approach
F. Z. Nouri (University of Annaba, Institut de Mathematiques, BP 12, Annaba 23000, Algeria).

A spectral element method for the Navier-Stokes equations with improved accuracy
Julia Novo (Universidad de Valladolid, Dpto. Matematica Aplicada y Computacion, Facultad de Ciencias, C/ Prado de la Magdalena s/n, 47011 Valladolid, Spain) & Javier de Frutos.

On some problems with the hyperbolic Givens transformation
Gerhard Opfer (University of Hamburg, Institute of Applied Mathematics, Bundesstrasse 55, D-20146 Hamburg, Germany).

Geometric theory of fully nonlinear parabolic equations under discretization
Alexander Ostermann (Universität Innsbruck, Institut für Mathematik und Geometrie, Technikerstrasse 13, A-6020 Innsbruck, Austria).

Coupled waves method of analysing boundary conditions
M. Parsaei (Tehran University, Dept. Math. & Comp. Science, Faculty of Science, Tehran, Iran).

Feature of the stress state determination of viscoelastic composites
Boris E. Pobedria (The Moscow State University, 119899 Vorobjevy Gory, Mechanics & Mathematics Department, Russia).

A fast two-grid banded matrix iterative scheme for electromagnetic scattering
Mizanur Rahman (Brunel University, Departments of Mathematics and Statistics, Uxbridge, Middlesex UB8 3PH, UK).

Making wave with multi-symplectic collocation methods
Sebastian Reich (University of Surrey, Department of Mathematics and Statistics, Guildford, GU2 5XH, UK).

The numerical solution of one-dimensional phase change problems using an adaptive moving mesh method
M. Robertson (University of Strathclyde, Department of Mathematics, Livingstone Tower, Richmond Street, Glasgow G1 1XH, UK) & J. A. Mackenzie.

A Fast Trend Transform—FTT
Carl H. Rohwer (University of Stellenbosch, Dept. of Mathematics, Stellenbosch 7600, South Africa).

Starting algorithms for some DIRK methods
Teo Roldan (Universidad Publica de Navarra, Departamento de Matematica e Informatica, Campus de Arrosadia, Pamplona, Spain) & Inmaculada Higueras.

Uniform convergence on layer adapted meshes
Hans-Görg Roos (TU Dresden, Institut für Numerische Mathematik, Mommsenstrasse 13, D O-8027 Dresden, Germany).

Divergence-free Raviart-Thomas finite elements for 2nd-order elliptic problems in 3D
R. Scheichl (University of Bath, Department of Mathematical Sciences, Bath BA2 7AY, UK).

Computation of evolutionary stable strategies for population models with chaotic states
S. W. Schoombie (University of the Orange Free State, Dept. of Mathematics and Applied Mathematics, P.O. Box 339, Bloemfontein 9300, South Africa) & M. H. Ellis.

Row ordering for frontal solvers
Jennifer A. Scott (Rutherford Appleton Laboratory, Computational Science and Engineering Department , Atlas Centre, Chilton, Didcot, Oxfordshire OX11 0QX, UK).

Nonconforming Galerkin methods based on quadrilateral elements for second order elliptic problems
Dongwoo Sheen (Seoul National University, Department of Mathematics, Seoul 151-742, Korea), J. Douglas, Jr., J. E. Santos & X. Ye.

Control operators and iterative algorithms for nonlinear data assimilation problems
V. Shutyaev (Russian Academy of Sciences, Institute of Numerical Mathematics, Gubkina str. 8, 117951 GSP-1 Moscow, Russia).

Fast solution techniques for convection diffusion problems
David J. Silvester (UMIST, Department of Mathematics, Manchester M60 1QD, UK), D. Kay & Syamsudhuha.

Approximate factorization in time-dependent PDEs
Ben Sommeijer (CWI, Centre for Mathematics and Computer Science, Kruislaan 413, 1098 SJ Amsterdam, The Netherlands) & P. J. van der Houwen.

Experiments on the edge of numerical stability
E. Sousa (University of Oxford Computing Laboratory, Numerical Analysis Group, Wolfson Building, Parks Road, Oxford OX1 3QD, UK) & I. J. Sobey.

Numerical solution of a nonlinear eigenvalue problem arising in resonance problems in electromagnetism
Alastair Spence (University of Bath, School of Mathematical Sciences, Claverton Down, Bath BA2 7AY, UK).

Simultaneous simulation with adaptive triangulation methods
Martin Staempfle (University of Glasgow, Centre for Systems and Control, Glasgow G12 8QQ, Scotland, UK).

A moving mesh method for hyperbolic conservation laws
John M. Stockie (Simon Fraser University, Department of Mathematics and Statistics, 8888 University Way, Burnaby, British Columbia, V5A 1S6, Canada), J. A. Mackenzie & R. D. Russell.

Non-smooth data error estimates for linearly implicit Runge-Kutta methods and applications to long-term behaviour
M. Thalhammer (University of Innsbruck, Department of Mathematics and Geometry, Technikerstrasse 13, A-6020 Innsbruck, Austria) & A. Ostermann.

Newton's method in floating point arithmetic and iterative refinement of generalized eigenvalue problems
Françoise Tisseur (University of Manchester, Department of Mathematics, Manchester, M13 9PL, UK).

The PDE coffee table book
Lloyd N. Trefethen (University of Oxford Computing Laboratory, Wolfson Building, Parks Road, Oxford OX1 3QD, UK) & Kristine Embree.

Exponentially-fitted Runge-Kutta methods
G. Vanden Berghe (Universiteit Gent, Toegepaste Wiskunde en Informatica, Krijgslaan 281 - S9, B9000- Gent, Belgium), H. De Meyer, M. Van Daele & T. Van Hecke.

Volterra discrete equations : summabilty of the fundamental matrix
Antonia Vecchio (CNR, Istituto per Applicazioni della Matematica, Via P. Castellino 111, 80131 - Napoli, Italy).

Stable numerical methods for non-linear neutral delay equations
R. Vermiglio (University of Udine, Department of Mathematics and Computer Science, via delle Scienze, 208, I-33100 Udine, Italy) & L. Torelli.

Asymptotic successive-corrections, compact operator methods for two-point boundary problems
Francisco R. Villatoro (Universidad de Malaga, Spain, E.T.S. Ing. Industriales, Dpto. Lenguajes y Ciencias de la Computacion, Plaza El Ejido s/n, E-29013, Malaga, Spain).

A symbolic algorithm for the reduction and numerical solution of higher index DAEs
J. Visconti (LMC-IMAG, 51, rue des Mathematiques, 38051 Grenoble Cedex 9, France).

Preconditioning for indefinite linear systems
Andy Wathen (Oxford University, Oxford University Computing Laboratory, Wolfson Building, Parks Road, Oxford OX1 3QD, UK), Malcolm Murphy & Gene Golub.

Use of potential vorticity as a variable in data assimilation
M. A. Wlasak (University of Reading, Department of Mathematics, Reading RG6 6AX, UK), N. K. Nichols & I. Roulstone.

An algorithm for continuous minimax problems with constraints
Stanislav Zakovic (Imperial College, London, Centre for Process Systems Engineering, London SW7 2BY, UK), Costas Pantelides & Berc Rustem.